NEUTRON SCATTERING IN LAYERED COPPER-OXIDE SUPERCONDUCTORS

T0224511

Physics and Chemistry of Materials
with Low-Dimensional Structures

VOLUME 20

The titles published in this series are listed at the end of this volume.

Neutron Scattering in Layered Copper-Oxide Superconductors

Edited by

Albert Furrer

ETH Zurich & Paul Scherrer Institute,
Laboratory for Neutron Scattering,
Villigen PSI, Switzerland

KLUWER ACADEMIC PUBLISHERS

DORDRECHT / BOSTON / LONDON

A C.I.P. Catalogue record for this book is available from the Library of Congress.

ISBN 978-90-481-5091-5

Published by Kluwer Academic Publishers,
P.O. Box 17, 3300 AA Dordrecht, The Netherlands.

Sold and distributed in the North, Central and South America
by Kluwer Academic Publishers,
101 Philip Drive, Norwell, MA 02061, U.S.A.

In all other countries, sold and distributed
by Kluwer Academic Publishers,
P.O. Box 322, 3300 AH Dordrecht, The Netherlands.

Printed on acid-free paper

TABLE OF CONTENTS

PHONON DISPERSIONS AND PHONON DENSITY-OF-STATES IN COPPER-OXIDE SUPERCONDUCTORS

L. Pintschovius and W. Reichardt

PREFACE

The phenomenon of superconductivity - after its discovery in metals such as mercury, lead, zinc, etc. by Kamerlingh-Onnes in 1911 - has attracted many scientists. Superconductivity was described in a very satisfactory manner by the model proposed by Bardeen, Cooper and Schrieffer, and by the extensions proposed by Abrikosov, Gorkov and Eliashberg. Relations were established between superconductivity and the fundamental properties of solids, resulting in a possible upper limit of the critical temperature at about 23 K. The breakthrough that revolutionized the field was made in 1986 by Bednorz and Müller with the discovery of high-temperature superconductivity in layered copper-oxide perovskites. Today the record in transition temperature is 133 K for a Hg based cuprate system. The last decade has not only seen a revolution in the size of the critical temperature, but also in the myriads of research groups that entered the field. In addition, high-temperature superconductivity became a real interdisciplinary topic and brought together physicists, chemists and materials scientists who started to investigate the new compounds with almost all the available experimental techniques and theoretical methods. As a consequence we have witnessed an avalanche of publications which has never occurred in any field of science so far and which makes it difficult for the individual to be thoroughly informed about the relevant results and trends.

Neutron scattering has outstanding properties in the elucidation of the basic properties of high-temperature superconductors. Neutron scattering has been widely applied to these systems and brought results which are crucial for the characterization of the properties of high-temperature superconductors as well as for the understanding of the mechanism of superconductivity. It is the purpose of the present volume to review and summarize the most important achievements obtained by neutron scattering which probes both the static and dynamical properties of these systems. As the key technique neutron diffraction has furnished insight into the atomic structure. The unparalleled precision and accuracy of subtle structural details were crucial to the final conclusions about physical properties. In particular, from a determination of the oxygen sites and their occupancy with concentrations deviating from the stoichiometric one, the superconducting properties could be related to the evolution of structural disorder. At the same time neutron diffraction turned out to be essential for the production of better quality samples. The relation between superconductivity and magnetism - and its coexistence at low temperatures - led to phase diagrams which are fundamental for any understanding of the mechanism of high-temperature superconductivity. Inelastic neutron scattering brought unique results with respect to the spin and lattice dynamics which yielded important information about the electron-phonon coupling, the extraordinary large energy scale of the spin fluctuations and the characteristics of the superconducting state through the investigation of the spin gap both in the normal and in the superconducting state. The disappearance of the flux-line lattice in high magnetic fields far below the critical temperature observed by small-angle neutron scattering might have important consequences for possible technological applications. Finally, the neutron has been used as a local probe to explore the inhomogeneous materials properties of high-temperature superconductors and the related effects of phase separation and charge-stripe order.

Copper-oxide superconductors are prototypes of layered structures, containing CuO_2 planes typical of the perovskite lattice. The parent compounds are all antiferromagnetic Mott insulators with quasi-two dimensionality. These two-dimensional features are essential for achieving high-temperature superconductivity by doping. The highly interdisciplinary topic of high-temperature superconductivity fits therefore well into the book series on the *Physics and Chemistry of Materials with Low-Dimensional Structures*. So far the achievements obtained by neutron scattering have never been reviewed and summarized in a comprehensive manner, although they are now fairly well established and understood. This is the main reason for editing the present volume which covers the relevant issues in the field accumulated during the last decade. The basic principles and methods of neutron scattering are summarized in the first chapter. The following nine chapters treat the most important applications of neutron scattering to copper-oxide high-T_c superconductors, i.e., the static and dynamical properties of the crystal lattice (oxygen site occupancies, charge transfer, charge-stripe order, phonon dispersions, phonon density-of-states), the static and dynamical magnetic correlations (phase diagrams, 2-d and 3-d magnetic ordering, spin waves, spin fluctuations, crystal-field excitations), and the structure of the flux-line lattice (field and temperature dependence). All the chapters are thoroughly introduced, methodically discussed and highlighted with many examples by acknowledged experts.

I am extremely thankful to all the authors who wrote a chapter for this volume. They have done an excellent job in providing concise, though comprehensive reviews in their fields of expertise. I acknowledge the encouragement received at all stages of the project by the editor-in-chief of this book series, Francis Lévy. Finally I am indebted to the publisher for the invaluable efforts towards a rapid and professional publication of this volume.

ALBERT FURRER   Villigen, Spring 1998

INTRODUCTION TO NEUTRON SCATTERING

PETER BÖNI AND ALBERT FURRER
Laboratory for Neutron Scattering
ETH Zürich & Paul Scherrer Institute
CH-5232 Villigen PSI, Switzerland

1. Introduction

Among most other methods, neutron scattering allows a detailed understanding of the static and dynamic properties on an atomic scale of materials that occur in our environment. Combined with x-ray scattering a very large range of momentum and energy transfers can be covered thanks to the high complementarity of both techniques. The most relevant, unique character of neutrons that cannot be matched by any other technique, can be summarized as follows:

- The neutron interacts with the atomic nucleus, and not with the electrons as x-rays do. This has important consequences: i) the response of neutrons from light atoms like hydrogen or oxygen is much higher than for x-rays, ii) neutrons can easily distinguish atoms of comparable atomic number, iii) neutrons distinguish isotopes: For example, deuteration of macromolecules allows to focus on specific aspects of their atomic arrangement or their motion.

- For the same wavelength as hard x-rays the neutron energy is much lower and comparable to the energy of elementary excitations in matter. Therefore, neutrons do not only allow the determination of the "static average" chemical structure, but also the investigation of the dynamic properties of the atomic arrangements that are directly related to the physical properties of materials.

- By virtue of its neutrality the neutron is rather weakly interacting with matter, which means that there is almost no radiation damage to living objects under study. Also, the rather weak interaction with matter results in a large penetration depth and therefore the bulk properties of matter can be studied. This is also important for the investigation of materials under extreme conditions such as very low

1

and high temperatures, high pressure, high magnetic and electric fields, etc.
- The neutron carries a magnetic moment that makes it an excellent probe for the determination of magnetic structures and magnetic excitations.

In this introduction we discuss a few general results from scattering theory, i.e. Fermi's golden rule and the first Born approximation, as well as some particular aspects of neutron scattering that may be relevant for the following contributions on high-T_c materials. This introduction is not supposed to replace any of the excellent text books on neutron scattering that have been published in the literature [1, 2].

2. Properties of the Neutron

Neutrons for scattering experiments are usually extracted from a moderator of a spallation source or reactor by means of beam tubes and guides. The probability of neutrons having a velocity between v and dv follows therefore closely a Maxwell-Boltzmann probability distribution

$$\mathcal{P}(v)\, dv = 4\pi \left(\frac{m}{2\pi k_B T} \right)^{3/2} v^2 e^{-\frac{mv^2}{2} \frac{1}{k_B T}}\, dv. \qquad (1)$$

Here, $m = 1.675 \cdot 10^{-27}$ kg is the mass of the neutron, $k_B = 1.381 \cdot 10^{-23}$ J/K is Boltzmann's constant and T is the temperature of the moderator. The maximum of $\mathcal{P}(v)$ occurs at a velocity v_n that corresponds to a kinetic energy of the neutron $E = \frac{1}{2}mv_n^2 = k_B T$. This relation explains the classifications "hot", "thermal" and "cold" neutrons, corresponding to neutrons being moderated by hot graphite ($T \simeq 2000$ K), water ($T \simeq 320$ K), or liquid deuterium ($T \simeq 30$ K), respectively.

The de Broglie wavelength λ of the neutron is defined by

$$\lambda = \frac{h}{mv}, \qquad (2)$$

where $h = 6.626 \cdot 10^{-34}$ Js. Therefore, momentum $\mathbf{p}$ and wavevector $\mathbf{k}$ are given by

$$\mathbf{p} = \hbar \mathbf{k} = \frac{h}{\lambda} \mathbf{n}, \qquad (3)$$

where $\mathbf{n}$ is a unit vector that defines the direction of propagation of the neutrons. Their kinetic energy ($\omega = 2\pi\nu$ is the frequency) is given by

$$E = k_B T = \frac{mv^2}{2} = \frac{p^2}{2m} = \frac{\hbar^2 k^2}{2m} = \frac{h^2}{2m\lambda^2} = \hbar\omega = h\nu. \qquad (4)$$

Inserting the values for m, $\hbar$, and k_B we obtain the following useful relations:

$$E = 0.08617\,T = 5.227\,v^2 = 2.072\,k^2 = 81.81\,\frac{1}{\lambda^2} = 0.0579\,gB$$
$$= 0.658 \cdot 10^{-12}\,\omega = 4.136\,\nu = 0.1239\,\nu_r, \tag{5}$$

where E is in meV, T in K, v in km/s, k in Å^{-1}, λ in Å, B in T, ω in s^{-1}, ν in THz, and ν_r in cm^{-1}. g is the gyromagnetic ratio of the unpaired electrons, i.e. $g = 2$ for spin only systems. The magnitudes of the units in Eq. (5) match nicely atomic spacings and excitation energies in most materials, explaining the uniqueness of neutron scattering. Accessible time scales are typically $10^{-9} - 10^{-15}$ seconds.

3. Neutron Scattering Cross Section

The principle of the scattering of particles by a sample is the same for most kinds of radiation. The geometry of a scattering experiment is shown in Fig. 1. Incident particles (neutrons, photons, etc.) having a wavevector k_i and a flux F (particles per cm^2 and s) are scattered by the sample into the element of solid angle $d\Omega = \sin\Theta d\Theta d\Phi$. Then the number of particles scattered per second into $d\Omega$ with wavevector k_f is given by

$$I[s^{-1}] = F[\text{cm}^{-2}\text{s}^{-1}]\left(\frac{d\sigma}{d\Omega}\right)[\text{cm}^2]\,d\Omega, \tag{6}$$

where $d\sigma/d\Omega$ is the differential cross-section with the dimension of an area.

Suppose that the detector measures also the energy of the scattered particles then the partial differential cross section

$$\frac{d^2\sigma}{d\Omega dE_f} \tag{7}$$

is measured. It provides the number of particles that are scattered into $d\Omega$ having an energy between E_f and $E_f + dE_f$. The total cross section is given by

$$\sigma = \int d\Omega\left(\frac{d\sigma}{d\Omega}\right) = \int d\Omega \int dE_f\left(\frac{d^2\sigma}{d\Omega dE_f}\right). \tag{8}$$

Because of conservation of momentum and energy, the momentum $\hbar Q$

$$Q = k_i - k_f \tag{9}$$

and the energy

$$\hbar\omega = E_i - E_f = \frac{\hbar^2}{2m}(k_i^2 - k_f^2) \tag{10}$$

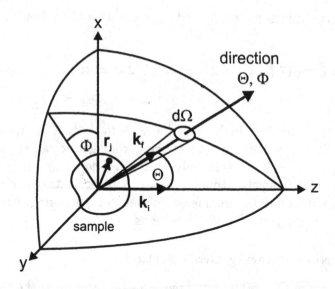

Figure 1. Geometry for a scattering experiment. A plane wave of neutrons that propagates along the z-direction is scattered by the sample into a solid angle $d\Omega$.

are transferred to the sample. $\mathbf{Q}$ is known as the scattering vector. Elastic scattering ($E_i = E_f \leftrightarrow |\mathbf{k}_i| = |\mathbf{k}_f|$) provides information about the position of the atoms or the size and distribution of inhomogeneities in a sample (Fig. 2). Inelastic scattering ($|\mathbf{k}_i| \neq |\mathbf{k}_f|$) provides information about the time and frequency dependent properties of the sample. If $E_i > E_f$, the neutrons transfer an energy $\hbar\omega$ (Stokes process) to the sample and vice versa (anti-Stokes process) thus creating or annihilating an excitation, respectively. The scattering vector $\mathbf{Q}$ is usually decomposed into $\mathbf{Q} = \boldsymbol{\tau} + \mathbf{q}$, where the reciprocal lattice vector $\boldsymbol{\tau}$ reflects the periodicity of the lattice and $\mathbf{q}$ is the wavevector of the excitation (Fig. 2).

The basic expression for the partial differential scattering cross section can be derived on the basis of Fermi's golden rule and is equivalent to the first Born approximation:

$$\frac{d^2\sigma}{d\Omega dE_f} = \left(\frac{m}{2\pi\hbar^2}\right)^2 \frac{k_f}{k_i} \sum_{\lambda_i} p_{\lambda_i} \sum_{\lambda_f} |\langle \mathbf{k}_f, \lambda_f | \breve{\mathbf{U}} | \mathbf{k}_i, \lambda_i \rangle|^2 \delta(E_{\lambda_i} - E_{\lambda_f} + \hbar\omega).$$

(11)

$|\lambda_i\rangle$ and $|\lambda_f\rangle$ denote the initial and final states of the sample, respectively. The averaging over the initial states is done on the basis of $|\lambda_i\rangle$ being occupied with the probability p_{λ_i}, and the summation over the final states is done by summing over the index λ_f [3]. Eq. (11) is only valid, if the scattering system remains in thermal equilibrium during the scattering event. Hence, $p_\lambda = \exp(-E_\lambda/k_B T)/Z$, where Z is the partition function. $\breve{\mathbf{U}}$ des-

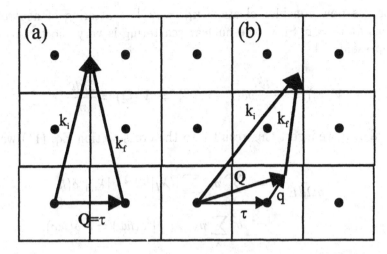

Figure 2.   (a) The reciprocal space for a quadratic lattice for elastic scattering. The condition $\mathbf{Q} = \tau$ is known as Bragg's law. (b) For inelastic scattering $|\mathbf{k}_i| \neq |\mathbf{k}_f|$. In a periodic lattice the scattering vector $\mathbf{Q}$ is usually decomposed into a reciprocal lattice vector τ and the wavevector $\mathbf{q}$ of the excitation. The lines indicate zone boundaries and the points indicate zone centres. Inelasticity deforms the scattering triangle significantly.

ignates the interaction operator between sample and neutrons and can be written as

$$\check{\mathbf{U}} = \sum_j V_j(\mathbf{r} - \mathbf{r}_j(t)), \tag{12}$$

where $\mathbf{r}_j$ is the position of the scattering objects in the sample. Inserting the (plane) wave functions for the neutrons, $|\mathbf{k}_i\rangle = \exp(i\mathbf{k}_i \cdot \mathbf{r})$ and $|\mathbf{k}_f\rangle = \exp(i\mathbf{k}_f \cdot \mathbf{r})$, one obtains

$$\langle \mathbf{k}_f, \lambda_f | \check{\mathbf{U}} | \mathbf{k}_i, \lambda_i \rangle = \langle f | \check{\mathbf{U}} | i \rangle = \langle \lambda_f | \int \sum_j e^{-i\mathbf{k}_f \cdot \mathbf{r}} V_j(\mathbf{r} - \mathbf{r}_j) e^{i\mathbf{k}_i \cdot \mathbf{r}} d\mathbf{r} | \lambda_i \rangle$$

$$= \sum_j \langle \lambda_f | \int e^{-i\mathbf{k}_f \cdot (\mathbf{x}_j + \mathbf{r}_j)} V_j(\mathbf{x}_j) e^{i\mathbf{k}_i \cdot (\mathbf{x}_j + \mathbf{r}_j)} d\mathbf{r} | \lambda_i \rangle, \tag{13}$$

where $\mathbf{x}_j = \mathbf{r} - \mathbf{r}_j$. Now, according to Eq. (9), $\mathbf{k}_i - \mathbf{k}_f = \mathbf{Q}$. Therefore

$$\langle f | \check{\mathbf{U}} | i \rangle = \sum_j \langle \lambda_f | \int e^{i\mathbf{Q} \cdot \mathbf{x}_j} V_j(\mathbf{x}_j) e^{i\mathbf{Q} \cdot \mathbf{r}_j} d\mathbf{x}_j | \lambda_i \rangle = \sum_j V_j(\mathbf{Q}) \langle \lambda_f | e^{i\mathbf{Q} \cdot \mathbf{r}_j} | \lambda_i \rangle,$$

$$\tag{14}$$

where

$$V_j(\mathbf{Q}) = \int V_j(\mathbf{x}_j) \exp(i\mathbf{Q} \cdot \mathbf{x}_j) \, d\mathbf{x}_j \tag{15}$$

is the Fourier transform of the interaction potential.

Let us now consider elastic ($|\mathbf{k}_i| = |\mathbf{k}_f|$) scattering of neutrons by a nucleus at $\mathbf{r}_j = \mathbf{r}$. Since the nuclear scattering is very short ranged it can be represented by

$$V(\mathbf{r}_j) = \frac{2\pi\hbar^2}{m} b \delta(\mathbf{r} - \mathbf{r}_j) \iff V(\mathbf{Q}) = \frac{2\pi\hbar^2}{m} b e^{i\mathbf{Q}\cdot\mathbf{r}_j} \qquad (16)$$

for real b. If we insert this result into the cross-section Eq. (11) we obtain

$$\begin{aligned}
\frac{d^2\sigma}{d\Omega dE_f} &= \sum_{\lambda_i} p_{\lambda_i} \sum_{\lambda_f} |\langle \lambda_f | b e^{i\mathbf{Q}\cdot\mathbf{r}_j} | \lambda_i \rangle|^2 \delta(\hbar\omega) \\
&= b^2 \sum_{\lambda} p_\lambda |\langle \lambda | \lambda \rangle|^2 \delta(\hbar\omega) = b^2 \delta(\hbar\omega). \qquad (17)
\end{aligned}$$

Therefore, the cross section is

$$\frac{d\sigma}{d\Omega} = b^2 \iff \sigma = 4\pi b^2. \qquad (18)$$

This result has an intuitive explanation: The scattering length b corresponds to the amplitude of the wavefunction of the scattered neutrons

$$\psi_{\text{sc}} = \frac{b}{r} \exp(ikr), \qquad (19)$$

when a nucleus is hit by a plane wave of neutrons $\psi_i = \exp(i\mathbf{k}_i \cdot \mathbf{z})$ (Fig. 1). b^2 defines the area for a scattering event to occur. b is roughly 3 orders of magnitude smaller than Bohr's radius ($5.3 \cdot 10^{-9}$ cm). Therefore the scattering of neutrons by nuclei is isotropic (s-wave scattering) in contrast to x-rays.

Let us consider now the cross section Eq. (11) for an assembly of nuclei at positions $\mathbf{r}_j$. According to Eq. (16) we can insert $\check{U} = (2\pi\hbar^2/m) \sum_j b_j \delta(\mathbf{r} - \mathbf{r}_j(t))$ in Eq. (14) and obtain

$$\frac{d^2\sigma}{d\Omega dE_f} = \frac{k_f}{k_i} \sum_{\lambda_i} p_{\lambda_i} \sum_{\lambda_f} |\langle \lambda_f | \sum_j b_j e^{i\mathbf{Q}\cdot\mathbf{r}_j} | \lambda_i \rangle|^2 \delta(E_{\lambda_i} - E_{\lambda_f} + \hbar\omega). \qquad (20)$$

In order to obtain a physically more transparent form for Eq. (20) we replace the δ-function in energy by the integral

$$\delta(E_{\lambda_i} - E_{\lambda_f} + \hbar\omega) = \frac{1}{2\pi\hbar} \int_{-\infty}^{\infty} e^{i(E_{\lambda_i} - E_{\lambda_f})t/\hbar} e^{-i\omega t} dt \qquad (21)$$

and obtain after some calculations, using the closure condition $\sum |\lambda_f\rangle\langle\lambda_f| = 1$, the result that the scattering cross section can be expressed by

$$
\frac{d^2\sigma}{d\Omega dE_f} = \frac{1}{2\pi\hbar}\frac{k_f}{k_i}\sum_{jj'} b_j b_{j'} \int e^{-i\omega t} dt \sum_{\lambda} p_\lambda \langle\lambda|e^{-i\mathbf{Q}\cdot\mathbf{r}_{j'}}e^{i\mathcal{H}t/\hbar}e^{i\mathbf{Q}\cdot\mathbf{r}_j}e^{-i\mathcal{H}t/\hbar}|\lambda\rangle
$$

$$
= \frac{1}{2\pi\hbar}\frac{k_f}{k_i}\sum_{jj'} b_j b_{j'} \int \langle e^{-i\mathbf{Q}\cdot\mathbf{r}_{j'}(0)}e^{i\mathbf{Q}\cdot\mathbf{r}_j(t)}\rangle e^{-i\omega t} dt. \tag{22}
$$

In the above expression, $\mathcal{H}$ is the Hamiltonian of the scattering system, i.e. $\mathcal{H}|\lambda\rangle = E_\lambda|\lambda\rangle$. The interpretation of Eq. (22) is as follows. The partial differential cross section is proportional to the Fourier transform of the expectation value to find a nucleus at position $\mathbf{r}_j$ at time t when it was at $\mathbf{r}_{j'}$ at time $t = 0$. Therefore it is convenient to write the cross section in the form

$$
\frac{d^2\sigma}{d\Omega dE_f} = \frac{\sigma}{4\pi}\frac{k_f}{k_i} NS(\mathbf{Q},\omega), \tag{23}
$$

where N is the number of nuclei in the scattering system. σ is essentially given by $\sum_{jj'} b_j b_{j'}$ (see Eq. (29)) in section 4) and

$$
S(\mathbf{Q},\omega) = \frac{1}{2\pi\hbar}\int G(\mathbf{r},t)e^{i(\mathbf{Q}\cdot\mathbf{r}-\omega t)} d\mathbf{r}\, dt. \tag{24}
$$

We see that the scattering function $S(\mathbf{Q},\omega)$ is simply the Fourier transform of the pair correlation function [4].

$$
G(\mathbf{r},t) = \left(\frac{1}{2\pi}\right)^3 \frac{1}{N}\int \sum_{jj'} e^{-i\mathbf{Q}\cdot\mathbf{r}}\langle e^{-i\mathbf{Q}\cdot\mathbf{r}_{j'}(0)}e^{i\mathbf{Q}\cdot\mathbf{r}_j(t)}\rangle d\mathbf{Q} \tag{25}
$$

with respect to space and time. One must be careful in evaluating the right hand side of Eq. (25) because, except when $t = 0$, the two operators $\exp(-i\mathbf{Q}\cdot\mathbf{r}_{j'}(0))$ and $\exp(i\mathbf{Q}\cdot\mathbf{r}_j(t))$ do not commute.

A closer investigation of the analytic properties of the scattering function shows that

$$
S(\mathbf{Q},\omega) = e^{\hbar\omega/k_BT}S(-\mathbf{Q},-\omega). \tag{26}
$$

This relation is known as *principle of detailed balance*. It expresses the physical result that the probability for the creation of an excitation is proportional to $\langle n+1\rangle$ and that the destruction of an excitation is proportional to $\langle n\rangle$ [5], where

$$
\langle n\rangle = \frac{1}{e^{\hbar\omega/k_BT} + 1}. \tag{27}
$$

For a proper data treatment, detailed balance has to be taken into account.

4. Coherent and Incoherent Scattering

The scattering lengths b do not only depend on the kind of element but also on the kind of isotope and on the quantum number of the angular momentum $T = I \pm \frac{1}{2}$ of the nucleus-neutron system. I and $\frac{1}{2}$ are the spin quantum numbers of the nucleus and the neutron, respectively. Therefore, during a scattering event in a sample, the outgoing partial neutron waves from the individual nuclei attain different phases and amplitudes and do not interfere completely anymore. Hence, one has to distinguish between coherent scattering and isotope and/or spin-incoherent scattering processes. It can be shown that the scattering length operator for a particular isotope α is given by [1, 2]

$$b_\alpha = A_\alpha + B_\alpha \boldsymbol{\sigma} \cdot \mathbf{I}_\alpha, \tag{28}$$

where $\boldsymbol{\sigma}$ is the Pauli spin operator for the neutron and the constants A_α and B_α are isotope-specific constants.

For coherent scattering, the cross section σ in Eq. (23) is given by the square of the average of the sum of the scattering lengths

$$\sigma_{\mathrm{c}} = 4\pi \left(\frac{1}{N} \sum_{i=1}^{N} b_i \right)^2 = 4\pi (\bar{b})^2, \tag{29}$$

and $S_{\mathrm{c}}(\mathbf{Q}, \omega)$ becomes proportional to the Fourier transform of the probability to find a particle at $\mathbf{r}$ at time t when there was another particle at $\mathbf{r} = 0$ at $t = 0$. Coherent processes are, for example, Bragg scattering and inelastic scattering by phonons or magnons.

The total incoherent scattering cross section has its origin in the "disorder" of the scattering lengths of chemically identical particles and is given by

$$\sigma_{\mathrm{i}} = 4\pi (\overline{b^2} - (\bar{b})^2). \tag{30}$$

$S_{\mathrm{i}}(\mathbf{Q}, \omega)$ provides information on self correlations, i.e. on the probability to find a particle at $\mathbf{r}$ at time t when the same particle was at $\mathbf{r} = 0$ at time $t = 0$. Diffusion and crystal-field excitations (see section 8.4) are typical examples for incoherent processes.

Hydrogen is an excellent example for a strong spin-incoherent scatterer because the scattering lengths for the triplet state ($T = 1$, $b_{\mathrm{trip}} = 1.085 \cdot 10^{-12}$ cm) and for the singlet state ($T = 0$, $b_{\mathrm{sing}} = -4.750 \cdot 10^{-12}$ cm) are very different leading to $\sigma_{\mathrm{i}} = 80.3$ barns $\gg \sigma_{\mathrm{c}} = 1.76$ barns. In contrast, the corresponding values for deuterium are $\sigma_{\mathrm{i}} = 2.05$ barns and $\sigma_{\mathrm{c}} = 5.59$ barns, respectively [6]. Therefore, it is possible to distinguish between coherent and incoherent processes by deuteration of a sample. In addition, structural studies in hydrogen containing samples can be facilitated by partial deuteration of particular groups in complicated molecules.

A strong isotopic-incoherent scatterer is Ni because the scattering lengths of the different isotopes vary over a wide range -8.7 fm $< b < +14.4$ fm [6]. For the investigation of coherent processes, like phonons (section 8.2) or magnons (section 8.3), it is usually necessary to use isotopically pure samples in order to reduce or even eliminate the incoherent scattering, for example ^{58}Ni or ^{60}Ni.

5. Magnetic Scattering

Due to the spin of the neutron, there is a strong interaction between the magnetic moment of the neutron and the magnetic field $\mathbf{B}$ created by the unpaired electrons in the sample. The magnetic interaction operator is given by

$$\check{U}_m = -\boldsymbol{\mu} \cdot \mathbf{B} = -\gamma \mu_N \boldsymbol{\sigma} \cdot \mathbf{B}, \tag{31}$$

where $\gamma = -1.913$ is the gyromagnetic ratio and $\mu_N = 5.051 \cdot 10^{-27}$ J/T is the nuclear magneton. Note that $\boldsymbol{\mu}$ is antiparallel to $\boldsymbol{\sigma}$. An unpaired electron at $\mathbf{r} = 0$ produces at the position $\mathbf{r}_j$ a magnetic field that is given by

$$\mathbf{B}_j = \nabla \times \left\{ \frac{\boldsymbol{\mu}_e \times \mathbf{r}_j}{|\mathbf{r}_j|^3} \right\} + \frac{(-e)}{c} \frac{\mathbf{v}_e \times \mathbf{r}_j}{|\mathbf{r}_j|^3}. \tag{32}$$

The first term ($\boldsymbol{\mu}_e = -2\mu_B \mathbf{S}$) describes the field due to the magnetic moment of the electron and the second term describes the magnetic field due to the orbital motion of the electron ($\mathbf{v}_e$ is its velocity). After Fourier transformation of Eq. (32) one obtains for the magnetic scattering length p of an electron

$$p = -\gamma r_0 \boldsymbol{\sigma} \cdot \left(\hat{\mathbf{Q}} \times (\mathbf{S} \times \hat{\mathbf{Q}}) + \frac{i}{\hbar |\mathbf{Q}|} (\mathbf{p}_e \times \hat{\mathbf{Q}}) \right) = -\gamma r_0 \frac{g}{2} \boldsymbol{\sigma} \cdot \left(\hat{\mathbf{Q}} \times (\mathbf{S} \times \hat{\mathbf{Q}}) \right). \tag{33}$$

$r_0 = 0.2818 \cdot 10^{-12}$ cm is the classical radius of the electron with momentum $\mathbf{p}_e$ and $\hat{\mathbf{Q}} = \mathbf{Q}/|\mathbf{Q}|$. On the right hand side of Eq. (33), the orbital contribution has been absorbed by means of the Landé splitting factor g [7]:

$$g = 1 + \frac{J(J+1) + S(S+1) - L(L+1)}{2J(J+1)}. \tag{34}$$

We see that $|p| \simeq 10^{-12}$ cm $\simeq |b|$. Therefore the magnetic scattering has the same order of magnitude as the nuclear scattering. The expression for p is more complicated than for its nuclear counterpart b because the dipole interaction between neutrons and electrons is a non-central force. Eq. (33) shows that only spin components perpendicular to the scattering vector $\mathbf{Q}$ contribute to the magnetic scattering cross section, providing an important selection rule for distinguishing between magnetic and nuclear scattering.

In order to obtain the cross section for magnetic scattering we proceed in a similar way as for nuclear scattering. We replace $\check{U}$ in Eq. (11) by

$$\check{U}_m = \frac{2\pi\hbar^2}{m} \sum_j p_j F_j(\mathbf{Q})\delta(\mathbf{r} - \mathbf{r}_j(t)) \tag{35}$$

(see also Eq. (16)), where $F_j(\mathbf{Q})$ is the magnetic form factor of the atom j at position $\mathbf{r}_j$. $F_j(\mathbf{Q})$ is given by the Fourier transform of the normalised spin density of the unpaired electrons, i.e. $F_j(\mathbf{Q} = 0) = 1$ [8]. Because the atomic orbitals are extended in space, $F(\mathbf{Q})$ is peaked near the forward direction, but not necessarily at $\mathbf{Q} = 0$.

The magnetic cross section for neutrons with the initial and final spin states σ_i and σ_f, respectively, becomes

$$\left(\frac{d\sigma}{d\Omega dE_f}\right)_{\text{mag}}^{\sigma_i \to \sigma_f} = \frac{1}{2\pi\hbar}\frac{k_f}{k_i}\int \langle \sigma_f| \sum_{jj'} p_j F_j(\mathbf{Q})p_{j'}F_{j'}(\mathbf{Q})|\sigma_i\rangle$$
$$\times \langle e^{-i\mathbf{Q}\cdot\mathbf{r}_{j'}(0)}e^{i\mathbf{Q}\cdot\mathbf{r}_j(t)}\rangle e^{-i\omega t}dt. \tag{36}$$

The comparison with the nuclear cross section Eq. (22) shows that apart from the vectorial nature of $p_j F_j$ the expressions are formally identical. For example, in a ferromagnet $p_j = p_{j'}$ and magnetic Bragg peaks appear at the same positions as the nuclear Bragg peaks. In a simple two-lattice antiferromagnet, the magnetic moments are staggered, i.e. $+p$ on one sublattice and $-p$ on the other sublattice. Therefore the periodicity of the magnetic unit cell is doubled with respect to the nuclear unit cell and magnetic Bragg peaks appear at zone boundaries of the nuclear reciprocal lattice.

The product $p_j p_{j'}$ can be expressed in the form

$$p_j p_{j'} = (\gamma r_0)^2 \frac{g_j g_{j'}}{4}\left(\left(\boldsymbol{\sigma}\cdot[\hat{\mathbf{Q}}\times(\mathbf{S}_j\times\hat{\mathbf{Q}})]\right)\right)\left(\boldsymbol{\sigma}\cdot[\hat{\mathbf{Q}}\times(\mathbf{S}_{j'}\times\hat{\mathbf{Q}})]\right)$$
$$= (\gamma r_0)^2 \frac{g_j g_{j'}}{4}\sum_{\alpha\beta}(\delta_{\alpha\beta} - \hat{Q}_\alpha\hat{Q}_\beta)S_{j\alpha}S_{j'\beta}\sigma_\alpha\sigma_\beta \tag{37}$$

$$\sigma_\alpha\sigma_\beta = \delta_{\alpha\beta} + i\sum_\gamma \epsilon_{\alpha\beta\gamma}\sigma_\gamma,$$

where α, β, and γ are Cartesian coordinates x, y, and z. The polarisation dependence of the cross section is caused by the Pauli spin matices σ_α [9]. For unpolarised neutrons we have to average over the initial states σ_i, and sum over the final states σ_f. In addition we assume identical magnetic moments and obtain the expression

$$\left(\frac{d\sigma}{d\Omega dE_f}\right)_{\text{mag}} = \frac{k_f}{k_i}(\gamma r_0 \frac{g}{2}F(\mathbf{Q}))^2 \sum_{\alpha\beta}(\delta_{\alpha\beta} - \hat{Q}_\alpha\hat{Q}_\beta)S^{\alpha\beta}(\mathbf{Q},\omega), \tag{38}$$

where $S^{\alpha\beta}(\mathbf{Q},\omega)$ is the magnetic scattering function:

$$S^{\alpha\beta}(\mathbf{Q},\omega) = \frac{1}{2\pi\hbar} \int \sum_{jj'} \langle S_{j'\alpha}(0)S_{j\beta}(t)\rangle \langle e^{-i\mathbf{Q}\cdot\mathbf{r}_{j'}(0)} e^{i\mathbf{Q}\cdot\mathbf{r}_j(t)}\rangle e^{-i\omega t} dt. \quad (39)$$

$S^{\alpha\beta}(\mathbf{Q},\omega)$ corresponds to the Fourier transform of the magnetic pair correlation function that gives the probability to find a magnetic moment at position $\mathbf{r}_j$ at time t with a component $S_{j\beta}(t)$ and a magnetic moment at position $\mathbf{r}_{j'}$ at time $t = 0$ with a component $S_{j'\alpha}(0)$. Therefore, the magnetic cross section depends on magnetic as well as on vibrational degrees of freedom.

In Eq. (39), $S^{\alpha\beta}(\mathbf{Q},\omega)$ is expressed in terms of expectation values of time-dependent operators (Heisenberg picture). Equivalently, we can write the scattering function in terms of time-independent operators (Schrödinger picture) by replacing the integral over time by a δ-function in energy, Eq. (21), and obtain the magnetic counter-part to Eq. (20)

$$S^{\alpha\beta}(\mathbf{Q},\omega) = \sum_{jj'} e^{i\mathbf{Q}\cdot(\mathbf{r}_j-\mathbf{r}_{j'})} \sum_{\lambda_i\lambda_f} p_{\lambda_i}\langle\lambda_i|\hat{S}_{j'}^\alpha|\lambda_f\rangle\langle\lambda_f|\hat{S}_j^\beta|\lambda_i\rangle\delta(E_{\lambda_i} - E_{\lambda_f} + \hbar\omega).$$

$$(40)$$

The magnetic scattering function is also directly related to the imaginary part ($\Im$) of the wavevector and frequency dependent susceptibility via the fluctuation-dissipation theorem [1, 2] by

$$S^{\alpha\beta}(\mathbf{Q},\omega) = \frac{\hbar}{\pi} \frac{1}{1 - e^{-\hbar\omega/k_BT}} \Im\chi^{\alpha\beta}(\mathbf{Q},\omega). \quad (41)$$

This theorem implies that the magnetic moment of the neutron acts on the sample as a frequency and wavevector dependent magnetic field $\mathbf{B}(\mathbf{Q},\omega)$, monitoring the response $\mathbf{M}(\mathbf{Q},\omega)$ of the sample

$$\mathbf{M}_\alpha(\mathbf{Q},\omega) = \sum_\beta \chi^{\alpha\beta}(\mathbf{Q},\omega)\mathbf{B}_\beta(\mathbf{Q},\omega), \quad (42)$$

where $\chi^{\alpha\beta}(\mathbf{Q},\omega)$ is the generalised susceptibility tensor. The fluctuation-dissipation theorem allows direct comparisons of $\chi_{\alpha\beta}$ as obtained by neutron scattering with bulk measurements of χ.

6. Polarisation Analysis

For the discussion of the polarisation dependence of the neutron scattering cross section [10] let us consider the matrix element $\langle\sigma_f\lambda_f|\sigma\cdot\check{\mathbf{M}}_\perp|\sigma_i\lambda_i\rangle$ in more detail. Here, we define the magnetic interaction operator by (see also

Eq. (33))

$$\check{\mathbf{M}}_\perp \equiv \sum_j e^{i\mathbf{Q}\cdot\mathbf{r}_j}\left(\hat{\mathbf{Q}} \times (\mathbf{S}_j \times \hat{\mathbf{Q}})\right). \tag{43}$$

It can be shown that $\check{\mathbf{M}}_\perp$ corresponds to the projection of the Fourier transform $\check{\mathbf{M}}(\mathbf{Q},\omega)$ of the magnetisation density $\mathbf{M}(\mathbf{r},t)$ on a plane perpendicular to the scattering vector $\mathbf{Q}$:

$$\check{\mathbf{M}}_\perp = \hat{\mathbf{Q}} \times (\check{\mathbf{M}} \times \hat{\mathbf{Q}}) \tag{44}$$

$$\check{\mathbf{M}} = -\frac{1}{2\mu_B} \int \mathbf{M}(\mathbf{r},t)e^{i\mathbf{Q}\cdot\mathbf{r}}d\mathbf{r} \tag{45}$$

Because $\lambda_{i,f}$ and $\sigma_{i,f}$ commute with each other we can consider the spin dependence of the matrix element $\langle\sigma_f|\boldsymbol{\sigma}\cdot\check{\mathbf{M}}_\perp|\sigma_i\rangle$ alone. Inserting the Pauli spin matrices σ_i [9] and using the relations $\sigma_x|\uparrow\rangle = |\downarrow\rangle$, $\sigma_x|\downarrow\rangle = |\uparrow\rangle$, $\sigma_y|\uparrow\rangle = i|\downarrow\rangle$, $\sigma_y|\downarrow\rangle = -i|\uparrow\rangle$, $\sigma_z|\uparrow\rangle = |\uparrow\rangle$, and $\sigma_z|\downarrow\rangle = -|\downarrow\rangle$, we arrive at the following selection rules with regard to the polarisations $\mathbf{P}_i$ and $\mathbf{P}_f$ of the incoming and scattered neutrons, respectively (the z-axis is the axis of quantisation):

$$\langle\uparrow|\boldsymbol{\sigma}\cdot\check{\mathbf{M}}_\perp|\uparrow\rangle = \check{M}_{\perp z} \tag{46}$$

$$\langle\downarrow|\boldsymbol{\sigma}\cdot\check{\mathbf{M}}_\perp|\downarrow\rangle = -\check{M}_{\perp z} \tag{47}$$

$$\langle\downarrow|\boldsymbol{\sigma}\cdot\check{\mathbf{M}}_\perp|\uparrow\rangle = \check{M}_{\perp x} + i\check{M}_{\perp y} = \check{M}^+ \tag{48}$$

$$\langle\uparrow|\boldsymbol{\sigma}\cdot\check{\mathbf{M}}_\perp|\downarrow\rangle = \check{M}_{\perp x} - i\check{M}_{\perp y} = \check{M}^-. \tag{49}$$

These matrix elements imply:

- spin densities $\check{M}_{\perp z}$ along $\mathbf{P}_i$ do not alter the spin eigenstate of the neutrons
- spin densities $\check{M}_{\perp x}$ and $\check{M}_{\perp y}$ perpendicular to $\mathbf{P}_i$ flip the neutron spin, i.e. $\mathbf{P}_f = -\mathbf{P}_i$.

In particular, if $\mathbf{Q} \parallel \mathbf{P}_i$ then $\check{M}_{\perp,z} = 0$ due to the selection rule for magnetic scattering that only spin components perpendicular to $\mathbf{Q}$ contribute to the magnetic scattering cross section. Hence, all magnetic scattering is spin flip. The scattering configuration $\mathbf{Q} \parallel \mathbf{P}_i$ is convenient for separating magnetic from nuclear scattering because coherent nuclear and isotopic-incoherent scattering do not flip the spin of the neutrons. Note, however, that spin-incoherent scattering contributes also to the spin flip cross section.

7. Instrumental Aspects

The determination of $S(\mathbf{Q},\omega)$ by neutron scattering techniques requires a controlled access to the variables $\mathbf{Q}$ and ω, which can be done in various

ways. A very efficient experimental method is the three-axis crystal spectrometer developed by Brockhouse [11]. As sketched in Fig. 3, an incident
beam of neutrons with a well defined wave vector k_i is selected from the
white spectrum of the neutron source by the monochromator crystal (first
axis). The monochromatic beam is scattered from the sample (second axis).
The intensity of the scattered beam with the wavevector k_f is reflected by
the analyser crystal (third axis) onto the neutron detector thereby defining
the energy transfer $\hbar\omega$ as well.

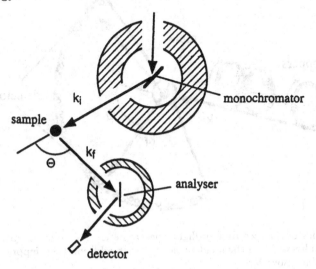

Figure 3. Schematics of a three-axis spectrometer. Monochromator (axis 1) and analyser
(axis 3) define the energy of the reflected neutrons.

The outstanding advantage of this spectrometer is that the data can
be taken at a pre-determined point in reciprocal space (constant-Q scan)
or for a fixed energy transfer (constant-E scan) along a particular line
in reciprocal space, so that measurements of dispersion relations in single
crystals can be performed in a controlled manner. Of course, general scans
in Q and ω are also possible.

For structural studies, the scattering law $S(Q, \omega)$ is integrated in ω-
space to yield the structure factor $S(Q)$, thus there is usually no need to
perform an energy analysis by the third spectrometer axis. Neutron diffractometers are therefore called two-axis instruments. Advanced instruments
often take advantage of position-sensitive detectors that speed up the rate
of data collection tremendously.

For neutron scattering experiments on polycrystalline, liquid and amorphous materials various types of time-of-flight spectrometers are usually
more appropriate. In the time-of-flight method the neutron beam is mono-
chromated by a series of choppers that produce pulses of neutrons with

the desired wavelength and that eliminate higher-order neutrons and frame overlap of pulses from different repetition periods as well (see Fig. 4). The monochromatic neutron pulses are then scattered from the sample and are detected by arrays of neutron counters covering a large solid angle. The energy transfer $\hbar\omega$ and the modulus of the scattering vector $\mathbf{Q}$ are then determined by the flight time of the neutrons from the sample to the detector and the scattering angle at which the detector is positioned, respectively.

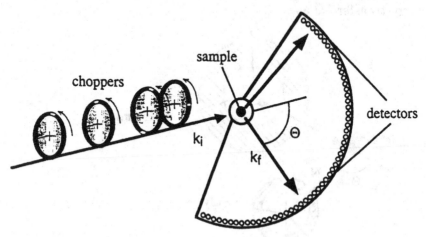

Figure 4. Schematics of a direct time-of-flight spectrometer. The four choppers define the energy and the pulse-width of the incident neutrons. In addition they suppress higher order neutrons and frame-overlap.

In the inverted time-of-flight method the white neutron beam is pulsed, for example by a spallation source itself and the energy of the scattered neutrons is analysed by means of banks of analyser crystals or choppers. Recently, the time-of-flight method has also been successfully used for the measurement of excitations in single crystals.

8. Examples

In this section we discuss four simple applications of neutron scattering that show some salient properties of the neutron scattering cross section. The examples have been selected from the field of high-T_c superconductivity.

8.1. DIFFRACTION

One of the most important tasks in a neutron scattering experiment is the determination of the position of the individuel atoms and magnetic moments in a sample. This information is contained in the elastic cross section. For its evaluation we start from Eq. (22) and obtain by setting

$\omega = 0$

$$\left(\frac{d^2\sigma}{d\Omega}\right)_{\text{nucl}} = \frac{1}{2\pi\hbar}\sum_{jj'} b_j b_{j'} \int \langle e^{-i\mathbf{Q}\cdot\mathbf{r}_{j'}(0)} e^{i\mathbf{Q}\cdot\mathbf{r}_j(t)}\rangle dt. \tag{50}$$

This expression is equivalent to [1, 2]

$$\left(\frac{d^2\sigma}{d\Omega}\right)_{\text{nucl}} = N\frac{(2\pi)^3}{v_0}\sum_{\tau} \delta(\mathbf{Q} - \tau)|F_N(\mathbf{Q})|^2, \tag{51}$$

where

$$F_N(\mathbf{Q}) = \sum_j b_j e^{i\mathbf{Q}\cdot\mathbf{d}_j} e^{-W_j}. \tag{52}$$

$F_N(\mathbf{Q})$ is the nuclear structure factor, $\exp(-W_j)$ is the Debye-Waller factor, $\mathbf{d}_j$ designates the position of atom j in the unit cell, and v_0 is the volume of the unit cell. The δ-function implies the well known fact that diffraction peaks appear if the scattering vector $\mathbf{Q}$ corresponds to a reciprocal lattice vector τ (Fig. 2).

In powder diffraction, the sample is composed of a random arrangement of single-crystalline grains. Therefore, the reciprocal space consists of spheres centred around (0 0 0). Fig. 5 shows as an example a powder diffraction pattern of the high-T_c compound $Y_2Ba_4Cu_7O_{15}$ [12]. From the position and the intensity of the peaks the structure of the material can be determined.

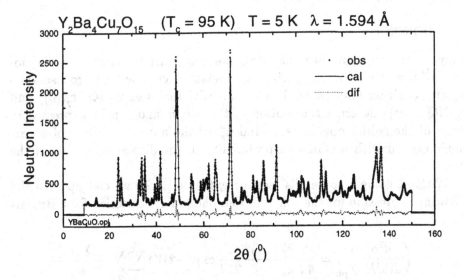

Figure 5.   Observed, calculated and difference neutron diffraction pattern for $Y_2Ba_4Cu_7O_{15}$ at 5 K ($T_c = 95$ K).

For magnetic Bragg scattering we obtain similarly (see Eq. (38))

$$\left(\frac{d^2\sigma}{d\Omega}\right)_{\text{mag}} = (\gamma r_0 \frac{g}{2} F(\mathbf{Q}))^2 N_m \exp(-2W) \sum_{\alpha\beta} (\delta_{\alpha\beta} - \hat{Q}_\alpha \hat{Q}_\beta) S^{\alpha\beta}(\mathbf{Q}, \omega = 0).$$

$$(53)$$

To be more specific, for a single domain Bravais ferromagnet, one obtains

$$\left(\frac{d^2\sigma}{d\Omega}\right)_{\text{mag}} = N \frac{(2\pi)^3}{v_0} \sum_{\tau} \delta(\mathbf{Q} - \tau) |F_M(\mathbf{Q})|^2, \qquad (54)$$

where

$$|F_M(\mathbf{Q})|^2 = \left(\gamma r_0 \frac{g}{2} \langle S \rangle F(\mathbf{Q}) e^{-W}\right)^2 \left(1 - (\hat{\mathbf{Q}} \cdot \frac{\mathbf{S}}{S})^2\right) \qquad (55)$$

showing that the magnetic Bragg peaks coincide with the nuclear Bragg peaks.

In more complicated magnetic structures, like incommensurate helical magnets, the magnetic Bragg peaks appear as satellites that are displaced from the nuclear Bragg peaks by the modulation wavevector $\mathbf{K}$ that defines the periodicity of the arrangement of the moments, i.e. peaks appear at positions $\mathbf{Q} = \tau \pm \mathbf{K}$.

The major difference between the nuclear and magnetic cross section is the vectorial dependence of the magnetic scattering on the relative orientation between $\mathbf{Q}$ and the direction of the magnetic moments allowing often a distinction between the nuclear and magnetic scattering contributions without using polarisation analysis.

8.2. PHONONS

Knowing the structure of a material, one may start to investigate the motions of the atoms by measuring the inelastic, coherent neutron scattering cross section (Eq. (22)). The exponential terms $\exp(-i\mathbf{Q} \cdot \mathbf{r}_{j'}(0))$ and $\exp(i\mathbf{Q} \cdot \mathbf{r}_j(t))$ describe the motion of the atoms and can be expressed in terms of the ladder operators $\hat{a}_s$ and $\hat{a}_s^+$ of the normal modes s of a harmonic crystal with a Gaussian probability of the displacements $\mathbf{u}_j$ of the atoms.

Without going into details (see for example [1, 2]) we end up with the following expression for the coherent one-phonon cross section for a Bravais crystal:

$$\left(\frac{d^2\sigma}{d\Omega dE_f}\right)_{\text{ph}} = \frac{\sigma_c}{4\pi} \frac{k_f}{k_i} \frac{(2\pi)^3}{v_0} \frac{1}{2M} \exp(-2W) \sum_s \sum_{\tau q} \frac{(\mathbf{Q} \cdot \mathbf{e}_s)^2}{\omega_s}$$

$$\times \left((n_s + 1)\delta(\mathbf{Q} - \mathbf{q} - \tau)\delta(\omega - \omega_s) + \langle n_s \rangle \delta(\mathbf{Q} + \mathbf{q} - \tau)\delta(\omega + \omega_s)\right). \quad (56)$$

$\mathbf{e}_s$ are the eigenvectors of the phonon belonging to the branch s defining the movement of the atoms around their equilibrium position. The δ-functions at $\omega \pm \omega_s$ give rise to Stokes (phonon creation) and anit-Stokes (annihilation) lines, respectively. Due to the polarisation factor $\mathbf{Q} \cdot \mathbf{e}_s$, only lattice vibrations along the scattering vector $\mathbf{Q}$ contribute to the scattering cross section. This selection rule allows an experimental discrimination between different eigenmodes, for example between longitudinal and transverse acoustic or optic phonons. The δ-functions are responsible for the conservation of energy and momentum during the scattering event and define the dispersion curve of the excitations, $\omega = \omega_s(\mathbf{q})$.

Fig. 6 shows as an example the temperature dependence of the soft optic phonon at the X-point in La_2CuO_4 that is responsible for the phase transition from the tetragonal to the orthorhombic phase at $T_0 = 503$ K. Above T_0, the transverse optic phonon softens, evolving into an overdamped mode near T_0. Below T_0, the degeneracy of the soft phonon is lifted and two modes are condensing out [13].

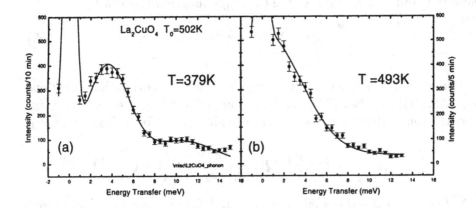

Figure 6. Energy scans showing phonon peaks in undoped La_2CuO_4. At 493 K, the damping of the phonons is so large that the two degenerate modes cannot be resolved. At 379 K, two distinct peaks from both the condensed zone-center mode and the uncondensed zone-boundary mode can be seen.

A quick overview about the lattice dynamics of a material can be obtained by measuring the (generalized) phonon density-of-states from polycrystalline samples using the time-of-flight technique. This technique is extremely useful when large single crystals are not available as was the case in the early days of high-T_c-superconductivity [14]. Peaks in the spectra occur at those energies, where the dispersion curves are flat. By measuring the T-dependence of such spectra, information can be obtained on the dynamics of phase transitions or on the electron-phonon coupling.

8.3. SPIN EXCITATIONS

The basic exciations in a magnetically ordered system are spin waves. If the unpaired electrons that carry the magnetic moment can be considered to be localised, the dispersion curve for spin waves can be calculated using the Heisenberg Hamiltonian (in zero magnetic field)

$$\mathcal{H} = -\sum_{jj'} J_{jj'} \mathbf{S}_j \cdot \mathbf{S}_{j'}. \tag{57}$$

The quantities $J_{jj'}$ are known as exchange constants arising from the requirement that the wave function of the electrons must be antisymmetric. Depending on the signs of $J_{jj'}$ different magnetic orderings can occur.

Within linear spin wave theory being only valid at low T and for reasonably large $\mathbf{S}$ one obtains for the spin wave dispersion

$$\hbar\omega_q = 2S(J(0) - J(\mathbf{q})), \tag{58}$$

where the Fourier transform of the exchange function is given by

$$\mathcal{J}(\mathbf{q}) = \sum_{jj'} J_{jj'} e^{i\mathbf{q}\cdot(\mathbf{r}_j - \mathbf{r}_{j'})}. \tag{59}$$

Also within the linear approximation, the spin correlation functions $\langle S_{j'\alpha}(0)S_{j\beta}(t)\rangle$ (see Eq. (39)) can be calculated yielding the inelastic magnetic neutron scattering cross section for spin waves (for a single domain)

$$\left(\frac{d^2\sigma}{d\Omega dE_f}\right)_{sw} = (\gamma r_0 \frac{g}{2} F(\mathbf{Q}))^2 \frac{k_f}{k_i} \frac{(2\pi)^3}{v_0} \frac{1}{2} S \left(1 + (\frac{Q_z}{Q})^2\right) e^{-2W}$$

$$\times \sum_{\tau q} \left(\langle n+1 \rangle \delta(\mathbf{Q} - \mathbf{q} - \tau)\delta(\omega_\mathbf{q} - \omega) + \langle \mathbf{n} \rangle \delta(\mathbf{Q} + \mathbf{q} - \tau)\delta(\omega_\mathbf{q} + \omega) \right). \tag{60}$$

The last two terms correspond to spin wave creation and annihilation, respectively, similarly as for phonons (Eq. (56)). By measuring the energy spectrum of scattered neutrons the dispersion relation of the spin waves can be measured and the magnetic interactions between the magnetic moments can be determined. For a multi-domain sample, $1 + (Q_z/Q)^2$ has to be replaced by 4/3.

As an example, Fig. 7 shows a high-energy spectrum of magnetic excitations in La_2CuO_4 as measured at the zone boundary using the time-of-flight spectrometer HET at ISIS [15]. The dispersion shows a typical sine-like q-dependence as expected for a simple Heisenberg antiferromagnet with nearest-neighbour interactions.

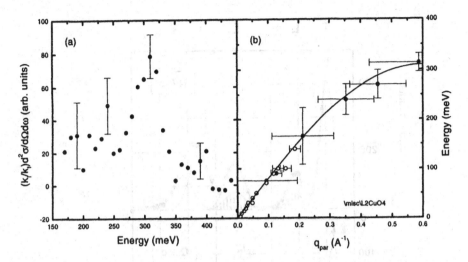

Figure 7. (a) Spin wave excitation as measured near the magnetic zone boundary $\mathbf{Q} \simeq (1, k, 0.5)$ at $T = 296$ K. The dashed line is a nearest-neighbour spin wave model. (b) Dispersion relation of the in-plane magnons in La_2CuO_4. Horizontal bars represent the range in q parallel to (100) over which the detectors integrate.

8.4. CRYSTAL FIELD EXCITATIONS

If the coupling between the magnetic ions is weak, we are left with a single-ion problem, thus the energy of the magnetic excitations will be independent of the scattering vector $\mathbf{Q}$. Typical examples are rare-earth compounds that exhibit very low magnetic ordering temperatures or do not order at all. In this case, the dominant mechanism is the crystal-field interaction.

The effect of an electric field on a rare-earth ion is to partially or totally remove the $(2J+1)$-fold degeneracy of the ground-state J-multiplet, giving rise to a sequence of crystal-field states Γ_n. In neutron scattering experiments, transitions between these states show up as resonance peaks in the observed energy spectra, so that one can determine directly the sequence of crystal-field levels. This is exemplified in Fig. 8 for some low-energy crystal-field excitations in a grain-aligned sample of the high-T_c superconductor $HoBa_2Cu_3O_7$ with a very low magnetic ordering temperature $T_N = 200$ mK associated with the Ho sublattice [16].

In evaluating the cross-section for the crystal-field transition $\Gamma_n \to \Gamma_m$ one starts with the scattering function Eq. (40) and neglects the lattice dynamical degrees of freedom. Since we are dealing with single-ion excitations, we have $j = j'$, and the $\mathbf{Q}$-dependence disappears. For N identical magnetic ions we can even drop the index j. Then $S^{\alpha\beta}(\mathbf{Q}, \omega)$ reduces to

$$S^{\alpha\beta}(\omega) = N p_{\Gamma_n} \langle \Gamma_n | \hat{J}^\alpha | \Gamma_m \rangle \langle \Gamma_m | \hat{J}^\beta | \Gamma_n \rangle \delta(\hbar\omega + E_{\Gamma_n} - E_{\Gamma_m}), \qquad (61)$$

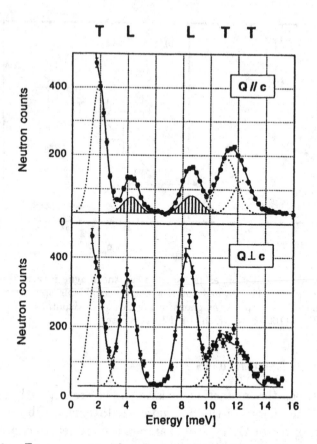

Figure 8. Energy spectra of neutrons scattered from grain-aligned $HoBa_2Cu_3O_7$ at $T = 1.5$ K with the scattering vector $\mathbf{Q}$ parallel and perpendicular to the c axis ($Q = 1.72$ Å^{-1}). The lines denote Gaussian fits to the observed crystal-field transitions. T and L refer to the dominant transverse and longitudinal character of the transitions, respectively. The shaded area corresponds to the transition strength due to the misorientation of the sample.

where p_{Γ_n} is the Boltzmann population factor for the initial state. From the symmetry relations associated with the matrix elements we find the cross section

$$\left(\frac{d^2\sigma}{d\Omega dE_f}\right)_{cef} = N\left(\gamma r_0\frac{g}{2}F(\mathbf{Q})\right)^2\frac{k_f}{k_i}e^{-2W}$$

$$\times p_{\Gamma_n}\sum_\alpha\left(1 - (\frac{Q_\alpha}{Q})^2\right)|\langle\Gamma_m|\hat{J}^\alpha|\Gamma_n\rangle|^2\delta(\hbar\omega + E_{\Gamma_n} - E_{\Gamma_m}). \qquad (62)$$

The polarisation factor permits a discrimination between transverse ($\alpha = x, y$) and longitudinal ($\alpha = z$) crystal-field transitions by measuring for different orientations of $\mathbf{Q}$. For example, for $\mathbf{Q} \parallel c$ only transverse transitions are observed, whereas for $\mathbf{Q} \perp c$ the transverse transitions lose half their intensities, and in addition longitudinal transitions appear. This is demonstrated in Fig. 8, however, the polarisation rules are not strictly fulfilled due to both, the rather large mosaic spread (FWHM $= 22^0$) and a fraction of 15% non-aligned powder of the grain-aligned sample [16].

For experiments on polycrystalline materials Eq. (62) has to be averaged in $\mathbf{Q}$-space, yielding

$$\left(\frac{d^2\sigma}{d\Omega dE_f}\right)_{\text{cef}} = N(\gamma r_0 \frac{g}{2} F(\mathbf{Q}))^2 \frac{k_f}{k_i} e^{-2W} p_{\Gamma_n} |\langle \Gamma_m | \hat{\mathbf{J}}_\perp | \Gamma_n \rangle|^2 \delta(\hbar\omega + E_{\Gamma_n} - E_{\Gamma_m}),$$
(63)

where $\mathbf{J}_\perp = \mathbf{J} - (\mathbf{J} \cdot \mathbf{Q})\mathbf{Q}/Q^2$ is the component of the total angular momentum perpendicular to the scattering vector $\mathbf{Q}$ and

$$|\langle \Gamma_m | \hat{\mathbf{J}}_\perp | \Gamma_n \rangle|^2 = \frac{2}{3} \sum_\alpha |\langle \Gamma_m | \hat{\mathbf{J}}^\alpha | \Gamma_n \rangle|^2.$$
(64)

9. Conclusions

In the preceeding sections we have derived some expressions for the most relevant scattering cross sections in neutron scattering. Because of the usually weak interaction between nuclei and neutrons, the cross sections are simply proportional to the Fourier transform in space and time of the pair correlation functions $G(\mathbf{r}, t)$ that describe position and dynamics of the nuclei and/or unpaired electrons in the sample. The real challenge of neutron scattering experiments is twofold, either to provide cross sections that give a hint to the underlying physical processes in the sample or to validate theoretical models.

References

1. G. L. Squires: *Thermal Neutron Scattering*, Cambridge University Press, Cambridge, 1978.
2. S. W. Lovesey: *Theory of Neutron Scattering from Condensed Matter*, International Series of Monographs on Physics 72, Oxford Science Publications, Oxford, 1987.
3. The final states must not be weighted with any probability p_{λ_f} because the probability for a transition from $|\lambda_i\rangle \longrightarrow |\lambda_f\rangle$ is given by the matrix elements of $\check{\mathbf{U}}$. If the states $|\lambda_i\rangle$ and $|\lambda_f\rangle$ form a continuum, then the sums have to be replaced by integrals and p_{λ_i} by a probability distribution for the initial states.
4. L. van Hove, Phys. Rev. **95** (1954) 249.
5. For comparison see the ladder operators for a harmonic oscillator.
6. Sears, V. F., Neutron News **3** (1992) 26.
7. For small $\mathbf{Q}$, $\mu = -\mu_B(\mathbf{L} + 2\mathbf{S})$.
8. We have used the property that the Fourier transform $\mathcal{F}(f_1(\mathbf{x}) \otimes f_2(\mathbf{x})) = \mathcal{F}(f_1(\mathbf{x})) \cdot \mathcal{F}(f_2(\mathbf{x}))$. Therefore, one can assume that f_1 describes the positions of the moments, $\delta(\mathbf{r} - \mathbf{r}_j)$, similar as for nuclear scattering, and f_2 describes the spacial extension of

the cloud of the unpaired electrons. The form factor for magnetic neutron scattering differs from the form factor for x-ray scattering. The former is due to the unpaired electrons and the latter is due to all electrons of an atom.

9. The elements of the Pauli spin matrix σ are defined as follows:

$$\sigma_x = \begin{pmatrix} 0 & 1 \\ 1 & 0 \end{pmatrix}, \sigma_y = \begin{pmatrix} 0 & -i \\ i & 0 \end{pmatrix}, \sigma_z = \begin{pmatrix} 1 & 0 \\ 0 & -1 \end{pmatrix}.$$

The polarisation of the neutron beam is given by $\mathbf{P} = 2\langle s \rangle = \langle \sigma \rangle$.

10. For reference see the excellent paper by R. Moon, T. Riste, and W. C. Koehler, Phys. Rev. **181** (1969) 920.

11. B. N. Brockhouse, Can. J. Phys. **33** (1955) 889.

12. P. Berastegui, P. Fischer, I. Bryntse, L.-G. Johansson, and A. W. Hewat, J. Solid State Chem. **127** (1996) 31.

13. T. R. Thurston, R. J. Birgeneau, D. R. Gabbe, H. P. Jenssen, P. J. Picone, N. W. Preyer, J. D. Axe, P. Böni, G. Shirane, M. Sato, K. Fukuda, and S. Shamato, Phys. Rev. B **39** (1989) 327.

14. B. Renker, F. Gompf, E. Gering, N. Nücker, D. Ewert, W. Reichardt, and H. Rietschel, Z. Phys. B: Cond. Matter **67** (1987) 15.

15. S. M. Hayden, G. Aeppli, R. Osborn, A. D. Taylor, T. G. Perring, S-W. Cheong and Z. Fisk, Phys. Rev. Lett. **25** (1991) 3622.

16. F. Fauth, U. Staub, M. Guillaume, J. Mesot, A. Furrer, P. Dosanjh, H. Zhou, and P. Vorderwisch, J. Phys.: Condens. Matter **7** (1995) 4215.

STRUCTURAL ANOMALIES, OXYGEN ORDERING AND SUPERCONDUCTIVITY IN $YBa_2Cu_3O_{6+x}$

P.G. RADAELLI[†]
Rutherford Appleton Laboratory
Chilton, Didcot, Oxfordshire
OX11 0QX, UNITED KINGDOM

1. Introduction

$YBa_2Cu_3O_{6+x}$ is perhaps the most heavily studied compound among the superconducting cuprates. Since its discovery in 1987[1], thousands of scientific papers, describing studies on this compound, have been published. One of the reasons is that this compound, due to the high T_c (93 K), the relatively simple synthesis route, and the very good intrinsic superconducting behavior in an applied field, has been viewed as being promising for practical applications. Also, $YBa_2Cu_3O_{6+x}$ became the aristotype of a family of compounds with the same basic structure, obtained by substitutions of one or more ionic species, which allowed the effect of ionic sizes, point and extended defects etc. to be explored in detail. From the structural point of view, one of the most interesting properties of this compound is its ability to support a large oxygen non-stoichiometry. In fact, its oxygen content can be varied from six (x=0) to seven (x=1) in a continuous way· The maximum critical temperature is achieved close to the highest oxygen content. The oxygen non-stoichiometry occurs in the so-called basal plane, which contains copper and oxygen atoms and acts as a "charge reservoir" for the CuO_2 double layer, where superconductivity primarily occurs. As new families of superconducting copper oxides were discovered over the years, it has become increasingly clear that, due to this specific structural features, $YBa_2Cu_3O_{6+x}$ is far from being a typical compound, and that it possesses a number of peculiarities, including the complex self-doping mechanism associated with oxygen ordering in the basal plane. In fact, the definition of a "typical" copper oxide superconductor is in itself questionable, as every system was found to possess peculiar features that appear to affect the superconducting properties in a significant way. Therefore, the search for fundamental correlations between the structural and superconducting properties of the copper oxide superconductors has not yet reached its objective. In particular, the factors determining the maximum critical temperature achievable in any given system by varying the electronic doping are far from being claified. Nevertheless, the detailed understanding of

---

[†] Formerly at the Institute Max von Laue - Paul Langevin, Grenoble (FRANCE).

the chemistry, physics and materials science that we now possess on the $YBa_2Cu_3O_{6+x}$-based family of compounds can be considered as one of the most significant achievement in this field in the last decade.

The contribution given by neutron techniques to the research on copper oxide superconductors in general and on $YBa_2Cu_3O_{6+x}$ in particular is of an impressive magnitude. Neutron studies, complemented by other techniques, are at the basis of most of what we know about the crystal structure, the oxygen doping mechanism, the lattice dynamics and the magnetic structure and dynamics of $YBa_2Cu_3O_{6+x}$. Some of the aspects related to the magnetic properties and to the magnetic and lattice excitations of this compound will be discussed in other chapters of this book. The purpose of the present chapter is to give an overview of the contribution given by neutron diffraction to the $YBa_2Cu_3O_{6+x}$ research. Reference will also be made to a number of other techniques, including electron and x-ray diffraction, which have played a very important complementary role in this field, as well as to some computational studies of oxygen ordering in $YBa_2Cu_3O_{6+x}$, since the phase diagrams obtained by these methods have been consistently compared to those obtained by neutron diffraction and other experimental techniques. Due to the enormous number of papers published on this subject in the last ten years, this review cannot be complete. Also, it is understandably rather difficult to work out priorities from the literature, based on submission or publication dates. We therefore apologize in advance for possible errors or omissions.

2. Structure Determination

The $YBa_2Cu_3O_{6+x}$ structure determination is in itself an interesting piece of scientific history, and it is worth summarizing in some detail. We shall, however, first describe the structure as we know it now, and then outline the work by which it was determined. The structures of $YBa_2Cu_3O_{6+x}$ are shown in Figure 2.1a ($x \geq 0.4$, orthorhombic) and Figure. 2.1b ($x \leq 0.4$, tetragonal).

Atom Labeling. For $YBa_2Cu_3O_{6+x}$, there is yet no universally accepted atom labeling scheme. The convention we adopt here, following Capponi et al.[2], is widely used, but we remark that an equally popular scheme exists, following Beno et al.[3], in which atoms O1 and O4 are interchanged. In both schemes, the labeling sequence is based on the orthorhombic structure (the first to be determined). In the tetragonal structure, the atoms O3 and O5 are missing, since they are equivalent by symmetry to O2 and O4, respectively.

Structural Building Blocks. The structure of $YBa_2Cu_3O_{6+x}$, like that of many other layered compounds, is best described in terms of 2-dimensional building blocks stacked on top of each other along one of the crystallographic axes (usually chosen to be the c axis). There are three main building blocks in the $YBa_2Cu_3O_{6+x}$ structure:

orthorhombic Pmmm   tetragonal P4/mmm

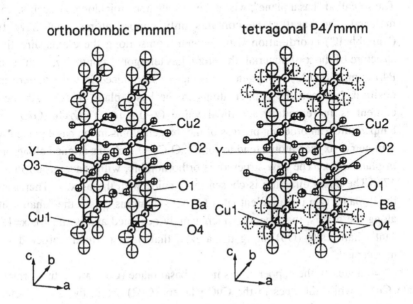

Figure 2.1. Crystal structures of the orthorhombic (left) and tetragonal (right) forms of YBa$_2$Cu$_3$O$_{6+x}$. The O5 (nearly empty) site of the orthorhombic structure (not shown) is between two Cu1 atoms along the a axis.

I).   The CuO$_2$ double-layer, with yttrium as a spacing ion. It is now known that 2-dimensional CuO$_2$ layers are the structural feature common to all copper oxide superconductors, and the structural element that supports superconductivity. The copper atoms in these layers, labeled Cu2, have 5-fold square-pyramidal coordination with oxygen atoms. Four of these 5 neighbors, known as *planar* oxygens (O2 and O3), are within these CuO$_2$ layers, while the fifth (*apical* oxygen O1) constitutes the vertex of the pyramid. The Cu2-O2/O3 bond lengths are in the range 1.92-1.96 Å, while the Cu2-O1 bond length is much longer (2.30-2.45 Å). One important structural feature of YBa$_2$Cu$_3$O$_{6+x}$, shared with a number of other copper oxide superconductors, is the fact that the CuO$_2$ are "puckered", meaning that the Cu2 atoms do not lie in the same plane with the oxygens, but are moved about 0.2 Å away from the Y-layer. This "puckering" has the effect that the Cu2-O2/O3-Cu2 liaisons are no longer straight, resulting in bond angles of ~ 164°. Yttrium has 8-fold coordination with oxygen, since no oxygen is contained in the Y layer.

II). The BaO layers (formed by Ba and the apical oxygen O1), that limit the double CuO_2 layer in the c-axis direction. Barium has coordination varying from 8-fold to 10-fold as x varies from zero to one.

II). The so-called "basal plane", where the oxygen non-stoichiometry occurs. In the x=0 material, the basal plane contains only copper atoms, that have two-fold ("dumbbell") coordination with oxygen atoms along the c-axis direction. The structure of the x=0 material, therefore, has tetragonal symmetry, with space group P4/mmm (No. 123). Oxygen atoms can be introduced in the structure in the 1b position, between the copper atoms in the basal plane (Cu1). As the oxygen content is increased to a critical value (x~0.4 for $YBa_2Cu_3O_{6+x}$ at room temperature), an ordering process of the oxygen atoms in the basal plane lowers the symmetry: oxygen atoms tend to form ...O-Cu-O... chains aligned along one of the in-plane axes. The new structure is orthorhombic, with space group Pmmm (No. 47)· The chain direction is chosen to coincide with the b-axis. Therefore, in the basal plane, two inequivalent sites for oxygen atoms exist: the "chain" site (O4), along the b-axis, that becomes increasingly occupied and is full for x=1, and the "anti-chain" site (O5) along the a-axis, that, in a perfectly ordered sample is unoccupied.

In the x=0 material, the copper atoms in the basal plane (Cu1) are in the formal valence state Cu^{+1}, while the ones in the CuO_2 layers (Cu2) are in the Cu^{+2} state. When oxygen is added to the basal plane, the largest change in oxidation state occurs for the Cu1 copper atoms adjacent to the added oxygen atoms, which take up the square planar coordination with 4 Cu-O bonds, characteristic of Cu oxidation states between +2 and +3. However, the charge carriers (holes) created by the oxidation process do not remain entirely confined to the basal plane, but are in part *transferred* to the CuO_2 layers. The basal plane, therefore, acts as a *charge reservoir* for the CuO_2 layers. This *charge transfer* process, indicated by a shortening (strengthening) of the Cu2-O bonds, constitutes the basic mechanism for electronic doping in $YBa_2Cu_3O_{6+x}$.

History of the $YBa_2Cu_3O_{6+x}$ structure determination. The first $YBa_2Cu_3O_{6+x}$ samples synthesized by Wu et al.[1] were polyphasic, and the stoichiometry of the superconducting compound was not correctly identified. Cava and coworkers[4] were the first to produce a reasonably single-phase sample of $YBa_2Cu_3O_{6+x}$ (x~0.9) and to present an x-ray powder diffraction pattern from it. Shortly afterwards, the first small single crystals (actually large grains from a powder heated near the melting point) became available, and a determination of the crystal structure of the oxidized material (obviously the most interesting one, since it was superconducting) was attempted. In fact, three paper describing single-crystal x-ray diffraction measurements on $YBa_2Cu_3O_{6+x}$ (x~0.9) appeared in the same issue of Physical Review B[5-7]. All three papers correctly described the metal ion positions, and the potential sites for oxygen, and recognized that no oxygen was present in the Y-layer. However, only one[6] established the correct space group Pmmm (the other two space groups were tetragonal), and none of them identified the presence of the ...O-Cu-O... chains in the basal plane. In retrospect, the reason of this inaccuracy is easy to understand: the crystals, as Hazen and coworkers suspected[5], were more or less heavily twinned, inducing averaging of non-equivalent reflections and, in the worst cases, a tetragonal pseudosymmetry. Therefore, the first correct structural solution for YBa$_2$Cu$_3$O$_{6+x}$ (x~1) was obtained using neutron powder diffraction (NPD), which is insensitive to macroscopic twinning. Between the end of March and the middle of May 1987, at least 7 groups submitted papers describing NPD measurements of oxygen-rich YBa$_2$Cu$_3$O$_{6+x}$ (references [2,3,8-12]). *All* of them correctly identified both the space group and the presence of ...O-Cu-O... chains, irrespectively of the fact that the measurement was performed on a high-flux reactor[2,12], medium flux reactor[9-11] or spallation source[3,8]. This success is now considered one of the major achievements of the NPD technique.

Structure of tetragonal YBa$_2$Cu$_3$O$_{6+x}$ (x ≈ 0.1). Early work has established that the oxygen stoichiometry of YBa$_2$Cu$_3$O$_{6+x}$ is not fixed, x being variable between ~0.10 and ~0.95 depending on the annealing conditions[13,14]. The oxygen-poor material was found to be metrically tetragonal by x-ray powder diffraction. The structure of tetragonal YBa$_2$Cu$_3$O$_{6+x}$ (x ≈ 0.1) was solved independently by Santoro and coworkers[15], Hewat and coworkers[16] and Jorgensen and coworkers [17], once again by NPD. It was readily established that, in the de-oxygenation process, oxygen was removed from the O4 basal plane site, which was found to be nearly but not exactly empty.

Temperature Dependence. The first low- and high-temperature neutron diffraction studies of YBa$_2$Cu$_3$O$_{6+x}$ were performed by the Grenoble group[2,16]. No structural phase transition was found to occur below room temperature, the only unusual feature in the 5 K data being the relatively large O4 Debye-Waller factors *perpendicular* to the direction of the chains (this issue will be the subject of further discussion: see section 6.). Hewat *at al.* heated *in situ* a sample of oxygen-rich YBa$_2$Cu$_3$O$_{6+x}$ (x≈1) in vacuum while taking NPD patterns at several temperatures. The O4 chain site was gradually depleted of oxygen with increasing temperature. Near 700 °C, an orthorhombic-to-tetragonal phase transition was detected. The sample started to decompose near 750 °C. After cool-down in vacuum, the authors determined the structure of the de-oxygenated tetragonal material at room temperature.

The Valence of Copper in YBa$_2$Cu$_3$O$_{6+x}$. The stoichiometry of YBa$_2$Cu$_3$O$_{6+x}$, as determined by thermogravimetric analysis (TGA), iodometric titration or other means, readily provided the *average* valence of copper in the compound: $<V_{Cu}>$ = 2.333 and 1.667 for x=1 and x=0, respectively. A more difficult problem was to determine the valence of copper for the two sites (Cu1 and Cu2) independently. These values can be obtained from crystallographic data by combining the information about Cu-O bond lengths and Cu coordination numbers. The fundamental concept was already proposed in the paper by Capponi and coworkers[2], who employed the so-called Zachariasen formula. A more precise determination of the Cu valences in oxygen-rich $YBa_2Cu_3O_{6+x}$ ($x\sim1$) was obtained by David and coworkers[8], who applied the so-called *bond valence sum* (BVS) method by Brown and Altermatt[18]. This method relies on assigning a "partial valence" V_b to each bond, which is exponentially related to the bond length r through the formula:

$$V_b = e^{\frac{r_0 - r}{B}}$$

(2.1)

where B is a universal parameter (= 0.37 Å), and the values of r_0 were determined experimentally and tabulated for a large number of ion pairs[18]. The sum of the "charges" over all the bonds around one atom is taken to correspond to the valence of that atom. David and coworkers concluded that Cu^{+3} preferentially occupies the chain Cu1 site: the Cu^{+3} : Cu^{+2} ratio was found to be of the order of 70(15):30(15) ($V_{Cu1} = 2.70(15)$), the formal valence of the planar Cu2 being $V_{Cu2} \approx 2.15$. Another important albeit qualitative observation was made by Santoro and coworkers about the de-oxygenated $YBa_2Cu_3O_6$ material: the Cu1 "dumbbell" coordination in this compound is similar to the one found in copper (I) oxide (Cuprite,Cu_2O) and in delafossites $Cu^{+1}M^{+3}O_2$ (M = La, Y, Ga, Al, Fe, Co, Cr), where copper is known to be monovalent. This observation strongly suggested that the formula of $YBa_2Cu_3O_6$ could be written as $YBa_2Cu^{+1}(Cu^{+2})_2O_6$. In the course of their *in situ* study (see previous paragraph)[16], Hewat and coworkers calculated the valences of the two copper sites using the previously described method by Brown and Altermatt[18]. They also found that the Cu1 valence was the one most affected by the de-oxygenation process, while the Cu2 valence varied only slightly (within their error bars) and always stayed above 2. These very early results contained *in nuce* the essential elements of the relationship between oxygen uptake and electronic doping in $YBa_2Cu_3O_{6+x}$: the main effect of the oxygen uptake is to oxidize the basal plane copper Cu1 from Cu^{+1} to Cu^{+3}. However, this oxidation process is not complete, since a fraction of the charge is *transferred* to the CuO_2 layers, increasing the Cu2 valence from 2 to ~2.15.

Structure-Property Relations. The peculiar structural features of $YBa_2Cu_3O_{6+x}$ generated enormous interest in connection with its superconducting properties. In fact, the simultaneous occurrence of 2-dimensional (the CuO_2 planes) and 1-dimensional (the ...O-Cu-O... chains) structural elements in the same structure are quite unusual, and so is the possibility of accommodating a large oxygen non-stoichiometry. In 1987, only two structural types ($La_{2-x}(Sr,Ba)_xCuO_{4+\delta}$ and $YBa_2Cu_3O_{6+x}$) were known to support superconductivity. The much higher value of T_c in $YBa_2Cu_3O_{6+x}$ was immediately correlated with the presence of chains, and the destruction of superconductivity in de-oxygenated sample (where ...O-Cu-O... chains are absent) further supported the hypothesis that electrical conduction may occur in the chains themselves[13], rather than in the CuO$_2$ layers. We now know that this hypothesis was not correct: the ...O-Cu-O... chains do participate in the electrical conductivity of YBa$_2$Cu$_3$O$_{6+x}$ (reference [19]), and do become superconducting by proximity effect, but the main structural element supporting superconductivity are the CuO$_2$ layers. As we shall see in the remainder, the primary role of the ...O-Cu-O... chains is to provide electronic doping, in the form of holes, to the CuO$_2$ layers.

3. Structural Properties as a Function of Oxygen Content

3.1. OXYGEN STOICHIOMETRY

Stoichiometry Determination by Physical and Chemical Methods.
The early work has established the extreme sensitivity of the YBa$_2$Cu$_3$O$_{6+x}$ stoichiometry and structure to the annealing atmosphere and the need of an oxidizing environment to obtain superconducting samples. For instance, Tarascon and coworkers established that the resistivity of samples annealed in vacuum was semiconducting in character, and did not show any trace of a superconducting transion[13]. These observations prompted a number of systematic studies (see for instance [14,20]), having several different goals:

1) Determine the exact range of existence of the YBa$_2$Cu$_3$O$_{6+x}$ structure as a function of x.
2) Study the evolution of x as a function of temperature in several different atmospheres.
3) Establish reliable and reproducible routes to produce YBa$_2$Cu$_3$O$_{6+x}$ samples with different values of x, for further studies.
4) Synthesize samples having optimal superconducting properties (sharp transitions and high Meissner fractions).
5) Determine the boundary between of the tetragonal and orthorhombic phases and the nature of the phase transition.

In these studies, the oxygen stoichiometry was typically varied by heating samples of known stoichiometry (previously determined by iodometric titration[20] or hydrogen reduction[14]) in a TGA apparatus, while monitoring the weight loss at the same time to determine the value of Δx (with respect to the initial state). Once the desired stoichiometry was attained, the samples were cooled to room temperature more or less rapidly in an inert atmosphere (to prevent oxygen pick-up). The samples were further characterized by x-ray diffraction and measurements of the superconducting properties.

As far as point 1) is concerned, it was established that YBa$_2$Cu$_3$O$_{6+x}$ existed for $0+\delta_1 \leq x \leq 1-\delta_2$, with $\delta_1 \approx 0.1$, $\delta_2 \approx 0.05$. Annealing in very low oxygen partial pressure (PO$_2$) and high temperatures, aimed at decreasing the oxygen stoichiometry below 6.0, resulted instead in decomposition[14]. Likewise, $YBa_2Cu_3O_{6+x}$ decomposed upon treatment in high PO_2 (10-5000 bars at 500°C), whereas annealing in a few bars of pure oxygen did not increase the oxygen content above 7[14].

For point 2) and 3), Kishio and coworkers performed a systematic study of the $YBa_2Cu_3O_{6+x}$ oxygen stoichiometry as a function of temperature and PO_2, by heating the samples in a TGA apparatus in flowing gasses containing different concentrations of oxygen (Figure 3.1) Samples with different values of x were recovered after quenching, allowing the authors to establish the remarkable effect of the oxygen stoichiometry on transport properties: as x decreased, the electrical resistivity increased and the value of T_c decreased. The samples with the best superconducting properties ($T_c = 92$ K and a sharp transitions) were obtained by slow-cooling the samples in oxygen between 500 °C and 400°C, since the oxygen uptake kinetics are very slow below these temperatures. Regarding point 5), a tetragonal-orthorhombic (T-O) phase boundary was found to be present both for samples studied *in situ* in various atmospheres[14,16] and for samples with variable x recovered after cooling or quenching and studied at room temperature. In the first case, the T-O transition temperature was found to *increase* as a function of increasing PO_2. For the recovered samples, the T-O phase transition was found to occur for $x \approx 0.4$-0.5. However, some of these samples were found to be mixtures of tetragonal and orthorhombic phases. The phase coexistence was probably due to oxygen inhomogeneities generated during the quenching process, and somewhat obscured the real nature of the T-O phase transition. Furthermore, the superconducting transitions for these samples were found to be broad, making it difficult to determine T_c exactly.

Oxygen Deficiency and Thermodynamic Equilibrium. The problems just outlined obviously had to do with the difficulty of preparing oxygen-deficient samples in true thermodynamic equilibrium. The isotherms in Kishio's phase diagram[20] (Figure 3.1) clearly illustrate the nature of the problem: the 450 °C isotherm barely reaches x=0.7 even in $PO_2 = 10^{-4}$, a pressure where it is already quite difficult to control concentration fluctuations. To reach x-values near the O-T transition, one needs to heat to higher temperatures (e.g., at 650 °C in 1% oxygen) and then quench. The quenching process needs to be very rapid in order to prevent further oxygen pick-up.

In order to avoid these problems, and to prepare oxygen-deficient samples thermodynamically equilibrated down to room temperatures, a number of alternative techniques were explored. Among them, it is worth mentioning near-room-temperature plasma oxidation[21] and zirconium gettering[22,23]. The latter technique was employed by Cava and coworkers[22,23] to synthesize a series of $YBa_2Cu_3O_{6+x}$ with variable oxygen content. This technique consist in sealing samples of oxygen-rich $YBa_2Cu_3O_{6+x}$ ($x \sim 1$) in quartz ampoules together with a variable amount of degassed Zr foil and annealing them for 48 h at temperatures between 360 °C and 520 °C. The amount of oxygen removed from the samples, determined *a posteriori* by TGA, depends both on the annealing temperatures and on the amount of Zr foil: the minimum value of x reached in this study was x=0.30. In any case, the oxygen-deficient samples were more

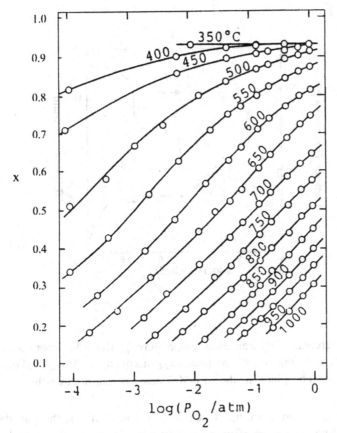

Figure 3.1. Excess oxygen concentration x in YBa$_2$Cu$_3$O$_{6+x}$ as a function of PO$_2$ at different temperatures, from reference [20].

homogeneous than previously prepared samples, as proven by the relatively sharp transition temperatures. Surprisingly, besides showing an improved homogeneity, this series of samples displayed structural and superconducting properties quite different from those of previous studies. In fact, all samples were found to be orthorhombic and superconducting, even the one with x=0.30, probably due to extended long-range ordering of the orthorhombic domains. Furthermore, T_c displayed pronounced "plateaus" at 60 K and 90 K for x≈0.5 and x ≈ 1, respectively, while the superconducting transition width and the room-temperature resistivity showed minima at the same values of x (Figure 3.2). These features, which first indicted the presence of intermediate phases between x=0 and x=1, have subsequently been reproduced with various techniques. Cava's work was therefore extremely important, because it anticipated some of the main issues to be discussed in the following years. In particular:

1) The "single-phase" nature of the 60 K phase. In many ways, the YBa$_2$Cu$_3$O$_{6+x}$ behaved as if only two well-distinct superconducting phases were present: the 90 K and

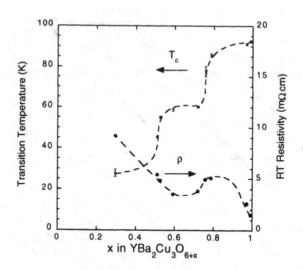

Figure 3.2. Superconducting transition temperature T_C (left) and room-temperature resistivity (right) as a function of the total oxygen content in $YBa_2Cu_3O_{6+x}$, from reference [22]. Vertical bars indicate the 10%-90% resistive transition widths.

the 60 K phases ($x \approx 1$ and $x \approx 0.6$). Outside these compositions, the transitions are broad, and could be suggestive of phase coexistence. However, this model was not corroborated by the diffraction evidence, since the samples appeared to be single-phase for all values of x. Cava suggests that the two "phases" may be the result of short-range oxygen ordering in the form of superstructures made of alternating full and empty chains, a speculation that was fully confirmed later on.

2) The difference between equilibrated and quenched samples. Cava suggests that this difference could go beyond the homogeneity issue, and proposed that the extended "plateau" of his series of samples could result from improved short-range ordering due to the low-temperature annealing, and was, therefore, dependent on sample history even at constant x. This second hypothesis was also to be later confirmed.

The Nature of the T-O Phase Transition: Neutron Diffraction Studies. Up to this point, the relationship between the tetragonal and orthorhombic phases had not yet been fully explored. On purely group-theoretical ground, it was known that the space group of the orthorhombic phase (Pmmm) is a subgroup of the one for the tetragonal phase (P4/mmm), and, in addition, one for which a second-order phase transition was possible. Furthermore, it was understood how the symmetry lowering could occur for the specific case of YBa$_2$Cu$_3$O$_{6+x}$: all occupied sites had the same multiplicity in the two structures, except for O2/O3 and O4/O5. O2, which has 4-fold multiplicity in the tetragonal phase, splits into O2 and O3 at the phase transition, both sites having 2-fold multiplicity in the orthorhombic phase. Likewise, O4 has 2-fold multiplicity in the tetragonal phase, and splits into O4 and O5 (1-fold) in the orthorhombic phase. No atomic displacement is involved in the phase transition except for those associated with the metric distortion. However, the actual evolution of the occupancies of these sites at the phase transition had not yet been determined. The first studies of these issues were once again performed by NPD, which is ideally suited to determine oxygen site occupancies with great accuracy. Interestingly, the T-O phase transition was initially studied as a function of temperature, not as a function of x, due to the previously discussed difficulty to synthesize homogeneous oxygen-deficient YBa$_2$Cu$_3$O$_{6+x}$ samples (systematic studied of oxygen-deficient samples by NPD were performed later on, and will be discussed in Section 3.2). The *in situ* study by Hewat and coworkers[16] had already determined the presence of a high-temperature T-O phase transition and studied it by neutron diffraction. However, this first study was performed under dynamic vacuum, due to the need to protect the vanadium heating elements of the furnace from oxidation. Under these conditions, YBa$_2$Cu$_3$O$_{6+x}$ is not in true thermodynamic equilibrium, and the de-oxygenation process is dominated by kinetics. Jorgensen and coworkers performed a detailed *in situ* study of YBa$_2$Cu$_3$O$_{6+x}$ as a function of temperature and PO$_2$ (reference [17]). The data were collected using the Special Environment Powder Diffractometer at the Intense Pulsed Neutron Source at Argonne National Laboratory. Jorgensen used a different furnace design, which took advantage of the fixed-angle geometry of the time-of-flight technique and did not require the use of heating elements with small coherent cross section for neutrons. This type of furnace can work in oxidizing environment, and allowed data to be collected for 3 values of PO$_2$ (100%, 20% and 2%) at temperatures between 25 °C and 900 °C. At every temperature, the sample was allowed to equilibrate for 30 minutes before data collection began. Furthermore, the structural parameters were monitored as a function of time, and found to be constant except for the 2% PO$_2$ run at 321 °C. For all the other runs, it was estimated that the sample was close to thermodynamic equilibrium conditions. The T-O phase transition was found to occur at different temperatures, depending on PO$_2$ (700 °C in 100% oxygen, 670 °C in 20% oxygen and 620 ° in 2% oxygen), but always at values of x close to 0.5 (x_{T-O} was determined by Rietveld refinement of the NPD data). Furthermore, the fractional site occupancy of the two chain sites O4 and O5 was found to vary continuously through the phase transition (Figure 3.3). As previously mentioned, for the fully oxygenated material at room temperature the O4 and O5 occupancies are close to 1 and 0, respectively. On heating above ~350 °C, the O4 occupancy f(O4) starts to decrease, indicating an overall loss of oxygen from the sample. Above 600 °C, however, f(O5) starts to *increase* and, at the T-O phase transition, f(O4) = f(O5) $\approx$ 0.25. This important observation clearly indicated that the T-O phase transition, at least at these high temperatures, is second-order, and can be classified as an order-disorder phase transition

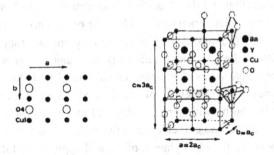

Figure 3.4. The ab plane (**Left**) and complete crystal structure (**Right**) of the fully-ordered Ortho-II phase (x=0.5), showing the arrangement of full and empty chains, from reference [27].

Oxygen-Ordered Superstructures by Electron Diffraction. Transmission electron microscopy (TEM) and electron diffraction (ED) have been widely employed to study YBa$_2$Cu$_3$O$_{6+x}$, starting immediately after its discovery. The contribution given by these techniques to our present knowledge of this compound stems from understanding its microstructure (e.g., twinning patterns in the orthorhombic phases) to unveiling some of the most subtle details of short-range oxygen ordering, and is far too broad to be summarized here. We are therefore compelled to focus only on perhaps the most significant aspect of the TEM-ED work, that is the discovery of the oxygen-ordered superstructures. In the spring of 1987, Van Tendeloo and collaborators performed a series of studies on YBa$_2$Cu$_3$O$_{6+x}$ samples[26]. The oxygen content was initially close to 7, and was locally changed *in situ* by beam heating the crystallites. The initial aim of Van Tendeloo's work was to confirm the oxygen-ordering into chains in the basal plane, and to follow it through the T-O phase transition. However, in certain parts of the crystallites, the ED patterns evidenced the presence of weak and elongated "streaks" centered around the 1/2 0 0 and 0 1/2 0 reciprocal lattice points. This was indicative of a local doubling of the a axis (with unit cell 2a × b × c), which was poorly correlated both in the a-b plane and along the c axis. The doubling could also be evidenced in TEM images, and can be considered the first observation of what is now called the "Ortho-II" phase (as opposed to the Ortho-I phase, where the O4 occupancy is not spatially modulated). However, the technique employed in this early work gave no control of the actual oxygen content of the sample, which is crucial for the formulation of a correct model of oxygen ordering associated with the superstructure.

A few months later, Chaillout[27] and coworkers reached the right solution by studying YBa$_2$Cu$_3$O$_{6+x}$ samples with controlled oxygen content. The most intense superlattice reflections were observed close to x=0.6, strongly suggesting that the superstructure resulted from a sequence of full ...Cu-O4-Cu... and ...Cu-[ ]4-Cu... empty chains ([ ]4 = O4 vacancy) along the a axis (Figure 3.4). A perfect arrangement of ...full-empty-full... chains would correspond to a stoichiometry of x=0.5. The persistence of this type of superstructure for other stoichiometries was attributed by the authors to partial occupancy of either the O5 or the [ ]4 sites. This work was later on extended by the same group[28] to a more complete series of samples: in addition to the Ortho-II phase, it was possible to evidence the presence of additional superstructures, with $2\sqrt{2}a \times 2\sqrt{2}b \times c$, occurring both at higher (x~0.75) and at lower (x~0.25) oxygen contents. Chaillout *et al.* noticed the coincidence between the region of x where the Ortho-II phase was observed and the 60 K plateau of T_c, and hypothesized that the two might be correlated: in this scenario, the Ortho-II phase would have a well-defined T_c of ~60 K.

3.2. SYSTEMATIC STUDIES OF OXYGEN-DEFICIENT SAMPLES AT ROOM TEMPERATURE

Synthesis Issues. The previously cited work by Cava *at al.*[22] has focused the attention of the community on the importance of synthesizing samples under low-temperature equilibrium conditions, and has indicated the sharpness of the superconducting transition and the presence of two plateaus in the T_c vs. x curve as useful gauges of this equilibrium. This work prompted an intense research activity aimed at optimizing the synthesis route for oxygen-deficient samples with methods other than the Zr-gettering techniques. One of the goals was to synthesize in a reliable and reproducible way homogeneous samples in the relatively large amounts (1-5 g) required to obtain high-quality neutron diffraction data. Different groups adopted different strategies toward this objective. Cava and coworkers continued to employ the Zr-gettering technique[29-31]. Other low-temperature methods (annealing temperature below 500 °) included equilibration of oxygen rich and oxygen-deficient samples sealed together in the same ampoule[32] and vacuum-extraction of oxygen at 400 °C[33]. The Argonne group prepared samples by annealing in different PO_2's at a slightly higher temperature (520 °C), followed by quenching into liquid nitrogen[34]. All these methods yielded series of samples with sharp transition temperatures and displaying extended plateaus, similar to the ones previously presented by Cava. Farneth and coworkers made a direct comparison between samples prepared by vacuum extraction and samples quenched from high temperatures, by using x-ray diffraction and magnetic measurements[33]. For any given value of x, high-temperature and low-temperature samples were indistinguishable by x-ray diffraction, but had markedly different magnetic properties in the region 0.4≤x≤0.8, the low-temperature samples having higher T_c and a more pronounced plateau.

In the course of these studies, an intriguing correlation was observed between the observation of superconductivity and the crystallographic symmetry of the compound.

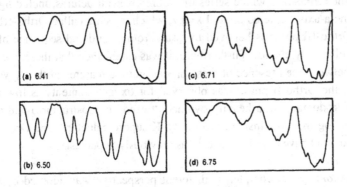

Figure 3.5. Electron diffraction pattern line traces along the a* direction, showing fundamental and superlattice reflections for various oxygen ordered superstructures in selected YBa$_2$Cu$_3$O$_{6+x}$ samples; from reference [35].

Obviously, for most sample, the appearance of superconductivity occurred for an oxygen content x_c in the vicinity of the T-O phase transition (x_{T-O}). Furthermore, all superconducting samples were found to be orthorhombic. Some authors went as far as speculating that the "..overall orthorhombic symmetry was crucial to any mechanism for superconductivity in this system"[33]. We now know that this is not true, either in general for copper oxide superconductors or in particular for the YBa$_2$Cu$_3$O$_{6+x}$ structure type. In fact, as we shall see, tetragonal superconducting samples can be made by appropriate doping.

Another very interesting method for preparing oxygen-deficient samples was proposed by Beyers and coworkers[3]: it made use of a solid-state ionic cell for coulometric titration[36]. The active element of this cell is a ceramic tube made of a fast oxygen conductor (yttrium-stabilized zirconia, YSZ), coated on the outside and the inside with porous platinum electrodes. The tube is inserted in a quartz chamber containing the sample, and held at fixed temperature (~850 °C). The sample temperature can be varied independently by means of a two-zone furnace. The inside of the YSZ tube is in a reference PO$_2$ (usually air), while the outside sees the same PO$_2$ as the sample. The open-circuit electromotive force between the two Pt electrodes is a direct measurement of PO$_2$ inside the quartz chamber. Furthermore, any flow of charge between the two electrodes can be mapped (in a 2:1 ratio) to a flow of oxygen atoms from the air to the chamber. In this way, the cell acts at the same time very precisely as a PO$_2$ gauge and as an oxygen pump. This method was extensively used by the authors to study the high-temperature phase diagram of YBa$_2$Cu$_3$O$_{6+x}$ and related phases, but was also suitable to prepare series of oxygen-deficient samples with very precisely determined values of x. Furthermore, once the desired stoichiometry was established, the samples could be cooled down to low temperatures without further oxygen uptake, allowing an excellent equilibration to be attained. Beyers studied these samples by TEM-ED, and observed, in addition to the Ortho-II, a whole series of ordered superstructures, including one with *tripling* of the a-axis, centered around x=0.7, which is being called Ortho-III[35](Figure 3.5). The Ortho-III phase is believed to result from an ordered sequence of full-full-empty-full-full-empty chains. Furthermore, it was established that there was no *direct* correlation between the presence of these phases and the features of the T_c vs. x curve: for instance, the Ortho-II phase was observed for oxygen contents as low as x=0.28 (albeit in microdomains in a heavily-twinned "tweed" microstructure) and as high as x=0.65, spanning from the insulator to the 60 K superconductor. The Ortho-II phase, therefore, could not have a well-defined T_c as previously hypothesized.

The c-axis Anomaly. A completely different perspective was adopted by Cava and coworkers by studying the structural properties of samples prepared by the Zr-gettering technique[29-31]. First of all, Cava determined that the exact annealing temperature before quenching (in the range 415 °C - 470 °C) influenced both structural and superconducting properties. In particular, non-superconducting orthorhombic samples could be prepared by annealing at the lowest temperatures, leading Cava to conclude that no correlation existed between the crystallographic symmetry and the occurrence of superconductivity. The value of x where superconductivity first appeared, x_{onset}, also varied for different annealing temperatures. The most striking feature of these series of samples is the evolution of the c lattice parameter as a function of x. In every case, the c axis displayed a step-like discontinuity in the vicinity of the T-O phase transition (Figure 3.6), although the exact value of x where the discontinuity occurs is not fixed, but varies depending on the annealing temperature. Based on preliminary laboratory x-ray powder diffraction data, Cava and coworkers speculated that this c-axis anomaly was associated with an electronic transition, whereby electronic charge would be transferred between chains and planes[29,30]. This hypothesis was later confirmed by an extensive study performed using the D2B high-resolution powder diffractometer at the Institut Laue-Langevin (Grenoble, France)[31]. The c-axis anomaly was found to be caused by a sudden *shortening* of the Cu2-O1 bond length with increasing x. The bond valence sum technique confirmed that this shortening was associated with an *increase* of the positive charge on the planar Cu2 atoms (~0.05 holes/Cu atom). Interestingly, the behavior of the Cu2 bond valence sum vs. x closely mimicked the 2-plateau structure of the T_c vs. x curve, strongly suggesting that the second plateau may me also due to the same mechanism: the T_c increase from 60 K to 92 K at x~0.7 would be associated with a further transfer of ~0.03 holes/Cu atom from the chains to the planes (Figure 3.7(a)). Cava suggests that the sudden charge transfer occurring at the onset of superconductivity is associated with the appearance of the Ortho-II phase, which would transfer holes to the CuO2 planes much more efficiently that the tetragonal phase. Cava also concluded that the Ortho-II phase would not occur at a specific value of x, but rather in a range of values roughly centered around x=0.5, and would also display a range of T_c's. The impact of this work on subsequent literature is very significant, but not completely uncontroversial. The charge transfer mechanism proposed by Cava is now universally

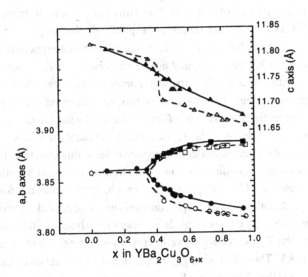

Figure 3.6. Lattice parameters as a function of total oxygen content in $YBa_2Cu_3O_{6+x}$, as determined by Cava et al. [31] (open symbols and dotted lines) and by Jorgensen, [34] (filled symbols and solid lines). Note how Cava's data evidence a step-like c-axis anomaly, which is absent in Jorgensen's data.

accepted to explain the onset of superconductivity in $YBa_2Cu_3O_{6+x}$. However, the structural properties of Cava's samples, and in particular the presence of a c-axis anomaly, cannot easily be reproduced by other techniques. Furthermore, it became increasingly clear that low-temperature annealing is not *per se* sufficient to produce samples with comparable *structural* properties. For instance, the method proposed by Jacobson (equilibration of oxygen-rich and oxygen-deficient samples sealed together in the same ampoule[32]) allows equilibration down to very low temperatures (see for instance [37]), but does not produce any c-axis anomaly. The typical behavior is the one described by Jorgensen and coworkers in their *in situ* study[17] (see also next paragraph), with the c axis displaying a change of slope. The absence of a c-axis anomaly would be reflected in a smoother increase of the hole concentration on the CuO_2 planes versus x, as measured, for instance, by bond valence sums.

On the contrary, most samples equilibrated at low temperatures display *superconducting* properties similar to Cava's, with a sharp onset of superconductivity and a broad first plateau, although the exact value of x_{onset} may vary. It is therefore clear that the Zr-gettering techniques produces somehow unique samples. In fact, Cava and coworkers already recognized this fact, by stating that the Zr-gettered samples do not correspond to any state of true thermodynamic equilibrium, but rather to quenching of metastable states. What this state exactly is has not completely been determined. One can only speculate that the fact that oxygen is mainly de-intercalated (and not vice versa) for Zr-gettered samples may play a role. The fact remains that the charge-transfer process, as established by Cava is a fundamental aspect of the doping mechanism in $YBa_2Cu_3O_{6+x}$, but one is forced to conclude that the correlation between T_c and bond valence sums shown in Figure 3.7(a) may not be universal. For Zr-gettered samples, the doping process occurs in a much narrower composition range than for other low-temperature techniques, but this has only a modest effect on the superconducting properties: superconductivity tends to be observed for somewhat lower values of x for Zr-gettered samples, but the shape of the T_c vs. x curve is notably the same. In the light of what we now know about superconducting copper oxides, this should not be surprising. In fact, the onset of superconductivity as a function of chemical substitution is very sudden even for materials, like $La_{2-y}Sr_yCuO_4$, for which the electronic doping is simply proportional to y. In other words, superconductivity is observed as soon as the hole concentration on the CuO_2 planes exceeds a critical threshold of ~0.05 holes/Cu atom, while the slope of the T_c vs. doping curve is very high at the onset of superconductivity (see also Section 8.) Therefore, changes in the doping mechanisms mainly affect the *value* of x_{onset} rather the *shape* of the T_c vs. x curve.

Structural Properties of Oxygen-Deficient $YBa_2Cu_3O_{6+x}$ by Neutron Powder Diffraction. In 1990, Jorgensen and coworkers published a detailed study of the structural properties of a series of $YBa_2Cu_3O_{6+x}$ samples, prepared by quenching from moderate temperatures (~520 °C)[34]. This extended work, complete with tables of crystallographic parameters and figures, is one of the most comprehensive to date. Furthermore, it was performed on a series of samples that were previously established to have attained thermodynamic equilibrium, as measured by the constancy of the lattice parameters as a function of time.

The evolution of the lattice parameters and of the O4 and O5 occupancies was found to be qualitatively similar to the one previously determined *in situ* as a function of temperature[17], and to be indicative of a second-order T-O phase transition occurring at $x \approx 0.4$. The "c-axis anomaly" manifests itself as a simple change of slope around x=0.4 (Figure 3.6). Jorgensen and coworker also focus their attention on the bond lengths along the c axis (Cu1-O1, Cu2-O1, Cu2-Cu2) and on the z fractional coordinate of Ba , since these parameters are the most sensitive to the electronic charge on the CuO_2 layers. The *slope* of all these parameters seems to change abruptly *at* the T-O phase transition: for instance, the copper-apical oxygen bond lengths are much more strongly dependent on x in the *orthorhombic* phase than in the tetragonal phase (Figure 3.7(b)). However, there is no indication of step-like discontinuities either in the c lattice parameter or in the internal parameters. Jorgensen and coworkers did not explicitly calculate bond valence sums (they can be easily obtained from their structural parameters). However, as shown later by Tallon[38], it is clear that, in the absence of c-

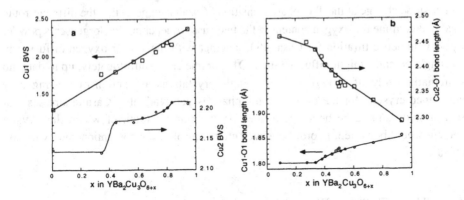

Figure 3.7. (a) Bond valence sums for the chain (Cu1) and plain (Cu2) copper atoms as determined by Cava[31]. Notice how the latter mimics the behavior of T_c, shown in *Figure 3.2.* (b) Copper-apical oxygen bond lengths as a function of x., as determined by Jorgensen *et al.* [34].

axis anomalies, the correlation between the T_c vs. x and BVS(Cu2) vs. x curves, as shown by Cava, is not present for Jorgensen's samples.

Single-Crystal Studies. Up to the end of 1990, most of the structural determinations on YBa$_2$Cu$_3$O$_{6+x}$ were performed by NPD. The reason, as we have already discussed, was the unavailability of twin-free single crystals of orthorhombic YBa$_2$Cu$_3$O$_{6+x}$ on one hand, and the difficulty of studying twinned crystals on the other. One of the few exceptions is represented by the single-crystal neutron diffraction work of McIntyre *et al.*, published in 1988[39]. McIntyre measured a series of 2-dimensional sections of reciprocal space by means of the position-sensitive detector of the instrument D19 at the ILL. For many reflections, the splitting due to crystal twinning was readily observable, and the peak multiplets could be fit by 2-dimensional Gaussians, consistent with a model accounting for microscopic and macroscopic twinning. A full structural refinement could also be obtained, albeit with some constraints made necessary by the pseudo-tetragonal symmetry induced by the twinning. After 1989, the first mechanically de-twinned orthorhombic YBa$_2$Cu$_3$O$_{6+x}$ became available. The de-twinning procedure is accomplished by placing the crystal under uniaxial stress along the a/b direction, and annealing it at ~450 °C[40]. Sullivan and coworkers studied at room temperature a de-twinned YBa$_2$Cu$_3$O$_{6+x}$ crystal with x=0.877 by x-ray diffraction, and carefully determined Debye-Waller ellipsoids and Fourier electronic density maps[41]. More recently, Casalta and coworkers performed a systematic study of YBa$_2$Cu$_3$O$_{6+x}$ single crystals as a function of x, by using both twinned and untwinned crystals in the orthorhombic phase[42]. Most of the structural parameters are in agreement with those determined by NPD[34]. One exception is the z coordinate of the apical oxygen O1 (z(O1)), which follows a smoother behavior for the single-crystal work, without the abrupt slope change at the T-O phase transition. Casalta suggests that the different route used to determine the oxygen content in the two cases (quenching for Jorgensen's powder work, volumetric titration for Casalta's), together with the slower oxygen diffusion in the single crystals, might influence the z(O1) parameter. Unfortunately, up to date, no systematic study of $YBa_2Cu_3O_{6+x}$ single crystals as a function of x using only untwinned crystals for the orthorhombic phase is yet available. Casalta stresses that such a study should be better performed on the *same* crystal, of which the oxygen content would be varied, in order to have better control of the absorption and extinction effects.

4. The x-T Phase Diagram

4.1. THERMODYNAMIC STUDIES

As we have seen in the previous paragraphs, since the discovery of the compound a very significant amount of experimental work has been devoted to determining the properties of $YBa_2Cu_3O_{6+x}$ as a function of x, both at ambient conditions and as a function of temperature and PO_2. Very early on the need was felt to unify all these diverse observations in a single phase diagram, on which the experimentally observed properties would be mapped out. This phase diagram would serve as a starting point to develop modeling theories, the ultimate goal being the correlations between oxygen ordering and transport properties. The very simple thermodynamic considerations needed for the basic understanding of these phase diagrams will be enunciated in the next few paragraphs.

Thermodynamic Phase Equilibria From the thermodynamic point of view, the complex interactions between oxygen molecules in the atmosphere, oxygen atoms in the basal plane and the rest of the $YBa_2Cu_3O_{6+x}$ crystal structure can be described in terms of a very simple two-component system. The two components are: 1) $YBa_2Cu_3O_6$ (i.e., all the metal ions plus all the oxygen ions except O4 and O5) and 2) Oxygen atoms in the lattice in equilibrium with oxygen molecules in the gas. This 2-component system can give rise to a variety of phases, the most obvious being $YBa_2Cu_3O_6$, a solid phase that contains only the first component, and gaseous oxygen, which contains only the second one. Elementary thermodynamics states that, in a multi-component system with c components, for any number p of coexisting phases in equilibrium with each other, the temperature, total pressure and chemical potentials μ of all the components must be the same:

$$T_I = T_{II} = \cdots = T_p$$

$$P_I = P_{II} = \cdots = P_p$$

$$\mu_I^1 = \mu_{II}^1 = \cdots = \mu_p^1 \tag{4.1}$$

$$\vdots$$

$$\mu_I^c = \mu_{II}^c = \cdots = \mu_n^c$$

This constitutes a system of (p-1)·c equations. Since the thermodynamic quantities of each phase are completely determined by temperature, total pressure and concentration of each component, the total number of variables in (4.1) is p·(c-1)+2. It follows that the number of degrees of freedom f is f = c+2-p, which is the well-known Gibbs' phase rule. In the specific case of YBa$_2$Cu$_3$O$_{6+x}$, c = 2 and p ≥ 2, hence f ≤ 2. Therefore, the phase relationships in this system can be described in terms of a *binary* phase diagram. Usually, the variables are chosen to be x (i.e., the oxygen concentration in the lattice phase(s) for the abscissa and T for the ordinate. We note that, if two or more solid phases coexist with the gas, the oxygen concentrations in the various phases will in general *not* be the same. Therefore, phase coexistence regions will appear as areas of the x-T phase diagram where the system is not allowed to be homogeneous (miscibility gaps). An exception is provided by second-order phase transitions: in this case, the coexisting phases are indistinguishable at equilibrium, and no miscibility gap will be present. We also notice that, if we consider oxygen as an ideal gas, the chemical potential can be obtained from the equation of state:

$$\mu_{gas}^O = \frac{1}{2}\mu_{gas}^{O_2} = \frac{1}{2}\left[RT\,ln\,PO_2 + \chi(T)\right] \tag{4.2}$$

where $\chi(T)$ is a known function that depends only on temperature. Equation (4.2) allows the changes of the thermodynamic quantities upon oxidation to be deduced from the knowledge of the PO$_2$ vs. x curves. In fact, in a generic oxidation process:

$$MO_{x-1} + \frac{1}{2}O_2 \Rightarrow MO_x \tag{4.3}$$

we know that the total free energy change must be zero:

$$\Delta G = G_{MO_x} - G_{MO_{x-1}} - \frac{1}{2}\left[RT\,ln\,PO_2 + \chi(T)\right] =$$
$$= \Delta G_0 - \frac{1}{2}RT\,ln\,PO_2 = 0 \tag{4.4}$$

where ΔG_0 is the oxidation free energy at standard conditions (1 Atm). From (4.4) we can also obtain the entropy and enthalpy of oxidation:

$$\Delta S_0 = -\frac{\partial}{\partial T}\Delta G_0 = \frac{1}{2}R\left[ln\,PO_2 + \frac{T}{PO_2}\frac{\partial}{\partial T}(PO_2)\right] \tag{4.5}$$

$$\Delta H_0 = \Delta G_0 + T\Delta S_0$$

where the partial derivative is taken at constant oxygen content.

Determination of PO₂ vs. x. We have already cited the early work by Kishio and coworkers[20], in which a set of PO_2 vs. x curves were determined by heating the samples in a TGA apparatus followed by quenching. A similar technique was employed later on by Specht and collaborators[25]. Tetenbaum *et al.* performed a series of *in situ* determinations of x at different partial pressures and temperatures, using an electrochemical cell similar to the one employed by Beyers *et al*[43,44]. Tetenbaum was also able to determine the changes of the thermodynamic quantities ΔG_0, ΔS_0 and ΔH_0 upon oxidation. Based on the shape of the PO_2 vs. x curves, he extrapolates the presence of a miscibility gap in the x-vs.-T phase diagram below 200 °C, where true thermodynamic equilibrium is impossible to obtain due to the slow kinetics.

In situ neutron diffraction studies. In the aforementioned the pioneering work by Jorgensen and coworkers, later extended by the same group[45], oxygen content was measured at selected PO_2's as a function of temperature by Rietveld-refining NPD data sets, while, at the same time, determining other important structural parameters, like the orthorhombic strain, the position of the O-T phase line and the oxygen site occupancies. NPD was also used by Andersen and coworkers[46,47], in a very systematic and beam-time-intensive study. Andersen, follows a different approach than Jorgensen: PO_2 and x are determined by a volumetric titration apparatus, in which a controlled amount of oxygen is injected into the sample chamber from a calibrated volume. NPD is used to determine the lattice parameters and, therefore, the position of the T-O phase transition. Andersen also monitored relaxation times for oxygen diffusion, finding an anomalous increase for x<0.5 and T<450 °C, similar to the previous findings by Jorgensen. Later on, the same group analyzed the evolution of the NPD peak widths through the phase diagram[48]. The increased width in the orthorhombic phase was attributed to the formation of finite-size twin domains. The results could be interpreted in terms of two typical sizes, which, as the authors speculated, could be associated with different orthorhombic phases.

4.2. THEORETICAL MODELS

By the middle of 1987, a very significant amount of experimental data already existed on the x-T phase diagram of $YBa_2Cu_3O_{6+x}$, and even more information became available in the subsequent years. All this needed a theoretical framework to be fit in. In principle, a satisfactory theoretical model should address both structural and thermodynamic aspects of the problem, and, in particular, should correctly describe:

1) All the observed $YBa_2Cu_3O_{6+x}$ phases, including the ordered phases (Ortho II, etc.)
2) The nature of the phase transition lines and (possible) miscibility gaps in the x-T phase diagram.
3) The exact position of the phase lines, and, in particular, of the T-O boundary at high temperature, to be compared with experimental observations.
4) The O4 and O5 oxygen occupancies as a function of x and T.

5) The oxygen chemical potential as a function of temperature and x, to be compared with the measured values of PO$_2$.

Various theoretical approaches has been used to tackle this very complex problem. In most cases, these models make the drastic simplification of describing the system as a lattice of oxygen atoms with interaction potentials (long- or short-ranged), which are usually considered to be temperature- and concentration-independent. This general approach goes under the name of *lattice gas model*. The equilibrium states are either calculated exactly in mean-field approximation, or obtained through a Monte Carlo simulation. Some of these models, as we shall see, met with considerable success, although some of the subtle details of the x-T phase diagram are still eluding even the more sophisticated descriptions. A complete review of this quite specialized field is not possible in the present article. We shall, however, mention some of the most significant achievements, since they are so closely connected to the experimental work in general and to neutron diffraction in particular.

Lattice Gas Models. At the beginning of 1988, Khachaturyan and coworkers discussed the possibility of calculating the x-T phase diagram in mean-field approximation, based on an isotropic long-range interaction between oxygen atoms in the basal plane[49]. This model was subsequently extended to explain the experimentally observed oxygen-ordered superstructures[50]. The most prominent feature of Khachaturyan's phase diagram is the presence of a miscibility gap between O and T phases at finite temperatures, which, however, is usually not observed experimentally.

A completely different approach was adopted by De Fontaine and coworkers. de Fontaine considers two sublattices of oxygen atoms α and β (equivalent to the O4 and O5 sublattices), and includes in the calculations only short-range interactions: a nearest-neighbor (NN) inter-sublattice interaction potential V1, and two next-nearest-neighbor (NNN) intra-sublattice interactions: V2 ("intra-chain", through the Cu atoms) and V3 ("inter-chain", through the "empty" site 1/2,1/2,0 (Figure 4.1). This is equivalent to an Ising model with asymmetric NNN interactions, and is often called ASYNNNI model. Initially, de Fontaine and coworkers performed stability[51] and ground-state analyses[52] of their model, with the aim of determining the sign and relative strengths of the potentials. The observation of chains in oxygen-rich YBa$_2$Cu$_3$O$_{6+x}$ strongly suggested that the inter-sublattice potential V1 should be taken as *positive* (i.e., repulsive). Furthermore, for the ground-state analysis, only the ratios between the potentials are relevant. Depending on the value of V2/V1 and V3/V1, different ordered phases appear at each value of x. A comparison with the experimentally observed phases Ortho-I and Ortho-II led to the conclusion that V1>V3>0>V2. In other words, oxygen atoms feel an attractive interaction through the copper atoms, but a repulsive one through the empty 1/2,1/2,0 sites. The latter, however, is not as strong as the NN repulsive interaction V1.

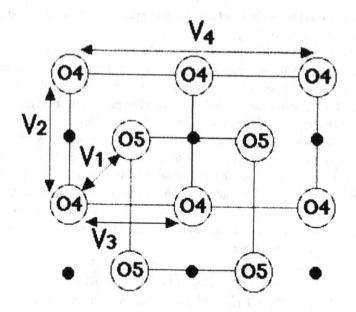

Figure 4.1. Interaction potentials for the ASYNNNI model in the
YBa$_2$Cu$_3$O$_{6+x}$ basal plane. The V$_4$ potential (see below), is associated
with a more sophisticated version of the model, capable of describing
higher-order superstructures.

Based on the ASYNNNI model, several authors calculated by various means the position
of the phase fields in a reduced x-T_R phase diagram, where the reduced temperature T_R is
T/k$_B$V1 and reasonable values were chosen for V2/V1 and V3/V1 (references [53-59]). All
these phase diagrams share some common features, like the presence of an T-O high-
temperature second-order phase transition for x > 0.5 and the existence of an Ortho-II
phase field at low temperatures, centered around x=0.5. However, there are significant
differences in the details of the phase diagrams and, in particular, in the nature of the
low-temperature phase transition. A significant breakthrough was obtained in 1990 by
Ceder and coworkers[60], who employed for the first time interaction potentials derived
from *ab initio* band structure calculations, and were able to present a phase diagram with
a real temperature scale (Figure 4.2). Qualitatively, Ceder's phase diagram is quite
similar to the one previously proposed by Kikuchi and coworkers[54]: the T-O and Ortho-
I to Ortho-II phase transitions are second-order, and a second orthorhombic phase field
(anti-Ortho-I) is found at low temperatures and low concentrations. Another interesting
feature, already identified by Kikuchi, is the presence of a miscibility gap between Ortho-
II and tetragonal phases, centered near x=0.36 and T=400 K. Ceder points out that there
is a good agreement between the position of the high-temperature T-O phase line (as

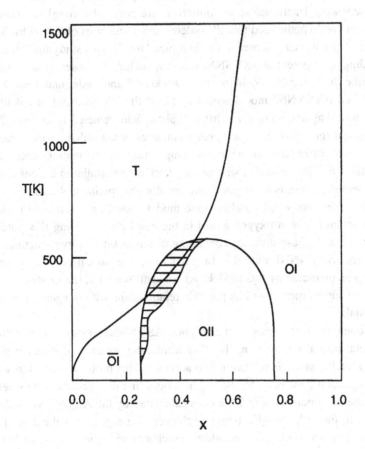

Figure 4.2. x-T phase diagram calculated on the basis of the 3-potential ASYNNNI model, with *ab initio* determination of the values of the interactions, from the work by Ceder *et al.*[60]

determined, for instance, by Specht and coworkers[25]) and that of the theoretical phase diagram.

Beyond the 3-Potential ASYNNNI Model. A detailed analysis of the zero-temperature phase diagram is discussed by de Fontaine, Ceder and Asta in a paper published in Nature[61]. At zero temperatures, only completely ordered phases are expected to be stable. However, only Ortho-I, Tetragonal and Ortho-II phases, which are partially disordered for all oxygen contents except for x=0.0, 0.5 and 1.0 respectively, are present in the x-T phase diagram calculated with the 3-potential ASYNNNI model. Higher-order phases can only be stabilized by a longer-range interaction potential, presumably a long-range extension of the inter-chain V3 potential. In fact, Monte Carlo simulations including an additional next-nearest chain (NNC) repulsive term V4 (Figure 4.1) yield a series of full-chain-ordered superstructures at zero temperatures[61-63]. These phases, having commensurate fractional stoichiometries, are reached through a branching process. de Fontaine hypothesized that the ordered superstructures observed by Beyers and coworkers[35] are in fact a subset of this complex "tree". This study underlined the need of extending the 3-potential ASYNNNI model to include the observation of higher-order phases like the Ortho-III. Zubkus and coworkers[64] and Ceder and coworkers[63] used the modified ASYNNNI model which included the V4 potential to calculate a modified x-T phase diagram. Indeed, an Ortho-III phase field, centered around x=0.7, was found to be present for T<400 K. The upper boundaries of the ordered superstructures and the superlattice diffraction intensities vs. temperature are the main concern of the work by Lapinskas and coworkers[65], who employ a technique similar to the one adopted by Ceder[63]. Lapinskas' results are in good agreement with experimental data.

Alagia and coworkers proposed an alternative model, based on a screened repulsive Coulomb interaction between oxygen atoms in the basal plane. Using this potential, they calculated an x-T phase diagram containing different set of superstructures with respected to the ASYNNNI model. In particular, the so-called "herringbone" superstructure was predicted by this model. As we shall see later, the existence of this particulat type of superstructure and its possible relationships with oxygen ordering are still controversials.

One of the recognized drawbacks of the original ASYNNNI model is that it totally ignores the metal-insulator transition. In other words, the presence of itinerant charge carriers in the metallic phase is not taken into account. This problem was addressed in detail by Schleger and coworkers[66]. Schleger notices that the agreement between the experimental chemical potentials and the ones calculated by the ASYNNNI model are especially poor in the orthorhombic phase. Schleger then argues that the discrepancy could be reduced by accounting for the additional charge and spin degrees of freedom associated with the itinerant carriers, and proposes a modified ASYNNNI model in which the entropic contribution of these effects is taken into account. The thermodynamic and structural parameters calculated by this model are in excellent agreement with experiments.

Relationships with Doping and Superconductivity. We have already mentioned that several authors have tried to correlate the observation of ordered superstructures with features of the superconductivity phase diagram of $YBa_2Cu_3O_{6+x}$. In particular, the association between the 60 K plateau and the Ortho-II phase has not gone unnoticed. All the experimental work, in particular the one by Cava and Beyers, suggested that this correlation had to do with charge transfer, rather than with a specific T_c of the Ortho-II phase. Clearly, the relationship between oxygen ordering and hole concentration on the CuO_2 planes is rather difficult to address. In fact, for oxygen-rich $YBa_2Cu_3O_{6+x}$ (x~1), electronic charge is transferred from the chains to the planes because both plains and chains give rise to highly dispersive bands crossing the Fermi surface[67]. These structures can be precisely determined by band structure calculations[67]. However, these calculations imply a periodic structure, which is obviously not present for states of partial order. The problem was cleverly addressed by Zaanen and coworkers in 1988[68]. Zaanen presents a band structure calculation of the electronic structure of YBa$_2$Cu$_3$O$_{6+x}$ as a function of x, by constructing a series of periodic structures aimed at mimicking oxygen ordering in the real material. First of all, for all values of x, Zaanen assumes an orthorhombic symmetry and an oxygen concentration modulation with period 2a perpendicular to the chain direction (i.e., the same as for the Ortho-II phase). Oxygen is therefore assumed to form sequences of ...-empty-partially full-empty-...chains for x≤0.5 and of -full-partially full-full-...chains for x≥0.5. In the partially filled chains, oxygen would be arranged as to form alternating chain fragments containing v copper and v-1 oxygen atoms, with x=1/2-1/(2v) for 0≤x≤1/2 and x = 1-1/(2v) for 1/2≤x≤1 (a mixture of different fragment lengths was used for incommensurate values). We shall note that this scheme yields the observed structures for x=0, x=1/2 (fully-ordered Ortho-II) and x=1 (fully-ordered Ortho-I), but it is otherwise a rather crude approximation of the realistic oxygen ordering in YBa$_2$Cu$_3$O$_{6+x}$ (which was not known in details at the time of Zaanen's work). However, the band structure calculation based on this model gave a deep insight in the relationship between oxygen ordering and doping. In fact, the chain fragments were found to be unable to transfer charge to the CuO$_2$ planes until they exceeded a critical length (v= 3), since, for v<3, the holes are trapped in localized chain-fragment states *above* the Fermi surface. Also, the doping-vs.-x curve constructed by Zaanen on the basis of this model displayed the 2-plateau structure of the real T$_c$ vs. x curve. The validity of this early work is still intact, in the sense that all the observed relationships between oxygen ordering and T$_c$, including the quenching and annealing effects to be later discussed (Section 5.) can be understood by the very simple concept that short chain fragments are ineffective dopants.

Later on, Poulsen and coworkers proposed a different model combining the charge-transfer hypothesis with the ASYNNNI model[69]. Their basic *ansatz* is rather simple: only Ortho-I- or Ortho-II-like domains are able to transfer charge to the CuO$_2$, planes. In the case of small chain fragments or disordered clusters in the basal plane, the holes are "trapped" into the basal plane and cannot contribute to superconductivity. Starting from this hypothesis, Poulsen and coworkers construct Monte Carlo simulations of the basal plane for several values of x, equilibrated at rather low temperatures (300-500 K). They then simply count the number of clusters of the two species (Ortho-I and Ortho-II), and assign a charge transfer to each of them, by imposing that the two fully-ordered states have the correct T$_c$ and assuming a linear T$_c$- vs. - charge-transfer relationship. The resulting T$_c$ - vs. - x curve is strikingly similar to the experimental one. However, Poulsen's work is not immune to criticism. In particular, the size and shape of the clusters is rather arbitrarily chosen, with Ortho-I clusters being 1/2 the size of Ortho-II clusters, and there is no guarantee that a different choice would lead to the same result. Furthermore, the need to introduce 2-dimensional clusters, with respect to the simple 1-dimensional chain fragments proposed by Zaanen is not completely justified. However, the merit of this work is to stress once again the need for a size threshold for ordered structures, below which no charge transfer can take place.

4.3. EXPERIMENTAL OBSERVATIONS OF OXYGEN-ORDERED SUPERSTRUCTURES (ORTHO-II AND ORTHO-III)

The Role of Single-Crystal Diffraction. Until the end of 1990, most of the information about the Ortho-II and other ordered superstructure had been obtained by TEM-ED. The reason for this fact is easily explained: the superstructures tend to have a short coherence length, and therefore yield weak and broad reflections. The short intrinsic coherence length of the ED probe is ideally suited for this type of problems, and is able to detect ordering over a few tenths of unit cells. On the contrary, ordering over hundreds of Angstroms is needed before an observable peak could be seen in x-ray or neutron diffraction. As the sample quality improved, however, neutron and x-ray single-crystal diffraction observations of ordered superstructures became possible. In this field, the single-crystal techniques proved to be far superior to powder diffraction, mainly due to the better signal-to background ratio and the ability to orient a whole macroscopic sample in Bragg reflection conditions for the weak superlattice peaks.

Single-Crystal Neutron and X-ray Diffraction of the Ordered Superstructures The first observation of superlattice reflections in $YBa_2Cu_3O_{6+x}$ with a technique other than ED was presented by Fleming and coworkers in 1988[70]. Fleming *at al.* observed a modulation of the x-ray diffuse scattering intensity in a single-crystal with x=0.7, with maxima centered at the (1/2,0,0) wavevector, corresponding to what expected for the Ortho-II phase. However, the coherence length of the oxygen ordering was very small (21, 16 and 9 Å for the a, b, and c directions, respectively). In 1991, Zeiske and coworkers presented the first observation of the Ortho-II superstructure by single-crystal neutron diffraction[71]. The measurement, on a $6 \times 3 \times 2$ mm³ twinned single crystal, was performed on a triple-axis spectrometer at Risø National Laboratory. The coherence lengths (40, 90 and 22 Å) were found to be much larger than in the measurement by Fleming. Later on, Burlet and coworkers provided a quantitative analysis of the superlattice intensities, allowing for a complete structural solution of the Ortho-II phase to be obtained[72-74]. It was found that, in addition to the formation of alternating Cu1-O4 and Cu1-vacancy chains, the apex oxygen O1 and the cations (Y, Ba, Cu2) are involved in the superstructure. In two adjacent unit cells these atoms have opposite displacements. More recently, Hadfield and coworkers presented a more accurate solution, obtained by simultaneously refining x-ray and neutron single-crystal diffraction data[75].

A different type of ordered superstructure was found by Sonntag and coworkers at the beginning of 1991[76]. Sonntag studied a $5 \times 5 \times 4$ mm³ *tetragonal* crystal with x=0.35, by employing the D10 diffractometer at the ILL, and observed a series of superlattice peaks, consistent with a $2\sqrt{2}a \times 2\sqrt{2}b \times c$, as previously observed by ED[28]. Subsequently, the same group presented a more precise determination of the superstructure by single-crystal neutron and x-ray diffraction[77,78]. The $2\sqrt{2}a \times 2\sqrt{2}b \times c$ cell turned out to be a superposition of two orthorhombic domains, having a smaller $2\sqrt{2}a \times \sqrt{2}b \times c$ unit cell. The intensity analysis suggests that the oxygen in the basal plane forms isolated Cu-O-Cu units which order in a "herringbone" fashion. However, subsequent work by Yakhou and coworkers[79-81] and TEM-ED[82] had refuted this model, arguing that oxygen ordering is not involved in this type of superstructure. In her Ph.D. thesis and related papers[79-81], Yakhou argues quite conclusively, on the basis of synchrotron and neutron difraction data, that the "herringbone" structure is actually an intergrowth of an extraneous phase with formula BaCu$_3$O$_4$.

Temperature Dependence. One of the important issues associated with the Ortho-II phase and other ordered superstructures is the rather short and anisotropic coherence length. This point was addressed by Schleger and coworkers by using single-crystal hard x-ray scattering. The shape and intensities of the Ortho-II reflections are carefully monitored as a function of temperature, and also as a function of time in quenching experiments (see below). Schleger noticed that the Ortho-II peaks at room temperature cannot be fit with simple Lorentzians, but require an additional contribution in the form of a Lorentzian square, typical of 2-dimensional domain wall scattering. Above 125°C, the simple Lorentzian description is adequate, consistent with pure 3-dimensional critical scattering *above* the phase transition . Schleger proposes, therefore, the following picture: as the crystal is cooled, critical scattering is observed, consistent with a second-order or weakly first-order phase transition. Below the phase transition (~125°C) the crystal would have the tendency to form the Ortho-II phase. However, true long-range order is never attained, since the antiphase boundaries between Ortho-II domains do not anneal out. Schleger speculates that the reason for this is the intrinsic quasi-1-dimensional nature of the system, possibly coupled with a random field.

More recently, Poulsen and coworkers studied the Ortho-II superstructure as a function of temperature, using x-ray synchrotron diffraction[83]. The most important issue addressed in this work is the nature of the phase transitions from the Ortho-II phase, ultimately leading to the tetragonal phase. Poulsen and coworkers chose the compositions x=0.35 and 0.36, quite close to the T-O phase boundary at room temperature, which is believed to provide a rather stringent test of the ASYNNNI model. In fact, in his region, de Fontaine's phase diagram displays a miscibility gap between Ortho-II and tetragonal phases around 400 K. In other words, upon heating, an x=0.35-0.36 sample, with Ortho-II room-temperature structure, should phase-separate into Ortho-II and tetragonal domains. On further heating, the Ortho-II domains should disappear. The experimental findings of Poulsen *et al.* for the x=0.36 composition were in fact quite different: the Ortho-II phase disappears into *single-phase* Ortho-I at T_{OII} = 358 K, followed by a nearly continuous OI-T phase transition at T_{OI} = 519 K, with critical exponent β = 0.35. No evidence of phase separation was observed. Corresponding measurements for x=0.35 gave similar results, with T_{OII} = 368 K, T_{OI} = 454 K, except for the presence of a small tetragonal component at room temperature. The latter, already observed by Radaelli *et al.* in a systematic study of $ErBa_2Cu_3O_{6+x}$ powder samples[37], is interpreted as a non-equilibrium feature.

Schleger and coworkers observed Ortho-III correlations in a x=0.77 single crystal by hard x-ray and neutron diffraction, and followed the temperature dependence of the peak intensity[84]. The disappearance of the peaks mark the position of the Ortho-III - to - Ortho-I phase transition. This and the previous pieces of experimental work, together by analyses performed by other techniques[85], were aimed at establishing the upper boundaries of the ordered superstructures. Their conclusions further confirmed the need of extending the ASYNNNI model to correctly describe the detailed features of the x-T phase diagram.

5.   Quenching and "Ageing" effects at Room Temperature

Superconducting Properties. The unusual dependence of the structural and superconducting properties of $YBa_2Cu_3O_{6+x}$ for samples having the *same* value of x but prepared in different ways (low-temperature annealing vs. quenching) has already been discussed in the previous paragraphs. The early work by Farneth *et al.* already stressed that significant differences in the T_c values could be obtained, especially in the region of the 60 K plateau, by adopting different quenching or annealing procedures[33]. However, no difference could be observed in conventional x-ray diffraction pattern for samples having the same oxygen content, indicating that the structural differences, if any, were rather subtle. Later on, the now famous controversy on the existence of a c-axis anomaly near the onset of superconductivity as a function of x (see Section 3.2) reflected the intrinsic *structural* differences between samples with the same x but subjected to different annealing conditions. Clearly, a need was felt for a more systematic study of these issues. Further attention was drawn on the subject by a study published in 1990 by Namgung *et al.*[86]. Namgung prepared a series of samples by quenching from various temperatures into liquid mercury, arguing that the better thermal conductivity of Hg with respect to liquid nitrogen (traditionally used in quenching experiments) would allow for much faster cooling rates. The oxygen content was calibrated by choosing the appropriated temperature and PO_2 for annealing before quenching, and was checked by iodometric titration for selected composition. The T_c's measured on samples prepared by this technique are spectacularly different from previous results: both the 60 K and the 90 K plateaus have now completely disappeared, and the T_c vs. x points lay on an almost strainght line.

Veal and collaborators[87-89] worked along the same line, but with a few substantial differences: first of all, their quenching experiments and measurements of T_c are performed mainly on small single crystals, which display much sharper transitions even after quenching, perhaps due to the absence of grain interaction stresses. Furthermore,

Veal optimized the quenching process in order to establish which is the *lowest* quenching temperature where effects are observed and what kind of "slow-cooling" or "annealing" is required to recover optimal superconducting properties. In fact, short-range oxygen mobility could be very high *within* the material at rather low temperatures, in spite of the fact that equilibrium with atmospheric oxygen can only be established above ~350 °C. Initially, the crystals were quenched from 520 °C in different PO$_2$, in order to establish the correct stoichiometry, which was demonstrated not to change in subsequent annealing-quenching processes. After quenching, the crystals were found to have T$_c$ reduced by as much as 27 K with respect to the "optimal" T$_c$, the largest T$_c$ changes being found near the superconductivity onset (x≈0.4) . Remarkably, Veal found that annealing (or "ageing") the crystals at *room temperature* for a sufficient amount of time is equivalent to slow-cooling, and is sufficient to recover the maximum T$_c$. Typical annealing time constants at room temperature are of the order of 600 min. It was also demonstrated that the maximum T$_c$ reduction can be obtained by quenching from as low as 230 °C (see Figure 5.1). Quenching from lower temperatures results in intermediate T$_c$'s between the 230 °C-quenched and the fully aged (or slow-cooled) values. These very low temperatures indicate that short-range oxygen diffusion must be responsible for the observed changes. Veal also measured the orthorhombic strain of the crystals in various quenching/annealing states, and found that the orthorhombicity *increased* with aging. This was an indication that oxygen *ordering*, which is coupled to the orthorhombicity, was the likely cause of the observed effects. This important work established two main points: 1) Oxygen ordering phenomena occur at temperatures as low as room temperature. 2) Low-temperature oxygen ordering has a dramatic effect on the superconducting properties. However, what kind of oxygen ordering was associated with the observed changes still had to be established. One possibility was that the relevant ordering occurred *between* the two sublattices O4 and O5. In other words, with ageing or slow-cooling, oxygen would be transferred from the "anti-chain" O5 sites to the "chain" O4 sites, thereby lengthening the average chain length. Another possibility was that oxygen atoms were already fully ordered onto the O1 sublattice immediately after quenching, but that, upon annealing, short-range order of the Ortho-II type would be established *within* the O1 sublattice.

Neutron Diffraction. In order to discriminate between these two models of oxygen ordering, a direct measurement of the sublattice occupancies as a function of ageing was needed. The experiment was performed by Jorgensen and coworkers by NPD[90]. A large powder sample was quenched from 520 °C in an appropriate N$_2$/O$_2$ mixture, in order to establish a x=0.41 composition. Magnetization measurements were performed as a function of time. The as-quenched sample did not display any measurable diamagnetic signal, but superconductivity with T$_c$ up to 20 K was recovered upon ageing. The largest portion of the sample, stored in liquid nitrogen, was transferred in a vanadium can and placed on the SEPD diffractometer. NPD data were collected at different time intervals up to 9000 min of room-temperature annealing.

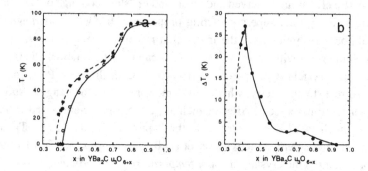

Figure 5.1. (a) T_c vs. x for oxygen-deficient $YBa_2Cu_3O_{6+x}$ single crystals "aged" at 25 °C (open circles) and quenched from 230 °C (closed circles). (b) ΔT_c, the difference between the two curves in (a), as a function of x. From reference [89].

All the lattice parameters and most of the internal parameters were found to evolve as a function of time with similar time constants with respect to the superconducting properties. In particular, the orthorhombicity *increased* with time, and the Cu2-O1 distance *decreased*, the latter signaling an increased amount of charge transfer of the CuO_2 layers[91] (Figure 5.2). However, the O4 and O5 occupancies were found to be constant, indicating that no oxygen transfer occurred between the two sublattices, at least within the time scale of the diffraction experiment. More recently, Shaked and coworkers performed a similar experiment on a *tetragonal, non-superconducting* sample with x=0.25[92]. The lattice parameters a and c were found to decrease as a function of time, evolving with a time constant similar to the one governing the evolution of the electrical resistivity upon ageing. Clearly, some form of oxygen ordering must occur also in the tetragonal phase, and must be associated with chain fragment lengthening, giving rise to a decrease of the electrical resistivity. This is a clear illustration of the subtle relationship between *local* ordering (which has an intrinsically orthorhombic symmetry, since is associated with the formation of chains) and *average* symmetry (allowed to be either tetragonal or orthorhombic), which results from an averaging of the chain orientations over hundreds of Angstroms.

Theory. Ceder analyzed ageing phenomena described above can be modeled by Monte Carlo simulations in the framework of the ASYNNNI model[93]. The initially random lattice gas was "annealed " (that is, allowed to reach equilibrium) at high temperature, and subsequently quenched to room temperature, where the time evolution was studied over several thousands of Monte Carlo cycles (each cycle was approximately equivalent to 1 minute. Some key parameters, like the fractions of copper atoms in various coordinations, were monitored as a function of time. It was also found that the largest "ageing" effects could be observed for oxygen concentrations in the range x=0.3-0.4.

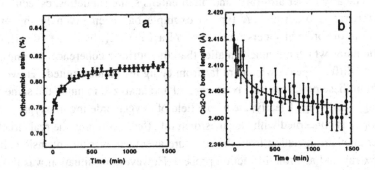

Figure 5.2. Orthorhombicity (a) and Cu2-O1 bond lengths (b) as a function of annealing time at room temperature as determined by NPD on a x=0.41 sample of YBa$_2$Cu$_3$O$_{6+x}$, quenched from 520 °C into liquid nitrogen. From reference [90].

This technique allowed to determine the presence of two distinct time constants: the first corresponds to oxygen ordering between the O4 and O5 and sublattices, and was too fast to be observed in Jorgensen's experiment. The second is associated with the Ortho-II-type ordering, and is likely to be the same as for the neutron diffraction experiments.

6.   "Local" Structural effects in oxygen-rich YBa$_2$Cu$_3$O$_{6+x}$ (x≈1)

Most of the structural work discussed up to this point was aimed a the determination of the *average* structure of YBa$_2$Cu$_3$O$_{6+x}$, where the word "average" is intended in the sense which is typical of Bragg diffraction: *time* average, since only elastic scattering is considered, and *lattice-space* average, since only scattering around the reciprocal lattice position is considered. In general, any real crystal will violate time invariance and perfect lattice periodicity because of dynamic fluctuations (phonons or local vibrational modes), static deviation from the average structure (e.g., ionic substitutions/vacancies with an associated local strain field) and, of course, finite crystal dimensions. However, it is useful to consider the total scattering intensity form a crystal as the sum of 2 components: the Bragg component, and the *diffuse* component. The Bragg scattering intensity is *by definition* equal to the square of the scattering amplitude calculated for a perfectly periodic and time-invariant structure, constructed by averaging the real nuclear or electronic density over all the unit cells and over time. The nuclear or electronic density lattice average (*coherent* average) is usually not performed over the whole crystal, but rather on smaller regions, the extent of which, called intrinsic *coherence volume*, is related to the resolution of the probe (the better the resolution, the larger the coherence length). For neutrons and x-rays, typical coherence lengths are of the order of a few thousand Angstroms. The Bragg scattering is associated with the sharp diffraction peaks, to be *diffuse* (elastic and inelastic). The distinction between Bragg and non-Bragg scattering is clearly rather arbitrary, and often entails some subtleties, especially in the case of short-range ordering. A typical example has been given in the previous paragraphs: in the ordered superstructures of $YBa_2Cu_3O_{6+x}$, like the Ortho-II, oxygen orders in domains which are much smaller than the intrinsic coherence volumes, i.e., it gives rise to "diffuse" peaks which are far from being resolution-limited. However, the intensity from these peaks can still be integrated and analyzed in much the same way as normal Bragg scattering. Clearly, the whole field of oxygen ordering in oxygen-deficient $YBa_2Cu_3O_{6+x}$ is concerned with "local" structural effects, whereby the real structure is, in a way or another, deviating from the structure averaged over the intrinsic coherence volume of x-ray and neutron diffraction probes. However, the situation was thought to be quite different for oxygen-rich $YBa_2Cu_3O_{6+x}$ ($x \approx 1$). In the ideal case of perfectly stoichiometric $YBa_2Cu_3O_7$, the structure can be thought as being perfectly ordered, with the O4 sublattice fully occupied and the O5 one completely empty. This is the exception rather than the rule for high-temperature copper oxide superconductors. In fact, doping of the CuO_2 layers is usually achieved by metal ion substitutions (e.g., $La_{2-x}Sr_xCuO_4$) and/or additional oxygen atoms in interstitial sites (e.g., $La_2CuO_{4+\delta}$ and $HgBa_2Ca_{n-1}Cu_nO_{2n+2+\delta}$), which inevitably lead to random or locally ordered deviations from the average structure. $YBa_2Cu_3O_7$ (together with the closely related $YBa_2Cu_4O_8$) was thought to be one of the few examples of a perfectly ordered high-T_c superconductor, due to the presence of a self-doping mechanisms from the chains. The idea of $YBa_2Cu_3O_7$ being a "perfect" superconductor was however variously challenged over the years. In the next few paragraphs we will discuss two issues related to this problem, which have had an ample resonance in the literature: the possibility of a "zig-zag" pattern for the ...Cu1-O4-Cu1... chains and the possible presence of a double-well potential for the apical oxygen O1.

Zig-zag Chain Pattern. The amplitude of the thermal ellipsoids, related to the Debye-Waller factors, has been discussed early on in the first structural determination of $YBa_2Cu_3O_{6+x}$, performed by NPD. Capponi and coworkers[2] determined the structure at various temperatures between 5 K and 300 K, and concluded that the only site for which the use of anisotropic Debye-Waller factors led to significant improvements of the fits was the oxygen chain site O4. The displacements *perpendicular* to the bonding direction (i.e., to the b axis) were much larger that parallel to it, and were found to have a significant residual component at low temperatures, suggesting that static displacements might be partly responsible for the observation. The issue was later re-examined by François and coworkers[94], by employing the high-resolution powder diffractometer D2B at the ILL, which allows a better accuracy to be attained for the structural parameters. François found that stable refinements could be obtained by employing a "split-site" model, in which the O4 oxygen is placed at the two-fold position 2k ($\pm x$,1/2,0), point symmetry 2mm (with 1/2 the occupancy), instead of the usual one-fold position 1e (0,1/2,0), point symmetry mmm. This model would simulate the situation in which oxygen atoms are randomly displaced by a distance a·x along the a axis. Therefore, the chains would not be straight, but would form a random zig-zag pattern. The Debye-Waller factors of the split site were refined *isotropically*, and found to be of the same order of magnitude than those of the other oxygen atoms in the cell. The split-site reliability factors were found to be comparable to those of the anisotropic model which has a higher number of refinable parameters (3 anisotropic B's vs. x and 1 isotropic B for the split-site model), which is a good indication of the plausibility of the model. The fractional displacements of the O4 atoms along the a axis , x(O4), were found to be slightly temperature dependent, but to be still fairly large at low temperatures, indicating the presence of a static component to the zig-zag pattern. The validity of this conclusion was more recently corroborated by a detailed single-crystal neutron diffraction work by Schweiss and coworkers[95]. Schweiss compared the values of the anisotropic Debye-Waller factors, measured as a function of temperature, with the ones calculated using an harmonic model based on lattice dynamics calculation, consistently checked against measured values of the phonon dispersion curves. The O4 displacements along the chain direction (b axis) was found to be comparable with the corresponding displacements for the CuO$_2$ planes and to be in good agreement with the calculated values. However, *perpendicular* to the chains (and especially along the a direction), the Debye-Waller parameters B were found to be extremely large but, at the same time significantly temperature-dependent. Moreover, the observed curve was found to be consistently higher than the calculated one by a temperature-independent amount, strongly suggesting that an additional static component had to be taken into account. Schweiss *at al.* also tried the split-site model proposed by François, and obtained a virtually temperature-independent shift of ~0.074 Å along the a axis, consistent with the previous findings by NPD. The zig-zag pattern may also affect other atoms beside O4: the measured a-axis displacements of the apical oxygen O1 and of the Ba atoms were found to be somewhat larger than the calculated ones. The origin of the zig-zag pattern of the ...Cu-O-Cu... chains is easily understood in terms of steric-coordination effects: the matching between Cu-O distances in the planes (with 5-coordinated copper having a formal valence of ~2.15) and in the chains (with 4-coordinated copper having a formal valence of ~2.70) is not perfect, the former having the natural tendency of being *shorter* than the latter. The mismatch is worsened by the buckling of the CuO$_2$ layers, which tends to further shorten the Cu-Cu distance in the CuO$_2$ layers. This puts the Cu1-O4 bonds in the basal plane under compressive strain, whence the tendency to forming zig-zag patterns.

Apical Oxygen Double-Well Potential. No special attention had been devoted by the early literature to the displacements along the c axis in the YBa$_2$Cu$_3$O$_{6+x}$ unit cell. When refined anisotropically, the Debye-Waller factors along the c axis are, for all atoms, either comparable or smaller than those in the ab plane. However, starting from 1990 Mustre de Leon and coworkers presented a series of c-axis-polarized extended x-ray absorption fine structures (EXAFS) measurements, taken at the Cu-K edge on c-axis-oriented powders, of the Cu2-O1 bond lengths, in which the possibility of a double-well potential for O1 was strongly put forth[96,97]. The experimental data could be fit with two equally populated Cu1-O1 distances, differing by 0.13 Å from each other. The separation between the two minima of the double-well potential was found to sharply decrease in a "fluctuation" region near T_c. In the original publications, a comparison with diffraction data is not presented. However, the obvious discrepancies with the diffraction results were usually dismissed citing a supposed "insensitivity" of the diffraction probe to the local structure. The measurements by Mustre de Leon initiated a controversy which is still alive today at least to a certain extent. In fact, subsequent diffraction experiments analyzed either with conventional structural refinements or with Fourier reconstruction of the nuclear/electronic density[41,98] all failed to evidence the proposed double maximum corresponding to the apical oxygen position. Furthermore, the presence of a double-well potential would require a lowest optical (infra-red) excitation energy below 150 cm^{-1}, which is not observed experimentally[99]. The sharp anomaly in the T_c region is also in contradiction with optical spectroscopy data, which clearly show a much broader step-like anomaly with onset at T_c. In general, the interpretation given by Mustre de Leon in terms of strong electron-phonon coupling with an anharmonic apical oxygen vibration is not supported by optical spectroscopy data[99].

Clearly, the information given by EXAFS is quite different in nature than the diffraction one. EXAFS is probing the *local instantaneous* structure, since the signals given by different sites are decorrelated from each other (incoherent average) and a full energy integration over the relevant vibrational energies is performed. Furthermore, the EXAFS technique allows a very high momentum transfer to be attained (15-20 Å^{-1}), making it ideally suited to study small atomic displacements. It is therefore not surprising that the "same" quantities, as measured by EXAFS and diffraction, can have different values. This is because the definition for apparently identical quantities (e.g., the bond lengths) is not the same whether a local or average probe is used. However, the diffraction and EXAFS results should be compatible with each other, in the sense that a unified structural model should be able to explain both. Clearly, the interpretation of the EXAFS data given by Mustre de Leon is not easily compatible with diffraction data. In fact, the centers of the two "minima" of the double-well potential, differing by 0.13 Å along the c axis, are well outside the average displacements from the average O1 position that can be calculated from the Debye-Waller ellipsoids. One could still argue that the apparent doubling of the Cu1-O1 distance may result from an anti-correlated displacement of copper and oxygen atoms. However, this would presumably entail a deviation from the harmonic model of the temperature dependence of the Debye-Waller factors for both Cu1 and O1. In the aforementioned paper[95], Schweiss and coworkers carefully measured the Debye-Waller ellipsoids, and compared them with the harmonic model. No significant discrepancies were found along the c axis, leading the authors to conclude that the double-well potential model should be excluded.

More recently, the EXAFS evidence for a double-well apical oxygen potential has been critically re-examined[100,101]. The "beat" structure around 12 Å^{-1} in the EXAFS data was found to be present in some samples but to be strongly sample-dependent. For instance, the EXAFS data from the highest T_c samples could be equally well fitted with either a double- or a single-potential model. Furthermore, the determination of the temperature dependence of the inter-well distance was found to be unreliable, due to strong correlations with the mean square disordered. In general the more recent experiments appear to confirm the experimental validity of Mustre de Leon's work, but to question the interpretation in terms on an on-site anharmonic potential. The actual origin of the "beat" structure, however, still remains a mystery. One possibility, suggested by Röhler[100,101], is that the extreme sample-dependency of the EXAFS results may critically depend on a small number of oxygen vacancies, which would create a local distortion field affecting a comparatively larger number of adjacent apical oxygen sites.

Structural Anomalies at T_c. Many attempts have been made to evidence the presence of structural anomalies associated with the superconducting transition temperature. One should stress that the presence of structural anomalies would not in itself be unexpected, since the quasiparticle condensation associated with superconductivity involves a coupled system of charge carriers and lattice vibrations (phonons). We have already mentioned that optical spectroscopy clearly evidences the presence of phonon anomalies at T_c, which are expected to yield, at the very least, a discontinuity in the thermal expansivity. The lattice parameter anomalies are detectable with dilatometry or careful diffraction experiments, but the effects on the integrated intensities (related to the internal parameters) are generally too small to be observed. Nevertheless, Schweiss and coworkers observe a very weak intensity change for some reflections at the superconducting transition[95], albeit at the limit of the statistical uncertainty.

The detection of structural anomalies at T_c in superconductors has been considerably more successful by using local probes. For instance, the pair distribution function (PDF) technique, based on a Fourier transform of NPD data (including the diffuse scattering), has evidenced significant anomalies at T_c in various compounds, including La$_{2-x}$Sr$_x$CuO$_4$, Ba$_{0.6}$K$_{0.4}$BiO$_3$ and YBa$_2$Cu$_3$O$_{6+x}$ (reference [102]). However, in the specific case of YBa$_2$Cu$_3$O$_{6+x}$, the PDF anomalies are rather small. More recently, Röhler and coworkers have published an extensive study of the temperature dependence of the YBa$_2$Cu$_3$O$_{6+x}$ structural parameters by EXAFS. Clear anomalies at T_c are observed for the Y-Cu$_2$ mean-square displacements, which depart significantly from a simple harmonic model.[103]. The relevance of these findings towards the general understanding of the structure-property relations is, however, still unclear.

7.   Substituted Compounds

The YBa$_2$Cu$_3$O$_{6+x}$ compound has become the aristotype of a large family of compounds, obtained by substitution of all the atomic species. Each substitution generates new series of compounds, which are, in general, describable on a two-dimensional phase diagram (oxygen content, x, and concentration of the substituent, y), and is often connected with an entirely new set of characteristics and "pathologies". Therefore, it is extremely difficult to make a thorough review of this vast field. We will restrict ourselves to citing some of the most important "themes" with some relevant references, especially in connection to the contribution given by neutron diffraction. For rather complete review of the early work see, for instance, the exhaustive book by Raveau and coworkers[104].

7.1. CU-SITE SUBSTITUTIONS

Location of the Substitution Sites. In the field of high-T_c superconductivity, Cu-site substitutions are of the utmost importance, both theoretically and experimentally. By substituting copper with other transition metals, T_c is invariably suppressed. The rate of the T_c suppression and the comparison between magnetic and non-magnetic substitutions serve as benchmarks against which all theories of superconductivity should be checked. For instance, for the 39 K superconductor, $La_{2-x}Sr_xCuO_4$, Ni and Zn substitutions both depress T_c dramatically by as much as 20 K for concentrations as low as 2% (for a review, see the aforementioned discussion by Raveau coworkers[104].) Interestingly, the effect on T_c is larger for non-magnetic Zn than for magnetic Ni. For $YBa_2Cu_3O_{6+x}$, the situation is quite different[105,106]: Zn and Ni were found to depress T_c rather drastically, much as for $La_{2-x}Sr_xCuO_4$. However, Co, Fe, Al and other substitutions have less of an effect: as much as 10% Co or Fe and 20% of Al are required before a significant effect on T_c can be observed. Once again, contrary to conventional BCS superconductors, no special correlation was observed with the intrinsic magnetic nature of the substituent species. The different effects on T_c of various substituents were already associated by early studies to the presence of two Cu sites. In fact, substitutions on the planar Cu2 site are expected to have a direct effect on T_c, much as in the case of $La_{2-x}Sr_xCuO_4$. Substitutions on the chain Cu1 have a subtler effect on T_c, since they affect the doping mechanism by distorting the oxygen coordination in the basal plane. The hypothesis that the cation having the *least* effect on T_c substitute preferentially the chain copper Cu1 was verified in 1988 by Tarascon and coworkers[105] in the case of cobalt, using NPD. Cobalt is ideally suited for this determination, since its neutron scattering length is quite different from that of copper ($b_{Cu} = 0.77 \times 10^{-12}$ cm, $b_{Co} = 0.25 \times 10^{-12}$ cm). Tarascon studied two samples of $YBa_2Cu_{3-y}Co_yO_{6+x}$, and concludes that cobalt only substitutes for the chain Cu1 for y=0.2, while, for higher values of the cobalt concentration, some Cu2 replacement is observed. A similar conclusion is reached by Siegriest and coworkers for Al substitutions, on the basis of an x-ray single crystal study[107]. Other substitutions are harder to probe, due to the smaller contrast with Cu. The problem was subsequently addressed by using anomalous x-ray diffraction[108] and a combination of the latter with neutron diffraction[109]. The results of these two studies demonstrated with fair precision that Ni and Zn were essentially randomly distributed, whereas Co and Fe occupied predominantly the chain sites. Other metal ion substitutions for copper (Ti, V, Cr, Mn) are discussed by Kilcoyne[110]. Rhenium substitutions were studied by NPD by Taylor and coworkers[111]. For a more recent experimental treatment of the Zn substitutions in YBa$_2$Cu$_3$O$_{6+x}$, including neutron diffraction data (see, for instance, the paper by Collin and coworkers[112]).

Effects on the Cu1 Substitutions on Chain Ordering. In all the early studies, a close correlation was found between Cu1 substitutions and orthorhombic strain. For instance, in the case of Fe, the orthorhombic strain was found to decrease quite rapidly, until an orthorhombic-to-tetragonal transition was encountered[113-117]. However, the exact value of y for which the compound became tetragonal was different by as much as a factor of 2 between different studies. Also, the behavior of the c axis and the T$_c$ variation were not consistently determined. The mechanism through which the orthorhombicity is reduced was addressed early on by Hodeau and coworkers[118] and by Xu and coworkers[119]. The ionic radius of the mixed-valence copper species on Cu1 is approximately 0.5 Å. As a consequence, the most likely valence states and coordinations for the substituents are Fe^{+3} in tetrahedral or pyramidal coordination and Co$^{+3}_{LS}$ or Al^{+3} in octahedral coordination. None of these cations has been reported in square-planar coordination. Therefore, substitutions of these species is bound to perturb the chain ordering, and, as a side effect, increase the maximum overall oxygen content of the compound, due to the increased average coordination in the basal plane. This latter effect was verified by Hodeau using NPD. On the basis of NPD and ED observations, Hodeau proposes that one of the likely mechanisms is Fe clustering in diagonal bands which would act as twin domain boundaries. As the average twin distance decreases below the intrinsic coherence length of the probe, an overall reduction of the orthorhombic symmetry and the appearance of diffuse scattering are observed, ultimately leading to an "average" tetragonal phase composed of "orthorhombic" microdomains, which is evident in TEM images as a typical "tweed" pattern. It should be kept in mind that, at the time of the elaboration of Hodeau's work (1988), the link between (local) orthorhombic symmetry and superconductivity was at the forefront of the discussion in the field of high-T$_c$. In fact, for both the best known compounds La$_{2-x}$Sr$_x$CuO$_4$ and YBa$_2$Cu$_3$O$_{6+x}$, superconductivity disappears near the O-T phase boundary. However, since the discovery of intrinsically tetragonal copper oxide superconductors (like the HgBa$_2$Cu$_n$Ca$_{n-1}$O$_{2n+2+\delta}$), this issue has been almost entirely superseded.

The reason for the discrepancy between various reports concerning the lattice parameter and critical temperature behaviors of Co- and Fe-substituted compounds remained unknown for quite a long time. Initially, it was hypothesized that the conflicting reports were due to differences in oxygen content[116]. However, this did not seem to be an entirely satisfactory explanation, since all the studies dealt with samples that were "fully oxygenated", i.e., that underwent a low-temperature annealing in oxidizing atmosphere. If differences in oxygen content and other properties were present, they must have been due to metal ion stoichiometry and/or distribution. Takayama-Muromachi[120] was among the first to systematically study the effect of the oxidizing conditions during the high-temperature annealing. It was found that, for "standard" samples prepared by cooling in oxygen or air from high temperatures, the orthorhombic-to-tetragonal transition occurs for y≈0.03. However, samples annealed in argon and then deoxidized near 400 °C remained orthorhombic for y up to 0.12. A similar effect was found in the case of cobalt. A number of conflicting interpretations of this effect were given, supported by various "local" experimental techniques like EXAFS and Mössbauer spectroscopy: among them, Fe or Co clustering, re-ordering within the basal plane, migration from the basal plane to the CuO_2 layers and simple re-ordering of the oxygen atoms. Suard[121] studied this subject in detail in her PhD thesis and, with other coworkers, in a number of subsequent publications[122-124]. By using NPD, they were able to unequivocally show that the high-temperature annealing induces a migration of the Co and Fe ions from the basal plane to the pyramidal CuO_2 layer sites. This was linked to the inability of these cations to assume the dumbbell coordination found in the basal plane of $YBa_2Cu_3O_6$. The NPD analysis of the Co/Cu site distribution is rather straightforward, due to the good scattering length contrast of the two species. The case of Fe is rather more delicate, but was successfully tackled[124] by preparing the samples with isotopic ^{57}Fe, which has a neutron scattering length (0.23×10^{-12} cm) comparable to that of cobalt.

Effects on the Doping Mechanism. The possibility of a tuneable double-site substitution makes the Fe and Co-substituted $YBa_2Cu_3O_{6+x}$ a very complex system indeed. In particular, understanding the doping mechanism in these compound is extremely difficult. At first sight, the replacement of Cu (which has an average valence of 2.333 in $YBa_2Cu_3O_7$) with Co^{+3}, Al^{+3} or Fe^{+3}, should result in a reduction of the remaining copper. However, this naive interpretation does not account for the additional oxygen atoms which could be introduced in the basal plane in the case of a Cu1 substitution. Furthermore, the creation of finely spaced twin domain boundaries ("tweed") is likely to result in a reduced charge transfer from the short chain fragments to the CuO_2 planes. An extensive experimental study of these effects has not yet been undertaken. However, it has been clarified that the net result of these competing effects is an overall reduction of the doping on the CuO_2 layers. In fact, near optimum doping can be restored by partially replacing trivalent yttrium with divalent calcium, which is known to increase the electronic charge on the CuO_2 layers (see Section 7.2). Suard and coworkers have shown[123] that the T_c of the series $Y_{1-y}Ca_yBa_2Cu_{3-y}Co_yO_7$ decreases only to ~80 K up to y = 0.4, whereas superconductivity disappears for the corresponding series without Ca. However, these high values of T_c can be attained only by assuring, by an appropriated high-temperature treatment in oxidizing environment, an almost complete substitution on the Cu1 site. Partial substitutions on the Cu2 site, confirmed by NPD[125], result in a much more pronounced T_c depression.

7.2. Y- AND Ba-SITE SUBSTITUTIONS

Our choice to discuss Y- and Ba-site substitutions together arises from the historical development of the work on rare-earth and alkali-earth metal substitutions in YBa$_2$Cu$_3$O$_{6+x}$. In YBa$_2$Cu$_3$O$_{6+x}$, Y and Ba are perfectly ordered upon the two respective sites because of the large size difference and the strong tendency of Y to prefer the 8-fold-coordinated interlayer site. In this respect, YBa$_2$Cu$_3$O$_{6+x}$ can be considered a triple oxygen-deficient perovskite with cation and oxygen vacancy ordering. As the size difference between the trivalent and the divalent species is reduced through substitutions, so is the tendency of the cations to order, which results in partial Y- and Ba-site occupancies, unless special synthesis routes are adopted. The poor understanding of the effects of double-site substitutions on the crystallographic symmetry and on the doping mechanism have generated in the literature a significant amount of confusion, which is only now beginning to be cleared. This is very unfortunate, because single-site, non-isovalent substitutions, in particular Ca-substitutions on the Y-site, could have offered a direct and powerful tool to understand the oxygen-doping mechanism: for instance, Y$_{1-y}$Ca$_y$Ba$_2$Cu$_3$O$_{6+x}$ would provide a link between self-doped compounds, like YBa$_2$Cu$_3$O$_{6+x}$, and direct-doped compounds like La$_{2-x}$Sr$_x$CuO$_4$. On the contrary, the learning process in this matter has been much more laboured. As a consequence, in spite of the success of crystallographic-based techniques like BVS calculations, a direct and reliable measure the planar copper valence in YBa$_2$Cu$_3$O$_{6+x}$ is still lacking.

Systematic Studies of Rare Earth Substitutions. Isovalent Y-site substitutions have been the object of numerous studies since very early on. Already in 1987, Tarascon and coworkers systematically studied the series REBa$_2$Cu$_3$O$_{6+x}$, with RE = Rare earths and lanthanum[13]. They determined that the structure could be formed with all these substituents, except for RE = Ce, Pm and Tb, probably due, at least in the case of Ce and Tb, to the strong tendency of these rare earths to have an oxidation state higher than +3. The REBa$_2$Cu$_3$O$_{6+x}$ family has therefore 13 members: Y, La and 11 of the 14 rare earths. When properly oxygenated, all of the compounds with the YBa$_2$Cu$_3$O$_{6+x}$ structure were found to be superconducting, with T$_c$'s between 87 and 96 K, except for RE = La and Pr (these two cases will be dealt with separately). This fact was immediately recognized to be remarkable, since, in most cases, the rare earth magnetism seems to have practically no effect on the superconducting properties. More recent systematic NPD studies of REBa$_2$Cu$_3$O$_{6+x}$ for x~0 and x~1 and almost all rare earths have been published by Guillaume and coworkers[126,127]. Structural data on the oxygen-rich compound and selected rare earths, as determined by NPD, have also been published by Currie *et al*[128]. In general, there is an increasing trend of T$_c$ vs. the ionic radius r$_{RE}$ of the rare earth, the highest T$_c$ belonging to NdBa$_2$Cu$_3$O$_7$. This may be due in part to the fact that small-r$_{RE}$ compounds could be slightly overdoped when fully oxygenated, whereas large-r$_{RE}$ compounds are optimally doped[129]. However, this only explains part of the effect (as we shall see Section 8.) T$_c$ = 93.5 K for optimally-doped

$YBa_2Cu_3O_{6+x}$, vs. 96 K for $NdBa_2Cu_3O_7$). The increasing trend is particularly intriguing, since substitutions with smaller cations could be expected to be equivalent to the application of external pressure, which, for most copper oxide superconductors including $YBa_2Cu_3O_{6+x}$, has the effect of *increasing* T_c (see Section 8). At present, the reason for this trend has not been clarified.

A systematic study of the properties of $REBa_2Cu_3O_{6+x}$ as a function of *both* r_{RE} *and* x entails a major research effort, which has not yet been undertaken for all rare earths by any group. When structural studies by neutron diffraction are required, and additional difficulty is represented by the need to use isotopically enriched rare earths, due to the high absorption cross section of the natural isotopic mixture for many rare earth elements. Transport and/or structural data as a function of x for short series of rare earths have been published by Veal[130] (Y, Gd, and Nd) and Kogachi[131] (Tm, Gd and Sm). Extensive data are available for selected rare earths: for example, for Er, see references 37,132-136. For Nd, see references 137,138. For Ho, see reference 139. These authors evidenced a remarkable correlation between r_{RE}, x_{T-O} and x_c: x_{T-O} was found to *increase* with increasing r_{RE}. However, this effect is correlated with an increase of x_c as well, so that, for most rare earths, only orthorhombic samples are superconducting (the apparent exception of La can be easily explained, and will be discussed later). Once again, this observation reinforced the belief that the orthorhombic symmetry, at least on the short range, could be somewhat necessary for superconductivity. Nowadays, the general validity of this view has been invalidated by the discovery of *intrinsically* tetragonal superconductors, like the mercury-based compounds. However, the correlation between orthorhombicity and superconductivity in $REBa_2Cu_3O_{6+x}$ remains intriguing. In our opinion, the best explanation is based on the charge-transfer argument. All $REBa_2Cu_3O_{6+x}$ phases require charge transfer from the basal plane to the CuO_2 layers to become superconducting, except, perhaps, for the heavily Ca-substituted compounds (which are difficult to synthesize as single-phases: see below). Oxygen insertion in the basal plane *does not* lead to charge transfer unless chain fragments of significant length are present (local orthorhombic symmetry). In defect-free compounds, the short-range ordering region is very narrow or non-existent, so that compositions with local orthorhombic symmetry are also orthorhombic in the average sense. As a consequence, only orthorhombic samples have significant charge transfer and can be superconducting. On the other hand, the presence of defects in the basal plane (e.g., Co or Fe substitutions) or near it (e.g., La-substitutions on the Ba site) tend to frustrate the average symmetry (e.g., by provide favourable conditions for microscopic twinning). By these substitutions, therefore, it is possible to synthesize superconducting samples with an average tetragonal symmetry.

Lanthanum Substitutions. Together with Pr, La represented an anomaly in the early work on the $REBa_2Cu_3O_{6+x}$ series. A $YBa_2Cu_3O_{6+x}$-type structure could be formed by using La_2O_3 instead of Y_2O_3, but the resulting samples did not have the same superconducting properties as the other members of the series. In their early systematic study of the REBa$_2$Cu$_3$O$_{6+x}$ series, Tarascon reported the absence of superconductivity for LaBa$_2$Cu$_3$O$_{6+x}$ (reference [13]). However, most other groups reported T_c's between 50 and 80 K with broad transitions. Furthermore, most samples were shown to contain impurity phases. The reason for this was clarified by a series of systematic studies, mainly by Japanese groups (see, for instance, reference [140] and references cited therein): LaBa$_2$Cu$_3$O$_{6+x}$ was in fact the end member of the series LaBa$_{2-y}$La$_y$Cu$_3$O$_{6+x}$, in which the Ba-site was partially substituted with La. In fact, the (non-superconducting) compound La$_3$Ba$_3$Cu$_6$O$_{14+\delta}$, corresponding to y=0.5, had been known for a number of years, and its crystal structure was studied in detail in the early eighties by Er-Rako and coworkers[141]. The double Y- and Ba- site substitution is not unique of La, and has been evidenced for other large rare earths, like Nd[142], Sm and Eu. However, for the latter elements, the end member of the series can be stabilized under the usual synthesis conditions, whereas for La this is not the case. Therefore, the early samples, with nominal composition LaBa$_2$Cu$_3$O$_{6+x}$ were in fact mixtures of LaBa$_{2-y}$La$_y$Cu$_3$O$_{6+x}$ (x~0.25) plus impurity phases that compensated the initial stoichiometry.

The partial substitution on the Ba-site has multiple effects:

i)   Since La^{+3} replaces Ba^{+2}, charge compensation requires a *reduction* of the average Cu valence for a given value of x: $<v_{Cu}> = 5/3 + 2/3x - 1/3y$.

ii)   NPD experiments[143,144] have shown that this effect is partially compensated by the insertion of additional oxygen in the basal plane. For instance, for y=0.5, the fully oxygenated samples contain 1.22(6) oxygen atoms per copper atom in the basal plane[144].

iii) The presence of point defects in a site which is quite close to the basal plane induces a reduction of the chain order parameter, and, as a consequence, of the orthorhombic strain. The mechanism driving this effect has not been clarified, but could possibly be related to the previous one: La ions could be associated with a local increase of the oxygen concentration in the adjacent basal plane, with one or more copper atoms having a coordination higher than 4. These sites, in turn, would promote a 90° rotation of the chains, thereby reducing the orthorhombicity. Incidentally, we will note that this mechanism, in order to be effective, would require La "clustering" along microtwin boundaries, similar to what proposed by Hodeau for Cu1-site substitutions[118].

The analysis of point i) suggested an effective method to prevent Ba-site substitutions, leading to the synthesis of stoichiometric LaBa$_2$Cu$_3$O$_{6+x}$. In fact, the average valence of Cu in the YBa$_2$Cu$_3$O$_{6+x}$ cannot be lower than 5/3, since Cu1^{+1} and Cu2^{+2} are the lowest valence states allowed by the respective coordinations. Therefore, a synthesis carried out under reducing condition will favour the formation of stoichiometric LaBa$_2$Cu$_3$O$_6$. Oxygen can be later introduced by a low-temperature annealing. This method was successfully employed by Maeda an coworkers[145], who were able to produce samples with T_c = 93 K and large orthorhombicity. The structural properties of this type of samples as a function of oxygen content were subsequently studied by Wada et al.[146].

The substitution of La on the Ba site can be perhaps better studied in the solid solution $Y(Ba_{2-y}La_y)Cu_3O_{6+x}$, where the 8-fold coordinated site is occupied by Y only. This system was extensively studied by several authors. For example, Cava and coworkers[147] studied a series of $Y(Ba_{2-y}La_y)Cu_3O_{6+x}$ samples with $0 \leq y \leq 1$. All the samples were "fully oxygenated", i.e., post-annealed in oxygen at low temperatures. Cava showed that a O-T phase transition occurred for $y \approx 0.4$. Charge compensation is achieved in a different way in the O and T phases: at low La contents, the total oxygen content of the samples remains constant ($x \sim 1$), while the average copper valence is progressively reduced. Above $y=0.4$, extra oxygen atoms are incorporated in the basal plane. After an initial increase of a few K at very low values of y the superconducting critical temperature T_c *decreases* as a function of y, since the reduction of hole doping is not completely compensated by the additional oxygen. Cava's samples were not superconducting for $y>0.5$. However, it has been shown[148] that different annealing schemes allow the synthesis of superconducting samples even for higher values of y, albeit with a reduced T_c.

Another interesting compound, in a sense "derived" from the previous ones, is $CaLaBaCu_3O_{6+x}$, in which the Y-site is occupied by calcium, while La and Ba share the Ba-site (see, for instance, references [149-151]). Note that the reducing effect of La on the Ba site is completely compensated by the replacement of Y with Ca. When properly oxygenated, this compound is and 80 K *tetragonal* superconductor, and beautifully illustrates the complex interplay between local symmetry, average symmetry and charge transfer. The average tetragonal symmetry, induced by the La substitution on the Ba site, does not prevent charge transfer, since chain fragment of sufficient length are still present in the basal plane (local orthorhombic symmetry). On the other hand, the La on the Ba-site is not sufficiently close to the basal plane to interfere with the charge transfer process in other ways, in contrast, for instance, with the analogous situation for the Al substitution on the Cu1 site. As a consequence, $CaLaBaCu_3O_{6+x}$ maintains a very high values of T_c even in the presence of a high degree of chain disorder.

Calcium Substitutions. Non-isovalent substitutions on the Y-site in the $YBa_2Cu_3O_{6+x}$ structure have been the subject of many studies since the discovery of the compound. In principle, these substitutions offer a powerful tool to vary the charge on the CuO_2 planes directly, and, therefore, study the behavior of structural and superconducting properties versus carrier concentration. In fact, at least in the oxygen-poor compound $YBa_2Cu_3O_6$ ($x=0$), one Ca atom replacing Y amounts to one half of a hole being doped into the CuO_2 layers, since, in this case, the copper in the basal plane is not mixed-valent (Cu^{+1}). Furthermore, a systematic study of the superconducting properties as a function of oxygen *and* calcium content would allow the charge transfer from chains to planes to be quantified. In addition, Ca substitutions in oxygen-rich $YBa_2Cu_3O_{6+x}$ ($x \sim 1$) allow the extension of the high-doping region with respect to pure $YBa_2Cu_3O_{6+x}$, well into the "overdoping" regime. Finally, the Y-site is the farthest from the basal plane, where the oxygen non-stoichiometry occurs. Therefore, substitutions on the Y-site are more likely to preserve, at least in a qualitative way, the features of the T-x phase diagram found in pure YBa$_2$Cu$_3$O$_{6+x}$. Unfortunately, the study of Ca-substitutions on the Y site has generated a significant amount of confusion, due essentially to two reasons: the narrow solubility range of Ca in the YBa$_2$Cu$_3$O$_{6+x}$ structure and the possibility of simultaneous substitutions on the Ba site. The latter effect is particularly insidious, since Ba-containing impurities are often elusive to x-ray and neutron investigations. Ca substitutions on the Ba site have several effects. First of all, they make the simple hole-counting argument, based on the overall sample stoichiometry, no longer valid, since the Ba-site substitution is isovalent. Secondly, similar to La substitutions on the Ba site, they reduce the chain ordering, and therefore the orthorhombic strain, thereby modifying the parameters of the x-T phase diagram. As a consequence of the latter effect, the oxygen-doping mechanism may also be perturbed.

The first attempt to study the influence of Ca on superconductivity in the "1-2-3" system was made by Manthiram and coworkers in the early 1988 [152]. They found the solubility limit for Ca to be y~0.3. These authors focused their attention on samples with high oxygen contents. For these samples, T_c was found to decrease with increasing y. The orthorhombic strain also diminishes, while the c-axis increases. Similar results were obtained by Tokiwa and coworkers [153].

The first extensive work on the (Y$_{1-y}$Ca$_y$)Ba$_2$Cu$_3$O$_{6+x}$ system, varying both calcium and oxygen content was published in 1988 by Tokura and coworkers [154], who were the first to clearly evidence an overdoping behavior for the Ca-substituted compound: samples with y$\geq$0.2 have a maximum T_c at intermediate oxygen content, followed by a decrease of the critical temperature. Also, the maximum oxygen uptake of the samples decreases with increasing y (x_{max}=0.8 for y=0.3).

In 1989, McCarron and coworkers presented the results of work on oxygen-deficient (Y$_{1-y}$Ca$_y$)Ba$_2$Cu$_3$O$_{6+x}$ [155]. Their main result was the observation of superconductivity at 50 K in tetragonal (Y$_{0.8}$Ca$_{0.2}$)Ba$_2$Cu$_3$O$_{6.2}$. The same authors subsequently analyzed the structural properties of the tetragonal superconductor by neutron diffraction [156,157]. Similar results were reported in an extensive study on chemical, structural and superconducting properties of tetragonal (Y$_{1-y}$Ca$_y$)Ba$_2$Cu$_3$O$_{6+x}$ by Liu and coworkers [158].

In 1990 Gledel and coworkers published the results of a systematic study on thermodynamic and structural properties of (Y$_{1-y}$Ca$_y$)Ba$_2$Cu$_3$O$_{6+x}$ where both calcium and oxygen contents were varied [159]. In a subsequent paper, they focused their attention on the y=0.3 samples, studying the transport and superconducting properties as a function of the oxygen content [160]. Contrary to the findings of Tokura et al., Gledel and coworkers found no evidence of overdoping behavior at high oxygen content in their samples: the critical temperature of all samples with oxygen contents above x=0.8 was almost constant (~80 K).

Most of the samples used for the aforementioned studies contained impurity phases. In particular, the BaCuO$_2$ phase was observed by the majority of authors. The presence of BaCuO$_2$ is an indication of metal-ion disorder: Ca partially substitutes both on the Y-

and on the Ba-sites, so that the composition of the "1-2-3" phase is $(Y_{1-y+z}Ca_{y-z})(Ba_{2-z}Ca_z)Cu_3O_{6+x}$, and $BaCuO_2$ is formed to satisfy the initial stoichiometry. Buckley and coworkers addressed the problem of metal-ion disorder in Ca-substituted $YBa_2Cu_3O_{6+x}$, and in the related compounds $YBa_2Cu_{3.5}O_{7.5}$, and $YBa_2Cu_4O_8$ [161]. They came to the conclusion that, depending on the value y of the calcium content , between 20% and 60% of the calcium can substitute for Ba. In samples with high values of y, calcium substitutes preferentially for barium. By adjusting the initial stoichiometry to compensate for the metal-ion disorder, Buckley and coworkers were able to obtain single-phase samples with y=0.5.

In his PhD thesis[162], Radaelli addressed the problem of the double Y- and Ba-site substitution and explored possible chemical venues to obtain single-site substituted samples. He found that one effective way to reduce the amount of Ca on the Ba site was to synthesize the samples at high temperatures under reducing conditions, which inhibited the formation of $BaCuO_2$. The solubility limit for single-site-substituted samples was found to be rather low (y≤0.2). A series of $(Y_{1-y}Ca_y)Ba_2Cu_3O_{6+x}$ samples with y=0.1 and varying oxygen contents (oxygen was later introduced in calibrated amounts at 400°C by a volumetric titration apparatus) was subsequently studied by NPD. The structural parameters determined by Radaelli were later used by Tallon and coworkers[163] in their systematic study of the hole concentration in the CuO_2 layers (see Section 8.) The structural and superconducting properties of Radaelli's samples are strikingly different from previous work. One particularly useful parameter to consider is the behavior of the orthorhombic strain σ as a function of x. Contrary to previous findings, at high oxygen contents the orthorhombic strain is *not* significantly reduced with respect to pure $YBa_2Cu_3O_{6+x}$ ($\sigma = 1.65 \times 10^{-2}$ in $Y_{0.9}Ca_{0.1}Ba_2Cu_3O_7$ vs. $\sigma \approx 1.67 \times 10^{-2}$ in $YBa_2Cu_3O_7$). Furthermore, the T-O phase transition is pushed to *lower* oxygen contents ($x_{T-O} = 0.28$ in $Y_{0.9}Ca_{0.1}Ba_2Cu_3O_{6+x}$). Superconductivity is observed only for orthorhombic samples, and the T_c vs. x curve clearly displays overdoping behavior, with $T_c^{MAX} = 90$ K for x=0.8 and $T_c = 68$ K for x=1.

Among the more recent systematic studies, we will cite the works by Berastegui and coworkers[164], Awana and coworkers[165,166] and Böttger and coworkers[167], all using NPD to calculate the behavior of internal parameters and BVS as a function of x and/or y. Berastegui studied 9 samples across the x-y phase diagram, while Awana and Böttger focus on selected oxygen contents and explore the behavior of the structural and superconducting properties as a function of y. In spite of the significant effort in this field, however, substantial differences can still be evidenced between samples with the same nominal values of x and y. For instance, the orthorhombic strain reported for y=0.1 and x≈1 by Berastegui and Böttger ($\sigma \approx 1.58 \times 10^{-2}$) is substantially *lower* than in the work by Radaelli, and, for the same samples, T_c is higher (85 K for Berastegui, 72 K for Böttger). Also, contrary to Radaelli's findings, Berastegui synthesized a *tetragonal superconducting* sample with x=0.1 ($T_c = 20$ K). A careful evaluation of these differences leads to the conclusion that they cannot be simply associated with slight stoichiometry variations, and must be attributed to different cation ordering

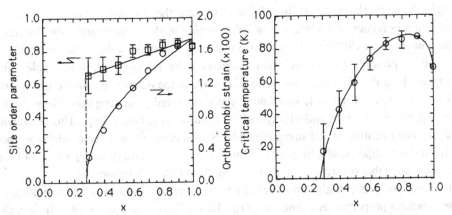

Figure 7.1. **Left**: Orthorhombic strain and site occupancy order parameter, defined as $(n(O4)-n(O5))/(n(O4)+n(O5))$ as a function of x in $Y_{0.9}Ca_{0.1}Ba_2Cu_3O_{6+x}$. **Right**: T_c vs. x in $Y_{0.9}Ca_{0.1}Ba_2Cu_3O_{6+x}$. The error bars indicate the width of the transitions.

arrangements. This implies that fundamental synthesis issues still represent an obstacle towards the comprehension of the intrinsic properties of $(Y_{1-y}Ca_y)Ba_2Cu_3O_{6+x}$.

Praseodymium and Pr/Ca Substitutions. $PrBa_2Cu_3O_{6+x}$ constitutes a remarkable exception in the $REBa_2Cu_3O_{6+x}$ series. After the explanation of the "apparent" anomaly of $LaBa_2Cu_3O_{6+x}$, the Pr compound is the only one of the 13-element family not to be superconducting. In fact, it has been shown that, in the series $Y_{1-y}Pr_yBa_2Cu_3O_7$, T_c is rapidly depressed, and reaches zero for y≈0.55 (see, for instance, [168]). This effect has generated a very significant amount of research, which still continues today, employing a number of different techniques. For a thorough summary of the state of the art in 1992, see, for instance, the review paper by Radousky[169]. Most of the discussion on the suppression of superconductivity in Pr-substituted $YBa_2Cu_3O_{6+x}$ centers around the valence of Pr. Magnetic susceptibility measurement indicates a value of nearly +4. X-ray absorption spectroscopy would indicate a valence state between +3 and +4. On the contrary, resonant valence-band photoemission studies indicate a value of +3. Clearly, the mechanism for the disappearance of superconductivity is different depending on the valence assigned to Pr. If the valence is higher than +3, the most logical explanation is hole filling: charge carriers would be trapped on the Pr ions instead of doping the CuO_2 layers. If the Pr valence is strictly +3, this mechanism cannot be effective, and an alternative explanation has to be sought, the most popular one being superconducting pair breaking by local moments. Obviously this creates a difficulty, since other rare earths, such as Gd, possess a much larger magnetic moment but do not destroy superconductivity. This discrepancy has been explained with the larger radial extent of the Pr-4f orbitals, leading to a significant hybridization with the CuO_2 valence band. A compromise view is that of hole *trapping*, in which the mobile holes would be localized in hybrid bands still largely retaining the original CuO_2 character, so that Pr would retain the +3 valence state. A strong indication in favour of the hole filling/hole trapping models was obtained by Neumeier and coworkers in 1989[170]. They argued that if T_c is depressed due to underdoping, it should be possible to recover superconductivity by injecting additional holes, for instance, by a double Y/Ca substitution. Indeed, in the doubly substituted compound $(Y_{1-y-z}Pr_yCa_z)Ba_2Cu_3O_7$, T_c could be partially recovered, indicating that at least part of the suppression of superconductivity had to be attributed to hole filling. This result was subsequently confirmed and strengthened by Norton et al.[171], who were able to prepare yttrium-free superconducting thin films of $Pr_{0.5}Ca_{0.5}Ba_2Cu_3O_7$ (T_c = 43 K), demonstrating that superconductivity is not confined to Y-rich regions.

Structural analysis can give an insight into this issue by the study of the evolution of the structural properties as a function of r_{RE} for the $REBa_2Cu_3O_{6+x}$ series. In fact, the ionic radii of Pr^{+3} (1.126 Å) and Pr^{+4} (0.96 Å) differ by 16%, which would locate the Pr compound in a completely different place in the rare earth series depending on its valence. With this in mind, Neumeier and coworkers performed a NPD experiment on $PrBa_2Cu_3O_7$ and the $Y_{1-x}Pr_xBa_2Cu_3O_7$ series[172], and compared the results with those of other rare earths. A particularly useful parameter is the Cu2-Cu2 separation, which is proportional to r_{RE} and is directly sensitive to the size of the rare earth. From this analysis, they concluded that Pr has a valence state of ~3.3. More recently, a particularly useful piece of information has been added by Guillaume et al., in their systematic NPD study of the rare earth series for x=0 and x=1. For the fully oxygenated compounds, Pr is, indeed, anomalous for many of the structural parameters, when the value of r_{RE} for Pr^{+3} is used. However, the structural parameters for the oxygen-deficient compound $PrBa_2Cu_3O_6$ are perfectly consistent with a valence state of +3. This strongly suggests that the valence of Pr is not constant, but varies as a function of the oxygen content, and is a very sensible result from the structural chemical point of view, since a higher Pr valence in $PrBa_2Cu_3O_6$ would imply unreasonably low oxidation states for Cu.

These studies constituted strong proofs of the importance of hole filling/trapping for the suppression of superconductivity, but did not rule out pair-breaking altogether. In fact, initially, Neumann indicated the presence of a strong pair breaking component to the T_c suppression, in addition to hole filling[170]. However, later on, the same group revised their statements, based on new measurements of the superconducting penetration depth λ by muon spin resonance (μSR) in Ca-free $Y_{1-y}Pr_yBa_2Cu_3O_7$ samples. The T_c vs. λ dependence appeared to follow the universal relationship previously established by Uemura and coworkers for underdoped compounds[173], strongly suggesting that no additional explanation beside underdoping is necessary. The apparent contradiction was later addressed by systematic studies of the structural properties of Pr/Ca doped compounds by NPD[174,175]. In particular, Lundqvist et al. have employed the BVS technique, evidencing a reduction of the charge on the CuO_2 layers as a function of y even for the "compensated" compound $(Y_{1-2y}Pr_yCa_y)Ba_2Cu_3O_7$, which should be optimally doped even for the extreme case of Pr^{+4}. Lundqvist suggests that small changes in interatomic distances could make the charge transfer mechanism less effective in these compounds. However, like in the case of $(Y_{1-y}Ca_y)Ba_2Cu_3O_7$, the possibility of Ca or even Ca/Pr double-site (Y/Ba) substitutions should not be ruled out.

8. Hole Count, and Phase Separation: Overdoping and Pressure Effects

Hole Count on the CuO$_2$ Layers. In the previous sections, we have outlined the main issues related to the determination of the hole count n_h on the CuO$_2$ layers in YBa$_2$Cu$_3$O$_{6+x}$ and related compounds. To summarize, if the overall stoichiometry is known, one can determine the average copper valence by simple charge balance considerations, assuming the valence of all the other cations and of oxygen to be known. However, the hole count on the CuO$_2$ layers is very difficult to determine exactly, since in YBa$_2$Cu$_3$O$_{6+x}$ mixed valence occurs both on the CuO$_2$ layers and in the basal plane. The determination of the hole count is of critical importance from the theoretical and experimental viewpoints, for several reasons:

i) It allows a direct comparison between YBa$_2$Cu$_3$O$_{6+x}$ and other systems, like La$_{2-x}$Sr$_x$CuO$_4$, where the CuO$_2$ layers are the only structural element believed to accommodate mixed valence. In the latter case, the hole count is simply proportional to x.

ii) The n_h determination would allow one to establish whether there is a general relationship between n_h and superconductivity. Most theoretical models of superconductivity in copper oxides assume more or less explicitly the existence of a universal T_c/T_c^{MAX} vs. n_h curve, where T_c^{MAX} is the value of T_c at optimum doping and it is system-dependent. If this assumption were to be disproved, this would have enormous consequences on our understanding of these systems.

iii) In the specific case of YBa$_2$Cu$_3$O$_{6+x}$, n_h is related to the charge transfer. A direct measure of n_h would be an important step toward refining the charge-transfer models based on oxygen configurations in the basal plane.

In principle, there are several ways to determine n_h. We have already discussed the potential benefits and the practical difficulties of non-isovalent substitutions. Also, since the first structural studies, the BVS method has been employed to determine the CuO$_2$ valence in YBa$_2$Cu$_3$O$_{6+x}$. The latter method, with significant modification mainly due to the contribution of Brown and Tallon and coworkers, is now widely used and is worth further discussion.

BVS for YBa$_2$Cu$_3$O$_{6+x}$. There are a number of obstacles to the quantitative determinations of n_h using BVS.

1) The BVS parameter r_0 depends on the valence of the species in question, which, on the other hand, is the parameter we want to determine.

2) The value of r_0 for Cu^{+3} is not well established, since very few Cu^{+3} compounds are known.

3) In calculating n_h from the valence of Cu2, one makes the implicit assumption that the holes reside on the Cu2 site. In fact, it is known that the holes on the CuO_2 layers have a significant oxygen 2p character.

4) In $YBa_2Cu_3O_{6+x}$, like in all other compounds, BVS are sensitive to internal (chemical) and external (pressure) strains. This is the reason why absolute determinations of valences by BVS are only reliable within 10-20%.

Issues 1) and 2) were addressed in a paper by Brown[176], who developed a self-consistent method to calculate BVS for $YBa_2Cu_3O_{6+x}$.

In 1990, Tallon simultaneously addressed 3) and 4)[38] and proposed a new method of calculating n_h. Instead of using V_{Cu2} to determine n_h, Tallon proposed to calculate BVS's for *both* planar oxygen *and* copper, and use a linear combination of these values as a measure of n_h. In particular, the following parameters are introduced:

$$V_- = 2 + V_{Cu2} - V_{O2} - V_{O3}$$
$$V_+ = 6 - V_{Cu2} - V_{O2} - V_{O3}$$

(8.1)

Note that, in (8.1), the absolute (positive) value of the oxygen valences are used. The parameter V_- is zero for Cu^{+2} and O^{-2}. Furthermore, its value increases by adding positive charge on either copper or oxygen. Therefore, V_- can be taken as a measure of n_h irrespectively of the hole distribution in the CuO_2 layers. Also, V_- has the additional benefit of being insensitive to either internal or external pressures, since their effect on the copper and oxygen valences would tend to cancel (they would cancel exactly for strains in the a-b plane). On the other hand, the ratio V_+/V_- could be taken as a measure of the hole distribution between copper and oxygen, varying between -1 and 1 for holes purely on copper/oxygen, respectively. However, V_+ is also very sensitive to strain, since its effects on copper and oxygen valences tend to add. Tallon calculated the values of V_- for $YBa_2Cu_3O_{6+x}$, based on NPD data by Jorgensen and Cava, and demonstrated that both the quantitative values of n_h (between 0 and 0.2 holes/Cu atom for $YBa_2Cu_3O_6$ and $YBa_2Cu_3O_7$, respectively) and the qualitative T_c vs. n_h dependence are very similar to underdoped $La_{2-x}Sr_xCuO_4$. Subsequently, Tallon and coworkers extended their study to the overdoped side by including Ca-doped samples[163], and proposed a universal parabolic relationship between n_h and T_c/T_c^{MAX} (Figure 8.1). Tallon also found the existence of a simple monotonic relationship between n_h and room-temperature thermopower, which he found to be valid for many systems. The calculation of V_- is now widely used in structural study to determine n_h in substituted compounds as well.

Overdoping in oxygen-rich $YBa_2Cu_3O_{6+x}$. The determination of the T_c vs. n_h curve (see 8.1) led to an apparently simple but very important conclusion: $YBa_2Cu_3O_7$ is very close to the "optimum doping" level. It is quite remarkable that, in this case, the nearly defect-free material has also the highest T_c, but we have no better option than attributing it to an extraordinary coincidence. This, in turn, brought many researchers to a striking realization: the 90 K-plateau of $YBa_2Cu_3O_{6+x}$ could be nothing but the

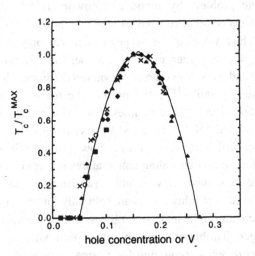

Figure 8.1. Critical temperature T_c of various copper oxide superconductors normalized to T_c^{MAX}, plotted as a function of hole concentration, p, as determined from either direct charge balance considerations or bond valence sums. From reference [163].

normal T_c saturation near optimum doping. Therefore, additional explanations in terms of complex oxygen ordering schemes near x=1 were no longer needed.

The details of the 90 K plateau region were studied by Loram and coworkers in 1991, using ceramic samples. In fact, Loram found that "fully oxygenated" YBa₂Cu₃O₇ is slightly overdoped (T_c = 92 K), and that the highest T_c (93.5 K) is obtained for x≈0.9. Similar conclusions were later obtained on single crystals by Claus and coworkers[177]. This work, together with a number of other studies suggestive of overdoping for fully oxygenated YBa₂Cu₃O₇, was later reviewed by Tallon[178]. Tallon analyzed the indication obtained from various techniques, including atomic substitutions, BVS, μSR, thermo-electric power, normal-state susceptibility, pressure dependence of T_c, etc., all leading to the same conclusion about the overdoping state of YBa₂Cu₃O₇.

Phase Separation for x≈1. One of the most intriguing phenomena associated with the x≈1 region of the YBa₂Cu₃O₆₊ₓ phase diagram is the possibility of segregation into two different phases, each having a slightly different value of T_c. This hypothesis was first put forth by Claus and coworkers, [179], as a possible interpretation of their x-ray diffraction and magnetic susceptibility data on high-quality single crystals. For x>0.92, the diffraction lines of Claus' crystals were sharply split, consistent with two different values of the c lattice parameter. This splitting was associated with a broadening of the resistive and diamagnetic transition, which would often reveal a two-step structure. The doubling of the superconducting transitions was also confirmed by specific heat measurement, confirming that both phases have bulk nature (for references, see the extensive review of the problem by Janod and coworkers[180]). Obviously, the possibility that these effects could arise from oxygen inhomogeneities had to be taken into account, since these high values of the oxygen content can only be reached for low annealing temperatures, where oxygen mobility is rather low. In an extended paper published in 1993, Janod and coworkers critically reviewed the susceptibility and specific heat evidence of phase separation[180], and present the results of a new series of experiments performed on polycrystalline samples annealed in pure oxygen at different temperatures between 300 and 550 °C, followed by quenching. Janod points out that both susceptibility and specific heat measurements show split transitions, but following a different pattern as a function of annealing temperatures (or, equivalently, of oxygen content). The specific heat data show double transitions in for *all* annealing temperatures, whereas the susceptibility curves are split only for annealing temperatures below 420 °C. All these data can be understood by assuming that the crystallites have an inhomogeneous oxygen distribution, with an oxygen-rich outer shell and a slightly oxygen-deficient inner core, each accounting for a large fraction of the total volume. Janot argues that a single susceptibility transition is observed as long as the outer shell has a higher T_c than the inner core, due to the shielding effect of the supercurrents. As soon as the outer shell reaches the overdoping region, double susceptibility transitions are observed, since the T_c of the core is higher than that of the outer shell. On the other hand, specific heat measures the true phase fraction and always sees double transitions. Janot, therefore, concludes that the apparent "phase separation" is not a thermodynamic property of the $YBa_2Cu_3O_{6+x}$ phase diagram, but is an extrinsic feature due to slow oxygen diffusion. Whether the actual oxygen diffusion rate is sufficiently slow to allow the formation of this "onion-like" structure, it still very much a controversial subject[181]. In the hope of clarifying the issue, Conder and coworkers performed very detailed NPD measurements in the overdoping region[181,182]. These measurements have revealed the presence of significant structural anomalies, later confirmed by EXAFS, which are extremely interesting in their own right, but have not provided clear evidence of a two-phase coexistence, perhaps due to the insufficient resolution of the experiments.

Structural and Electronic Effects of Applied Pressure. Understanding the effect of applied pressure on the evolution of the structural and superconducting properties is undoubtedly one of the keys towards the understanding of superconductivity in copper oxides, and is too much a wide subject to be thoroughly discussed were. In extreme synthesis, applied pressure affects superconductivity in copper oxides in three ways:

1) By affecting the hole count on the CuO_2 layers, mainly by promoting charge transfer from the charge reservoir layers.

2) By changing the amount of "puckering" or "buckling" of the CuO$_2$ layers, which is believed to reduce T$_c$. In perovskite-related compounds, the application of pressure usually reduces the distortion on the transition metal - oxygen bond angles.

3) By an intrinsic effect on T$_c$, which is usually *positive* (at least below or near optimum doping).

The actual evolution of T$_c$ as a function of pressure results from an often complex interplay of these effects, and is therefore dependent on a number of factors, including the nature of the charge reservoir, the doping state of the compound (under-, optimally- or under-doped) and by the pre-existing amount of distortion on the CuO$_2$ layers.

In the specific case of REBa$_2$Cu$_3$O$_{6+x}$, Chu and coworkers showed in 1998 that the value of dT$_c$/dP is very different for oxygen-deficient and oxygen-rich materials[183]. dT$_c$/dP was found to be 9K/GPa for a T$_c$=60 K, oxygen-deficient sample of EuBa$_2$Cu$_3$O$_{6+x}$, whereas the oxygen-rich material (T$_c$=92 K) dT$_c$/dP was only 1K/Gpa. With the aim of clarifying the issue, Jorgensen and coworkers studied the effect of applied pressure on the structure of oxygen-rich (x≈0.9) and oxygen-deficient (x≈0.6) samples of YBa$_2$Cu$_3$O$_{6+x}$ by NPD[184]. Jorgensen employed a helium gas cell, which can only reach modest pressures (P≤0.6 GPa), but operates in hydrostatic conditions, yielding extremely accurate and reproducible structural parameters. For both samples, the structural changes were consistent with an *increase* of the hole count on the CuO$_2$ layers. By applying a variant of the bond valence sum method, Jorgensen is able to quantify the change in valence on the Cu2 copper ions to be dV/dP ≈ 0.007 holes/GPa/Cu-atom in both cases. This result provides an explanation for the largely different values of dT$_c$/dP, and the strong T$_c$ increase of the strongly underdoped x=0.6 compound, but fails to explain why the optimally doped (or perhaps slightly overdoped) oxygen-rich compound still has a *positive* (albeit small) dT$_c$/dP. Clearly, the intrinsic T$_c$ increase and/or the reduction in the "puckering" of the CuO$_2$ layers as a function of pressure are the likely candidate to explain this effect. However, up to date, a complete, coherent and uncontroversial explanation of the pressure effects in this compound and in the other copper oxide superconductors has not yet been found.

Acknowledgements

This work has benefited from the help and the advice of a number of colleagues. My special thanks go to Alan Hewat and James D. Jorgensen, who extensively revised this manuscript. Also, I wish to mention J. Röhler for his help with the literature on EXAFS, P. Schleger for useful discussions on the theoretical and experimental aspects of the x-T phase diagram and H. Casalta and E. Suard for pointing out a number of important references. Finally, I acknowledge the Institute Max Von Laue - Paul Langevin, of which I was formerly staff member, for support during much of the elaboration of the present manuscript.

References

1 H.K. Wu, J.R. Ashburn, C.J. Torng, P.H. Hor, R.L. Meng, L. Gao, Z.J. Huang, Y.Q. Wang and C.W. Chu, Phys.Rev.Lett. **58**, 908 (1987).

2 J.J. Capponi, C. Chaillout, A.W. Hewat, P. Lejay, M. Marezio, N. Nguyen, B. Raveau, J.L. Soubeyroux, J.L. Tholence and R. Tournier, Europhys.Lett. **3**, 1301 (1987).

3 M.A. Beno, L. Soderholm, D.W. Capone II, D.G. Hinks, J.D. Jorgensen, J.D. Grace, I.K. Schuller, C.U. Segre and K. Zhang, Appl.Phys.Lett. **51**, 57 (1987).

4 R.J. Cava, B. Batlogg, R.B. van Dover, D.W. Murphy, S. Sunshine, T. Siegrist, J.P. Remeika, E.A. Rietman, S. Zahurak and G.P. Espinosa, Phys.Rev.Lett. **58**, 1676 (1987).

5 R.M. Hazen, L.W. Finger, R.J. Angel, C.T. Prewitt, N.L. Ross, H.K. Mao and V.G. Hadidiacos, Phys.Rev.B **35**, 7238 (1987).

6 T. Siegrist, S. Sunshine, D.W. Murphy, R.J. Cava and S.M. Zahurak, Phys.Rev.B **35**, 7137 (1987).

7 Y. LePage, W.R. McKinnon, J.M. Tarascon, L.H. Greene, G.W. Hull and D.M. Hwang, Phys.Rev.B **35**, 7245 (1987).

8 W.I.F. David, W.T.A. Harrison, J.M.F. Gunn, O. Moze, A.K. Soper, P. Day, J.D. Jorgensen, D.G. Hinks, M.A. Beno, L. Soderholm, D.W. Capone II, I.K. Schuller, C.U. Segre, K. Zhang and J.D. Grace, Nature (London) **327**, 310 (1987).

9 J.E. Greedan, A.H. O'Reilly and C.V. Stager, Physocal Review B **35**, 8770 (1987).

10 F. Beech, S. Miraglia, A. Santoro and R.S. Roth, Phys.Rev.B **35**, 8778 (1987).

11 S. Katano, S. Funahashi, T. Hatano, A. Matsushita, K. Nakamura, T. Matsumoto and K. Ogawa, Jpn.J.Appl.Phys. **26**, L1046 (1987).

12 D.E. Cox, A.R. Moodenbaugh, J.J. Hurst and R.H. Jones, J. Phys. Chem. Solids **49**, 47 (1988).

13 J.M. Tarascon, W.R. McKinnon, L.H. Greene, G.W. Hull and E.M. Vogel, Phys.Rev.B **36**, 226 (1987).

14 P.K. Gallagher, H.M. O'Bryan, S.A. Sunshine and D.W. Murphy, Mat.Res.Bull. **22**, 995 (1987).

15 A. Santoro, S. Miraglia, F. Beech, S.A. Sunshine, D.W. Murphy, L.F. Schneemeyer and J.V. Waszczak, Mat.Res.Bull. **22**, 1007 (1987).

16 A.W. Hewat Capponi J J, Chaillout C, Marezio M, Hewat E A, Solid State Communications **64**, 301 (1987).

17 J.D. Jorgensen, M.A. Beno, D.G. Hinks, L.Solderholm, K.L. Violin, R.L. Hitterman, J.D.Grace, I. K. Shuller, C.U. Segre, K. Zhang and M.S. Kleefisch, Phys.Rev.B **36**, 3608 (1987).

18 I.D. Brown and D. Altermatt, Acta Crystallographica **B41**, 244 (1985).

19 U. Welp, S. Fleshler, W.K. Kwok, J. Downey, Y. Fang, G.W. Crabtree and J.Z. Liu, Phys.Rev.B **42**, 10189 (1990).

20 K. Kishio, J. Shimoyama, T. Hasegawa, K. Kitazawa and K. Fueki, Jpn.J.Appl.Phys. **26**, L1228 (1987).

21   B.G. Bagley, L.H. Greene, J.-M. Tarascon and G.W. Hull, Appl.Phys.Lett. **51**, 622 (1987).

22   R.J. Cava, B. Batlogg, C.H. Chen, E.A. Rietman, S.M. Zahurak and D. Werder, Phys.Rev.B **36**, 5719 (1987).

23   R.J. Cava, B. Batlogg, C.H. Chen, E.A. Rietman, S.M. Zahurak and D. Werder, Nature **329**, 423 (1987).

24   J.D. Jorgensen, H. Shaked, D.G. Hinks, B. Dabrowski, B.W. Veal, A.P. Paulikas, L.J. Nowicvki, G.W. Crabtree, W.K. Kwok, L.H. Nunez and H. Claus, Physica C **153&155**, 578 (1988).

25   E.D. Specht, C.J. Sparks, A.G. Dhere, J. Brynestad, O.B. Calvin and D.M. Kroeger, Phys.Rev.B **37**, 7426 (1988).

26   G. Van Tendeloo and S. Amelinckx, Sol.St.Comm. **63**, 603 (1987).

27   C. Chaillout, M.A. Alario-Franco, J.J. Capponi, J. Chenavas, P. Strobel and M. Marezio, Sol.St.Comm. **65**, 283 (1988).

28   M.A. Alario-Franco, C. Chaillout, J.J. Capponi, J. Chenavas and M. Marezio, Physica C **156**, 455 (1988).

29   R.J. Cava, B. Batlogg, S.A. Sunshine, *et al.*, Physica C **152-155**, 560 (1988).

30   R.J. Cava, B. Battlog, K.M. Rabe, E.A. Rietman, P.K. Gallagher and J. L.W. Rupp , Physica C **156**, 523 (1988).

31   R.J. Cava, A.W. Hewat, E.A. Hewat, B. Batlogg, M. Marezio, K.M. Rabe, J.J. Krajewski, W.F. Peck Jr. and L.W. Rupp Jr., Physica C **165**, 419 (1990).

32   A.J. Jacobson, J.M. Newsam, D.C. Johnston, D.P. Goshorn, J.T. Lewandowski and M.S. Alvarez, Phys.Rev.B **39**, 254 (1989).

33   W.E. Farneth, R.K. Bordia, E.M. McCarron III, M.K. Crawford and R.B. Flippen, Sol.St.Comm. **66**, 953 (1988).

34   J.D. Jorgensen, B.W. Veal, A.P. Paulikas, L.J. Nowicki, G.W. Crabtree, H. Claus and W.K. Kwok, Phys.Rev.B **41**, 1863 (1990).

35   R. Beyers, B.T. Ahn, G. Gorman, V.Y. Lee, S.S. Parkin, M.L. Ramirez, K.P. Roche, J.E. Vazquez, T.M. Gur and R.A. Huggins, Nature **340**, 619 (1989).

36   B.T. Ahn, V.Y. Lee, R. Beyers, T.M. Gur and R.A. Huggins, Physica C **167**, 529 (1990).

37   P.G. Radaelli, C.U. Segre, J.D. Jorgensen and D.G. Hinks, Phys.Rev.B **45**, 4923 (1992).

38   J.L. Tallon, Physica C **168**, 85 (1990).

39   G.J. McIntyre, A. Renault and G. Collin, Phys. Rev B **37**, 5148 (1988).

40   U. Welp, M. Grimsditch, H. You, W.K. Kwok, M.M. Fang, G.W. Crabtree and J.Z. Liu, Physica C **161**, 1 (1989).

41   J.D. Sullivan, P. Bordet, M. Marezio, K. Takenaka and S. Uchida, Phys.Rev.B **48**, 10638 (1993).

42   H. Casalta, P. Schleger, P. Harris, B. Lebech, N.H. Andersen, LiangRuixing, P. Dosanjh and W.N. Hardy, Physica C **258**, 321 (1996).

43   M. Tetenbaum, B. Tani, B. Czech and M. Blander, Physica C **158**, 377 (1989).

44   M. Tetenbaum, L.A. Curtiss, B. Tani, B. Czech and M. Blander, *NATO Conference on High-Temperature Superconductors - Physics and Material Science, Bad Windsheim, Germany, 1989* (Kulver Academic Publisher, Norwell, MA, p. 279).

45   H. Shaked, J.D. Jorgensen, D.G. Hinks, R.L. Hitterman and B. Dabrowski, Physica C **205**, 225 (1993).

46   N.H. Andersen, B. Lebech and H.F. Poulsen, J.Less-Comm.Met. **164&165**, 124 (1990).

47   N.H. Andersen, B. Lebech and H.F. Poulsen, Physica C **172**, 31 (1990).

48   H.F. Poulsen, N.H. Andersen and B. Lebech, Physica C **173**, 387 (1991).

49   A.G. Khachaturyan and J.W. Morris_Jr., Phys.Rev.Lett. **61**, 215 (1988).

50   A.G. Khachaturyan, S.V. Semenovskaya and J.W. Morris_Jr., Phys.Rev.B **37**, 2243 (1988).

51   D. de Fontaine, L.T. Wille and S.C. Moss, Phys.Rev.B **36**, 5709 (1987).

52   L.T. Wille and D. de Fontaine, Phys.Rev.B **37**, 2227 (1988).

53   L.T. Wille, A. Berera and D. de Fontaine , Phys.Rev.Lett. **60**, 1065 (1988).

54   R. Kikuchi and J. Choi, Physica C **160**, 347 (1989).

55   N.C. Bartelt, T.L. Einstein and L.T. Wille, Phys.Rev.B **40**, 10759 (1989).

56   N.C. Bartelt, T.L. Einstein and L.T. Wille, Physica C **162-164**, 871 (1989).

57   T. Aukrust, M.A. Novotny, P.A. Rikvold and D.P. Landau, Phys.Rev.B **41**, 8772 (1990).

58   C.C.A. Gunther, P.A. Rikvold and M.A. Novotny, Phys.Rev.B **42**, 10738 (1990).

59   L.G. Mamsurova, K.S. Pigalskiy, V.P. Sakun, A.I. Shushin and L.G. Sherebakova, Physica C **169**, 11 (1990).

60   G. Ceder, M. Asta, W.C. Carter, M. Kraitchman, D. de Fontaine, M.E. Mann and M. Sluiter, Phys.Rev.B **41**, 8698 (1990).

61   D. de Fontaine, G. Ceder and M.Asta, Nature **343**, 544 (1990).

62   D. de Fontaine, M.E. Mann and G. Ceder, Phys.Rev.Lett. **63**, 1300 (1989).

63   G. Ceder, M. Asta and D. de Fontaine, Physica C **177**, 106 (1991).

64   V.E. Zubkus, S. Lapinskas and E.E. Tornau, Physica C **166**, 472 (1990).

65   S. Lapinskas, E.E. Tornau, A. Rosengren and P. Schleger, Phys.Rev.B **52**, 15565 (1995).

66   P. Schleger, W.N. Hardy and H. Casalta, Phys.Rev.B **49**, 514 (1994).

67   W.E. Pickett, R.E. Cohen and H. Krakauer, Phys.Rev.B **42**, 8764 (1990).

68   M.S. Islam and C. Ananthamohan, Phys.Rev.B **44**, 9492 (1991).

69   H.F. Poulsen, N.H. Andersen, J.V. Andersen, H. Bohr and O.G. Mouritsen, Nature **349**, 594 (1991).

70   R.M. Fleming, L.F. Scheemeyer, P.K. Gallagher, B. Batlogg, L.W. Rupp and J.V. Waszczak, Phys.Rev.B **37**, 7920 (1988).

71   T. Zeiske, R. Sonntag, D. Hohlwein, N.H. Andersen and T. Wolf, Nature **353**, 542 (1991).

72   P. Burlet, V.P. Plakthy, C. Marin and J.Y. Henry, Phys.Lett.A **167**, 401 (1992).

73   P. Burlet, L.P. Regnault, J. Rossat-Mignod, P. Bourges, C. Vettier, J.Y. Henry, V.P. Plakthy and C. Marin, *Workshop on Phase Separation in Cuprate Superconductors, Erice, Italy, 6-12 May 1992,* (World Scientific, Singapore, p. 208).

74   V. Plakhty, B. Kviatkovsky, A. Stratilatov, Y. Chernenkov, P. Burlet, J.Y. Henry, C. Marin, E. Ressouche, J. Schweizer, F. Yakou, E. Elkaim and J.P. Lauriat, Physica C **235-240**, 867 (1994).

75   R.A. Hadfield, P. Schleger, H. Casalta, N.H. Andersen, H.F. Poulsen, M. vonZimmerman, J.C. Schneider, M.T. Hutchings, D.A. Keen, RuixingLiang, P. Dosanjh and W.N. Hardy, Physica C **235-240**, 1267 (1994).

76   R. Sonntag, D. Hohlwein, T. Bruckel and G. Collin, Phys.Rev.Lett. **66**, 1497 (1991).

77   R. Sonntag, T. Zeiske and D. Hohlwein, Physica B **180-181**, 374 (1992).

78   T. Zeiske, D. Hohlwein, R. Sonntag, F. Kubanek and G. Collin, Z.Phys.B **86**, 11 (1992).

79   F. Yakhou. *L'ordre des oxygènes dans $YBa_2Cu_3O_{6+x}$: Etude physico-chimique et Cristallographie des Surstructures* (Université Joseph Fourier, Grenoble, 1996).

80   F. Yakhou, V. Plakthy, G. Uimin, P. Burlet, B. Kviatkovsky, J.Y. Henry, J.P. Lauriat, E. Elkaim and E. Ressouche, Sol.St.Comm. **94**, 695 (1995).

81   F. Yakhou, V. Plakhy, A. Stratilatov, P. Burlet, J.P. Lauriat, E. Elakim, J.Y. Henry, M. Vlasov and S. Moshkin, Physica C **261**, 315 (1996).

82   Krekels, Kaesche and Van Tendeloo, Physica C **248**, 317 (1995).

83   H.F. Poulsen, M. von Zimmermann, J.R. Schneider, N.H. Andersen, P. Schleger, J. Madsen, R. Hadfield, H. Casalta, R. Liang, P. Dosanjh and W. Hardy, Phys.Rev.B **53**, 15335 (1996).

84   P. Schleger, H. Casalta, R. Hadfield, H.F. Poulsen, M. VonZimmermann, N.H. Andersen, J.R. Schneider, LiangRuixing, P. Dosanjh and W.N. Hardy, Physica C **241**, 103 (1995).

85   S. Yang, H. Claus, B.W. Veal, R. Wheeler, A.P. Paulikas and J.W. Downey, Physica C **193**, 243 (1992).

86   C. Namgung, J.T.S. Irvine and A.R. West, Physica C **168**, 346 (1990).

87   B.W. Veal, H. You, A.P. Paulikas, H. Shi, Y. Fang and J.W. Downey, Phys.Rev.B **42**, 4770 (1990).

88   B.W. Veal, A.P. Paulikas, H. You, H. Shi, Y. Fang and J.W. Downey, Phys.Rev.B **42**, 6305 (1990).

89   H. Claus, S. Yang, A.P. Paulikas, J.W. Downey and B.W. Veal, Physica C **171**, 205 (1990).

90   J.D. Jorgensen, S. Pei, P. Lightfoot, H. Shi, A.P. Paulikas and B.W. Veal, Physica C **167**, 571 (1990).

91   J.T. Tallon, Physica C **176**, 547 (1991).

92   H. Shaked, J.D. Jorgensen, B.A. Hunter, R.L. Hitterman, A.P. Paulikas and B.W. Veal, Phys.Rev.B **51**, 547 (1995).

93   G. Ceder, R. McCormack and D. de Fontaine, Phys.Rev.B **44**, 2377 (1991).

94   M. François, A. Junod, K. Yvon, A.W. Hewat, J.J. Capponi, P. Strobel, M. Marezio and
     P. Fischer, Sol.St.Comm. **66**, 1117 (1988).

95   P. Schweiss, W. Reichardt, M. Braden, G. Collin, A. Heger, H. Claus and A. Erb,
     Phys.Rev.B **49**, 1387 (1994).

96   J. Mustre de Leon, S.D. Conradson, I. Batistic and A.R. Bishop, Phys.Rev.Lett. **65**,
     1675 (1990).

97   J. Mustre de Leon, S.D. Conradson, I. Batistic and A.R. Bishop, Phys.Rev.B **44**, 2422
     (1991).

98   M. Takata, E. Nishibori, T. Takayama, M. Sakata, K. Kodama, M. Sato and C.J. Howard,
     Physica C **263**, 176 (1996).

99   C. Thomsen and M. Cardona, Phys.Rev.B **47**, 12320 (1993).

100 E.A. Stern, M. Qian, Y. Yacoby, S.M. Heald and H. Maeda, Physical C **209**, 331
     (1993).

101 J. Röhler. in *Materials and crystallographic aspects of HT$_c$-superconductivity* (eds.
     Kaldis, E.) (Kluwer Academic Publishers, 1994), vol. 1, p. 353 .

102 T. Egami, B.H. Toby, S.J.L. Billinge, H.D. Rosenfeld, J.D. Jorgensen, D.G. Hinks, B.
     Dabrowski, M.A. Subramanian, M.K. Crawford, W.E. Farneth and E.M. McCarron,
     Physica C **185-189**, 867 (1991).

103 J. Röhler, P.W. Loeffen, S. Müllender, K. Conder and E. Kaldis, *Materials Aspects of
     High-T$_c$ Superconductivity: 10 Years after the discovery, Delphi, Greece, 1996* (NATO
     Advanced Study Institute) In Press.

104 B. Raveau, C. Michel, M. Herview and D. Groult. *Crystal chemistry of high-T$_c$
     superconducting copper oxides* (Springer-Verlag, Berlin, Heidelberg, New York, 1991).

105 J.M. Tarascon, P. Barboux, P.F. Miceli, L.H. Greene, G.W. Hull, M. Eibschultz and S.A.
     Sunshine, Phys.Rev.B **37**, 7458 (1988).

106 Y. Xu, R.L. Sabatini, A.R. Moodenbaugh, Y. Zhu, S.-G. Shyu and M. Suenaga, Physica
     C **169**, 205 (1990).

107 T. Siegrist, L.F. Schneemeyer, J.V. Waszczak, N.P. Singh, R.L. Opila, B. Batlogg,
     L.W. Rupp and D.W. Murphy, Phys.Rev.B **36**, 8365 (1987).

108 R.S. Howland, T.H. Geballe, S.S. Laderman, A. Fischer-Colbrie, M. Scott, J.M.
     Tarascon and P. Barboux, Phys.Rev.B **39**, 9017 (1989).

109 G.H. Kwei, R.B. Von Dreele, A. Williams, J.A. Goldstone, A.C. Lawson II and W.K.
     Warburton, J. Mol. Structure **223**, 383 (1990).

110 S.H. Kilcoyne, S.J. Hibble and R. Cywinski, Physica B **180-181**, 423 (1992).

111 C.R. Taylor, A. Das, C. Greaves, I. Zelenay and R. Suryanarayanan, Physica C **262**,
     135 (1996).

112 G. Collin, R. Villeneuve, Y. Sidis, P. Bourges, B. Hennion, F. Onufrieva, B. Gillon, F.
     Bouree, I. Mirebeau, P. Mendels, H. Alloul, F. Marucco and P. Schweiss, *Neutron
     Scattering in Materials Science II. Symposium, Boston, MA, USA, 1994* (Mater. Res.
     Soc, Pittsburgh, PA, USA, p. 513).

113 Y. Maeno and T. Fujita. in *Novel Superconductivity* (eds. Wolf, S.A. & Kresin, V.)
     (Plenum, New York, 1987), p. 1073 .

114 M. Ishikawa, T. Takabatake, A. Tohdake, Y. Nakazawa, T. Shibuya and K. Koga, Physica C **153-155**, 928 (1988).

115 B. Ullmann, R. Wordenweber, K. Heinemann, H. Krebs and H.C. Freyhardt, Physica C **153-155**, 872 (1988).

116 B. Raveau, F. Deslandes, C. Michel, M. Hervieu, G. Heger and G. Roth. in *High T$_c$ superconductors* (eds. Weber, H.A.) (Plenum, New York, 1988), p. 3 .

117 J. Bieg, J. Jing, H. Engelmann, Y. Hsia, U. Gonser, P. Gutlich and R. Jakubi, Physica C **153-155**, 952 (1988).

118 J.-L. Hodeau, P. Bordet, J.-J. Capponi, C. Chaillout and M. Marezio, Physica C **153-155**, 582 (1988).

119 Y. Xu, M. Suenaga, J. Tafto, R.L. Sabatini, A.R. Moodenbaugh and P. Zolliker, Phys.Rev.B **39**, 6667 (1989).

120 E. Takayama-Muromachi, Y. Uchida and K. Kato, Jpn.J.Appl.Phys. **26**, L2087 (1987).

121 E. Suard. *Etude structurale des phases Y$_{1-y}$Ca$_y$Ba$_2$Cu$_{3-x}$M$_x$O$_{7-\delta}$ (M=Fe, Co). Relations avec les propriétés supraconductrices et magnetiques* (Université de Caen, 1993).

122 E. Suard, V. Caignaert, A. Maignan and B. Raveau, Physica C **182**, 219 (1991).

123 E. Suard, A. Maignan, V. Caignaert and B. Raveau, Physica C **200**, 43 (1992).

124 E. Suard, I. Mirabeau, V. Caignaert, P. Imbert and A.M. Balagurov, Physica C **288**, 10 (1997).

125 E. Suard, V. Caignaert, A. Maignan, F. Bourée and B. Raveau, Physica C **210**, 164 (1993).

126 M. Guillaume, P. Allenspach, J. Mesot, B. Roessli, U. Staub, P. Fischer and A. Furrer, Journal of Alloys and Compounds **195**, 599 (1993).

127 M. Guillaume, P. Allenspach, W. Henggeler, J. Mesot, B. Roessli, U. Staub, P. Fischer, A. Furrer and V. Trounov, Journal of Physics: Condensed Matter **6**, 7963 (1994).

128 D.B. Currie and M.T. Weller, Physica C **214**, 204 (1993).

129 H. Luetgemeier, I. Heinmaa, D. Wagener and S.M. Hosseini, *2nd Workshop on Phase Separation in Cuprate Superconductors, Cottbus, Germany, 1993* (Springer-Verlag, Heidelberg, p. 225).

130 B.W. Veal, A.P. Paulikas, J.W. Downey, H. Claus, K. Vandervoort, G. Tomlins, H. Shi, M. Jensen and L. Morss, Physica C **162-164**, 97 (1989).

131 M. Kogachi, S. Nakanishi, K. Nakahigashi, H. Sasakura, S. Minamigawa, N. Fukuoka and A. Yanase, Jpn.J.Appl.Phys. **28**, L609 (1989).

132 R. Liang and T. Nakamura, Jpn.J.Appl.Phys. **27**, L1277 (1988).

133 H. Maletta, E. Porschke, B. Rupp and P. Meuffels, Z.Phys.B **77**, 181 (1989).

134 H. Maletta, E. Porschke and B. Rupp, *Neutron scattering for material science, Boston, MA, 1989* (Materials Research Society, Pittsburgh, PA, p. 181).

135 B. Rupp, E. Porschke, P. Meuffels, P. Fischer and P. Allenspach, Phys.Rev.B **40**, 4472 (1989).

136 H. Maletta, E. Porschke, T. Chattopadhyay and P.J. Brown, Physica C **166**, 9 (1990).

137 H. Shaked, B.W. Veal, J. Faber, R.L. Hitterman, U. Balachandran, G. Tomlins, H. Shi, L. Morss and A.P. Paulikas, Phys.Rev.B **41**, 4173 (1989).

138 B. Roessli, P. Allenspach, P. Fischer, J. Mesot, U. Staub, H. Maletta, P. Bruesch, C. Ritter and A.W. Hewat, Physica B **180-181**, 396 (1992).

139 U. Staub, P. Allenspach, J. Mesot, A. Furrer, H. Blank and H. Mutka, Physica B **180-181**, 417 (1992).

140 M. Izumi, T. Yabe, T. Wada, A. Maeda, K. Uchinokura, S. Tanaka and H. Asano, Phys.Rev.B **40**, 6771 (1989).

141 L. Er-Rakho, C. Michel, J. Provost and B. Raveau, J.Sol.St.Chem. **37**, 151 (1981).

142 M.J. Kramer, S.I. Yoo, R.W. McCallum, W.B. Yelon, H. Xie and P. Allenspach, Physica C **219**, 145 (1994).

143 F. Izumi, H. Asano, T. Ishigaki, E. Takayama-Muromachi, Y. Matsui and Y. Uchida, Jpn.J.Appl.Phys. **26**, L1153 (1987).

144 C.U. Segre, B. Dabrowski, D.G. Hinks, K. Zhang, J.D. Jorgensen, M.A. Beno and I.K. Schuller, Nature **329**, 227 (1987).

145 A. Maeda, T. Noda, H. Matsumoto, T. Wada, M. Izumi, T. Yabe, K. Uchinokura and S. Tanaka, J.Appl.Phys. **64**, 4095 (1988).

146 T. Wada, N. Suzuki, T. Maeda, A. Maeda, S. Uchida, K. Uchinokura and S. Tanaka, Phys.Rev.B **38**, 7080 (1988).

147 R.J. Cava, B. Battlogg, R.M. Fleming, S.A. Sunshine, A. Ramirez, E.A. Rietman, S.M. Zahurak and R.B. van Dover, Phys.Rev.B **37**, 5912 (1988).

148 A.S. Lebedev, V.G. Zubkov, V.L. Kozhevnikov, V.I. Voronin, A.P. Tyutyunnik and I.A. Kontsevaya, Superconductivity: Physics, Chemistry, Technology **4**, 2331 (1991).

149 D.M. de Leeuw, C.H.A. Musaers, H.A.M. van Hal, H. Verweij, A.H. Carim and H.C.A. Smoorenburg, Physica C **156**, 126 (1988).

150 Y. Tokura, J.B. Torrance, T.C. Huang and A.I. Nazal, Phys.Rev.B **38**, 7156 (1988).

151 T. Yagi, M. Domon, Y. Okajima and K. Yamaya, Physica C **173**, 453 (1991).

152 A. Manthiram, S.-J. Lee and J.B. Goodenough, J.Sol.St.Chem. **73**, 278 (1988).

153 A. Tokiwa, Y. Syono, M. Kikuchi, R. Suzuki, T. Kajitani, N. Kobayashi, T. Sasaki, O. Nakatsu and Y. Muto, Jpn.J.Appl.Phys. **27**, L1009 (1988).

154 Y. Tokura, J.B. Torrance, T.C. Huang and A.I. Nazzal, Phys.Rev.B **38**, 7156 (1988).

155 E.M. McCarron III, M.K. Crawford and J.B. Parise, J.Sol.St.Chem. **78**, 192 (1989).

156 J.B. Parise and E.M. McCarron III, J.Sol.St.Chem. **83**, 188 (1989).

157 J.B. Parise, P.L.Gai, M.K. Crawford and E.M.M. III, *High temperature superconductors: relationship between properties, structure and solid-state chemistry, San Diego, CA, 1989* (Materials Research Society, Pittsburgh, PA, p. 105).

158 R.S. Liu, R.J. Cooper, J.W. Loram, W. Zhou, W. Lo, P.P. Edwards, W.Y. Liang and L.S. Chen, Sol.St.Comm. **76**, 679 (1990).

159 C. Gledel, J.-F. Marucco and B. Touzelin, Physica C **165**, 437 (1990).

160 C. Gledel, J.-F. Marucco, E. Vincent, D. Favrot, B. Poummelec, B. Touzelin, M. Gupta and H. Alloul, Physica C **175**, 279 (1991).

161 R.G. Buckley, D.M. Pooke, J.L. Tallon, M.R. Prestland, N.E. Flower, M.P. Staines, H.L. Johnson, M. Meylan, G.V.M. Williams and M. Bowden, Physica C **174**, 383 (1991).

162 P.G. Radaelli. *Oxygen ordering and superconductivity in pure and Ca-substituted $REBa_2Cu_3O_{6+x}$ systems* (Illinois Institute of Technology, Chicago, IL, USA, 1992).

163 J.L. Tallon, C. Bernhard, H. Shaked, R.L. Hitterman and J.D. Jorgensen, Phys.Rev.B **51**, 12911 (1995).

164 P. Berastegui, S.-G. Eriksson, L.G. Johansson, M. Kakihana, M. Osada, H. Makaki and S. Tachihara, J.Sol.St.Chem. **127**, 56 (1996).

165 V.P.S. Awana, S.K. Malik and W.B. Yelon, Physica C **262**, 272 (1996).

166 V.P.S. Awana, S.K. Malik and W.B. Yelon, Modern Physics Letters B **10**, 845 (1996).

167 G. Bottger, I. Mangelschots, E. Kaldis, P. Fischer, C. Kruger and F. Fauth, Journal of Physics: Condensed Matter **8**, 8889 (1996).

168 L. Soderholm, K. Zhang, D.G. Hinks, M.A. Beno, J.D. Jorgensen, C.U. Segre and I.K. Schuller, Nature **328**, 604 (1987).

169 H.B. Radousky, J.Mat.Res. **7**, 1917 (1992).

170 Neumeier, T. Bjørnholm, M.B. Maple and I.K. Shuller, Phys.Rev.Lett. **63**, 2516 (1989).

171 D.P. Norton, D.H. Lownders, B.C. Sales, J.D. Budai, B.C. Chakoumakos and H.R. Kerchner, Phys.Rev.Lett. **66**, 1537 (1991).

172 J.J. Neumeier, T. Bjørnholm and M.B. Maple, Physica C **166**, 191 (1990).

173 Y.J. Uemura, G.M. Luke, B.J. Sternlieb, J.H. Brewer, J.F. Carolan, W.N. Hardy, R. Kadono, J.R. Kempton, R.F. Kiefl, S.R. Kreitzman, P. Mulhern, T.M. Riseman, D.L. Williams, B.X. Yang, S. Uchida, H. Takagi, J. Gopalakrishnan, A.W. Sleight, M.A. Subramanian, C.L. Chien, M.Z. Cieplak, G. Xiao, V.Y. Lee, B.W. Statt, C.E. Stronach, W.J. Kossler and X.H. Yu, Phys.Rev.Lett. **62**, 2317 (1989).

174 M. Andersson, O. Rapp and R. Tellgren, Physica C **205**, 105 (1993).

175 P. Lundqvist, C. Tengroth, O. Rapp, R. Tellgren and Z. Hegedues, Physica C **269**, 231 (1996).

176 I.D. Brown, J.Sol.St.Chem. **82**, 122 (1989).

177 H. Claus, M. Braun, A. Erb, K. Röhberg, B. Runtsch, H. Wühl, G. Bräuchle, P. Schweib, G. Müller-Vogt and H. Lohneysen, Physica C **198**, 42 (1992).

178 J.L. Tallon and N.E. Flower, Physica C **204**, 237 (1993).

179 H. Claus, U. Gebhard, G. Linker, K. Röhberg, S. Riedling, J. Franz, T. Ishida, A. Erb, G. Müller-Vogt and H. Wühl, Physica C **200**, 271 (1992).

180 E. Janod, A. Junod, T. Graf, K.-Q. Wang, G. Triscone and J. Muller, Physica C **216**, 129 (1993).

181 K. Conder, D. Zech, C. Krüger, E. Kaldis, H. Keller, A.W. Hewat and E. Jilek, *2nd Workshop on Phase Separation in Cuprate Superconductors, Cottbus, Germany, 1993* (Springer-Verlag, Heidelberg, p. 211).

182 K. Conder, D. Zech, C. Krüger, E. Kaldis, H. Keller, A.W. Hewat and E. Jilek, Physica C **235-240**, 425 (1994).

183 C.W. Chu, Z.J. Huang, R.L. Meng, L. Gao and P.H. Hor, Phys.Rev.B **37**, 9730 (1988).

184 J.D. Jorgensen, S. Pei, P. Lightfoot, D.G. Hinks, B.W. Veal, B. Dabrowski, A.P. Paulikas, R. Kleb and I.D. Brown, Physica C **171**, 93 (1990).

PHASE DIAGRAMS AND SPIN CORRELATIONS IN $YBa_2Cu_3O_{6+x}$

L.P. REGNAULT[1], Ph. BOURGES[2] AND P. BURLET[1]

[1]Département de Recherche Fondamentale sur la Matière
Condensée, SPSMS/MDN
CEA-Grenoble,17, Rue des Martyrs
8054 Grenoble cedex 9, France

[2]Laboratoire Leon Brillouin, CEA-CNRS
CEA-Saclay , 91191 Gif sur Yvette cedex, France

Since the first discovery in 1986 by J.G. Bednorz and K.A. Müller [1] of a superconductivity at about 30 K in the cuprate $(La,Ba)_2CuO_4$, a incredible amount of work has been carried out in order to establish and understand the fascinating properties of the new "high-T_C" superconductors. Very rapidly, it has been shown that conventional electron-phonon couplings could not be the only origin for the high-T_C superconductivity (HTSC) and that the electron-electron interactions were fundamental in explaining the "strange metal" properties of the new materials, giving rise for some of them to a clear non-Fermi-liquid behavior (for a comprehensive description of both the experimental and theoretical situations, see for example the review article by A. Kampf [2]). It is now well accepted that the low energy properties of high-T_C cuprates are essentially determined by the charge and spin dynamics within the CuO_2 layers. As early recognized, the existence of layered crystallographic structures build from a sequence of CuO_2 layers separated by block layers playing the role of charge reservoirs is at the origin of the charge transfer mechanism upon doping. Based on experimental as well as theoretical results, a generic phase diagram of High-T_C cuprates (shown in Figure 1) has been established. The quite interesting peculiarity of the cuprate superconductors is that they continuously evolve from a magnetic insulating ground state to a superconducting ground state by varying the charge-carrier concentration. Starting from an insulating antiferromagnetic (AF) state, the introduction of a small amount of charge carriers in the CuO_2 layers (in the following one will consider only hole doping as it is for YBCO, labeling the charge-carrier density n_h) suppress completely the initial long range order above a characteristic concentration $n_{c1}(\approx 2\%$ hole/Cu) and the CuO_2 layers becomes metallic. Superconductivity further occurs above a second characteristic concentration $n_{c2}(\approx 6\%$ hole/Cu) and persists up to n_{c3}(corresponding roughly to 30-35% hole/Cu), above which value a normal-

85

metal behavior is recovered. The range of doping between n_{c1} and n_{c2} defines the so called quantum-disordered-spin state, characterized by the existence of strong AF correlations and a non-Fermi liquid behavior. $T_c(n_h)$ reaches a maximum at some value defining the "optimal" doping n_{opt}.

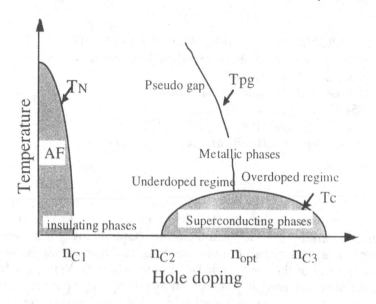

Figure 1. Schematic phase diagram of High-T_C cuprates as a function of hole doping in the CuO_2 layers showing the different doping states.

Doping rates smaller and larger than n_{opt} define the so called underdoped and overdoped regimes, respectively. In the underdoped regime, there is now a convergence of experimental works showing an anomalous behavior in many physical properties below some characteristic temperature $T_{pg}(n_h)$ [3-8]. Such a specificity has been attributed to the opening of a pseudo gap, a feature no more existing in the overdoped regime. The neutron-scattering technique, by probing the imaginary part of the spin susceptibility of CuO_2 layers as a function of momentum q and energy ω ($\chi''(q,\omega)$ has brought invaluable contributions to the characterization of this generic phase diagram, both from the static and dynamical viewpoints.

Among the about fifteen cuprate superconductors which have been successfully synthesized till now, the $YBa_2Cu_3O_{6+x}$ family has particularly attracted the neutronician's attention since the availability of large and good quality single crystals on which the doping could be quasi continuously tuned. This offered the rather attractive possibility to explore the evolution of the spin excitation spectrum as a function of doping on the same sample, an information which is of paramount importance for the theory. There are some questions left to be addressed by neutron scattering: what is the interplay between magnetism and superconductivity? How to understand the very different normal-state properties in the underdoped and overdoped regimes? what are the various length and energy scales of the AF spin fluctuations?

In this chapter we will review the main results obtained on the $YBa_2Cu_3O_{6+x}$ family by the neutron-scattering technique, focusing on the magnetic aspects of the T-x phase diagram. The first section will be devoted to the description of the T-x phase diagram. The insulating phases will be treated in section 2. We will first report on the magnetic ordering and the spin wave spectra in the antiferromagnetic undoped phase. This will allow to get a good description of all the exchange and anisotropy parameters in the undoped material. We will then study the effect of a small electronic doping both on the magnetic ordering and magnetic excitation spectra. The third part will be devoted to the metallic phases for which the magnetic fluctuations have been probed both in the normal and superconducting states. The temperature and doping dependencies of the main features of the magnetic fluctuation spectra will be presented. Finally, the effect of cationic substitutions will be presented in the last part for zinc-substituted compounds $YBa_2Cu_{3-y}Zn_yO_{6+x}$.

1. The temperature -concentration phase diagram

Since its first discovery in 1987 [9] the system $YBa_2Cu_3O_{6+x}$ continue to be one of the most extensively studied high-T_C materials. This exceptional status is mainly due to the fact that these compounds, which exist for oxygen contents $0<x<1$, cover a large range of doping state going from undoped up to slightly overdoped. The possibility to synthesize good powder and large single crystals with a rather good control of the oxygen content and thus of the doping allowed a quite comprehensive and accurate experimental investigation from a great variety of techniques, among which neutron scattering. The crystal structure is relatively simple [10,11] and the structural modifications occurring when x increases from 0 to 1 are well described and understood [12] (for more details see the chapter by P. Radaelli and A. Hewat and references therein). The crystallographic structures of the two extreme compounds $YBa_2Cu_3O_6$ and $YBa_2Cu_3O_7$ are given in Figure 2.

The main characteristic of importance for our purpose is that these structures consist of a stacking along the c axis of different planes. Among them one can distinguish two different types of planes containing the Cu-ions:
- the CuO_x planes (copper Cu1 site) at z=0 , which consist of a square lattice of Cu atoms in which the x oxygen occupy randomly the O4 and O5 sites for x<0.4 and nearly only the O4 sites for larger x , thus forming ...Cu-O-Cu-O-Cu-O... chain segments.
- the CuO_2 planes at crystallographic positions ±z (z ≈ 0.364 for x=0 and ≈ 0.355 for x=1) which form bilayers, defining two Bravais sublattices of copper Cu2 sites. In these planes the Cu^{2+} ions bear a magnetic moment.

The consensus temperature-versus-concentration phase diagram of $YBa_2Cu_3O_{6+x}$ is summarized in Figure 3, regarding the crystallographic, magnetic and transport aspects. It has been established from the results of an extreme variety of experimental techniques and bears all the characteristic features of the generic phase diagram of high-T_C cuprates previously given in Figure 1.

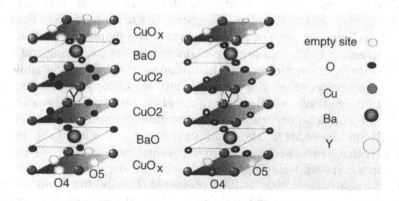

Figure 2. Schematic view of the crystal structure of $YBa_2Cu_3O_6$ (tetragonal) and $YBa_2Cu_3O_7$ (orthorhombic) showing the CuO_2 plane and CuO chain subsystems. The main difference is the additional oxygen site O4 occupied in the latter and empty in the former.

Two main different concentration ranges can be first distinguished above and below a critical concentration $x_C=0.4$ at which drastic changes occur in the physical properties. At low oxygen content, the samples have a tetragonal crystallographic structure (space group P4/mmm). They are insulators and exhibit a long-range ordered antiferromagnetic ground state at sufficiently low temperature. Above x_C a crystallographic transition takes place and the crystal structure becomes orthorhombic (space group Pmmm) while the resistivity takes a clear metallic character. No long range magnetic ordering exists but superconductivity develops at low temperature.

Inside these two main ranges several subdivisions can be done which concern the structural aspect (superstructures Ortho-I, Ortho-II, Ortho-III), the level of electronic doping (undoped ($0<x<0.2$) or doped ($0.2<x<0.4$) AF-state, slightly($0.4<x<0.5$), strongly ($0.5<x<0.94$) and overdoped ($x>0.94$) superconducting states, following the terminology defined in [4]) and the metallic character (superconducting or normal). In fact all these aspects are strongly connected one to the other. As an example the structural and charge transfer aspects result directly from the intercalation mechanism of the x extra oxygen, giving rise to the quite exceptional doping dependence of the magnetic and transport properties of this cuprate [12-15].

Starting from the reduced $YBa_2Cu_3O_6$ (usually classified as a ionic compound with well defined valence states Y^{3+}, Ba^{2+}, Cu^+, Cu^{2+} and O^{2-}), the inserted oxygen atoms occupy the vacant sites (O4, O5) in the square lattice of the Cu1 atom (see Figure 2). At low concentration ($x<0.15$-0.2) only isolated and randomly oriented Cu^{2+}-O^{2-}-Cu^{2+} units are created which maintain the tetragonal symmetry and, by changing two monovalent copper ions into bivalent ones, do not induce any charge transfer. The pure AF-state is conserved. On further increasing the oxygen concentration ($0.2<x<x_C$), longer and longer randomly oriented ...Cu-O-Cu-O...chain segments are formed, maintaining the tetragonal structure. Because the charge equilibrium can no more be satisfied in the Cu1 planes, some charge transfer with the rest of the structure (and in particular the CuO$_2$ planes) starts. Around $x=x_C \approx 0.4$ the chain fragments orient along the a crystallographic axis of the structure forming long ...-Cu-O-Cu-O-... chains and a cooperative crystallographic phase transition takes place, yielding a tetragonal phase in which only the O4 sites are occupied. The instability of long chain segments with respect to p-hole doping makes that a large amount of holes are transferred from the Cu1-planes to the CuO$_2$-planes, inducing the insulator-metal transition and finally the superconductivity. For $x \approx 0.5$, the repulsive interaction existing between the chains gives rise to a superstructure, the so called Ortho II phase, which consists in an alternating sequence of one empty and one full chain [16]. For $0.5<x<1$, the empty chains are progressively filled following the same mechanism, giving rise to the characteristic ("60 K" and "90 K") two-plateau behavior of T$_C$ clearly visible in Figure 3.

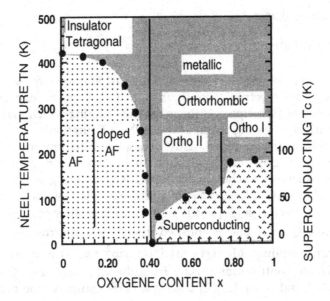

Figure 3. T-x phase diagram of YBa$_2$Cu$_3$O$_{6+x}$ as a function of the oxygen content x in the Cu1 planes, showing the insulating-metallic, magnetic-non magnetic and structural borderlines, and the two-plateau superconducting regime.

Thus, in YBCO$_{6+x}$, the hole doping of the CuO$_2$ planes increases continuously from x=0 to x=1, and the difficulty in this material is to determine the hole density n_h for a given x. For $x \approx 1$, XAS studies of untwinned samples yield $n_h \approx 0.25$ hole/Cu [17], whereas an appreciably lower value of $n_h \approx 0.16$ hole/Cu was speculated from thermoelectric and structural studies [18,19]. At lower oxygen content (x~0.5-0.8), n_h depends in a crucial way on the quality of the oxygen ordering in the Cu1-O chains. Consequently, samples with nominally the same oxygen content can exhibit slight discrepancies depending on the sample-preparation procedure. This is particularly true for the large single crystals suitable for neutron inelastic scattering (NIS). On the theoretical side, a microscopic approach has been proposed by G. Uimin [14,15], suggesting a realistic correspondence $n_h(x)$. Another relevant parameter able to influence the magnetic and transport properties concerns the micro inhomogenous character of the hole distribution in the CuO_2 planes, inherent to the charge separation phenomena due to the proximity of the Mott localization [20,21] (for more details, see the chapter by J. Tranquada).

The magnetic and superconducting behavior of $YBa_2Cu_3O_{6+x}$ are also sensitive to cationic substitution which can occurs on the Y (see the chapters by P. Fisher and by J. Mesot and A. Furrer, and references therein for more information), Cu and Ba sites (see the chapter by P. Radaelli and A. Hewat). The Zn-substituted compounds have been quite extensively studied from the point of view of their magnetic (both static and dynamic) properties, as we shall report later.

2. The insulating phases

2.1. MAGNETIC ORDERING

In the fully reduced compound $YBa_2Cu_3O_6$ the Cu1 ions are monovalent with a spinless (S=0) $3d^{10}$ electronic configuration and thus, as experimentally proved, they do not participate to the magnetic ordering which involves only the bivalent Cu2 ions bearing a spin S=1/2 ($3d^9$ configuration).

The magnetic structure of $YBa_2Cu_3O_6$ has been determined by neutron diffraction, first on powder [22,23] and further on single crystal samples [24,25]. Figure 4 shows a scheme of the magnetic ordering which builds up below T_N. Inside a CuO_2 plane, the spin arrangement is described according to the propagation vector $k_{AF}=[1/2,1/2,0]$. From the absence of intensity at scattering vector $Q = k_{AF}$ one established firstly the antiferromagnetic nature of intrabilayer couplings ($I_M(Q) \propto m_0^2 . f^2(Q). \sin^2(\pi z Q_1)$) and secondly the absence of sizable contribution arising from the Cu1 sites. The AF-direction lies in the (a,b) basal plane [22,24] and an accurate study of the evolution of $I_M(Q)$ under magnetic field has proved that the spontaneous moment direction is along the [100] direction [26], implying a small orthorhombic anisotropy in the spin Hamiltonian describing the system. From the structure given in Figure 4, one will also note the existence of antiferromagnetic interbilayer couplings. At low temperature the magnetic moment, assuming an isotropic form factor, amounts to $m_0 \approx 0.64$ μ_B [27]. A value somewhat smaller (≈ 0.52 μ_B) is found when considering a more realistic anisotropic magnetic form factor [28], much lower than the mean field prediction $g\mu_B S \approx 1.06 \mu_B$. This strong reduction is well explained from both quantum and covalence effects [4,28]. In increasing temperature, the moment value decreases continuously (see Figure 5) and finally vanishes at $T_N \approx 410$ K, a value which implies extremely large in-plane nearest-neighbor Cu-Cu superexchange interactions, taking into account for the layered character of this system.

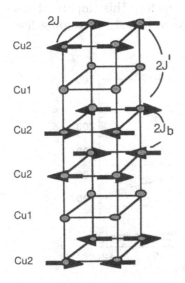

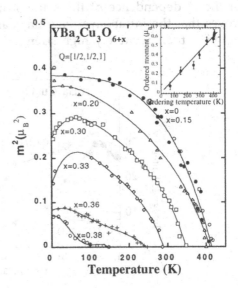

Figure 4. antiferromagnetic structure of $YBa_2Cu_3O_6$. The main exchange interactions are also indicated.

Figure 5. Temperature dependencies of the antiferromagnetic intensity at various oxygen contents.

An identical magnetic ordering is observed for oxygen contents up to 0.15, with very weak change in both T_N and m_0. In fact, such a behavior is not very surprising since it is a direct consequence of the weakness of the number of transferred holes for this concentration range.

Above 0.2 and up to x_C, the AF-ordering appears much more affected by the hole doping. Figure 5 summarizes the temperature dependence of the square of the magnetic moment $m^2(T)$ ($\propto I_M(T)$) for different oxygen contents between 0 and 0.38 [27,29,30]. If the main features of the spin ordering are preserved (propagation vector, spin couplings, absence of Cu1 contribution ....), the Néel temperature and the ordered moment are progressively reduced and both vanish abruptly around $x_C \approx 0.4$ [29] (as mentioned above, the critical concentration value is slightly dependent on the sample preparation, particularly on the quenching temperature). Interestingly, T_N and m_0 obey the linear relation $T_N(x)/m_0(x) \approx$ constant, whatever x (see the inset in Figure 5). We will come back to this point later. The decrease in intensity of the magnetic Bragg peaks ($\propto m^2(T)$) is correlated with the appearance of an increasing quasi 2D diffuse scattering centered around reciprocal-space positions $Q_{AF}^{2D}=(n+1/2,n+1/2)$, with n integer. Transverse scans across the 2D rods show a finite Q-width larger than the instrumental resolution, indicating a finite correlation length characteristic of short-range magnetic correlations. Typically, one has $\xi_a/a \approx 8$ for a sample with $x \approx 0.37$ [4]. Interestingly, scans along the rods give evidence for a clear sine modulation of the diffuse magnetic intensity reflecting the $\sin^2(\pi z Q_1)$ variation expected for the Q_1-dependence of the static structure factor. This important result proves that the intrabilayer couplings are preserved at this doping levels, implying their relatively high strength.

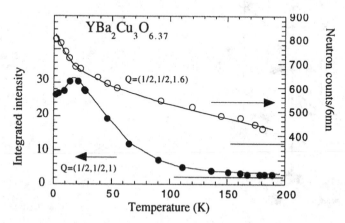

Figure 6.   Temperature dependence of the Bragg intensity (closed symbols) and of the 2D scattering (open symbols) in YBa₂Cu₃O₆.₃₇ showing the reentrant behavior at low temperature.

The data depicted in Figure 5 reveal another interesting feature of the doped AF-state. For the intermediate concentration range $0.30 < x < 0.37$, the ordered magnetic moment raises a maximum at some temperature $T_r(x)$ and further decreases on lowering the temperature. This reentrant behavior is correlated with an enhancement of the 2D component, as shown in Figure 6 for a sample with $x=0.37$.

All these facts can be qualitatively explained by considering that a hole transferred in a CuO_2 plane creates a defect in the network of antiferromagnetic bonds of the square lattice. To accommodate the resulting frustration an orientational disorder occurs around the defect (a magnetic polaron is created), implying a reduction of the 3D ordered moment (which corresponds to the average component along the **a**-direction) and a renormalization of the stiffness constant of low-q magnetic excitations [31,32], the latter allowing to define an effective in-plane coupling constant $J^{eff}(x)$. The linear relation between T_N (proportional to $J^{eff}(x)$) and m_0 can be understood by assuming that $J^{eff}(x) \propto m_0(x)$ [4]. The 2D diffuse part is associated with the spin components inside the magnetic polarons. It is therefore expected to increase directly as the hole density $n_h(x)$. Due to the strong intrabilayer couplings, each magnetic polaron within a CuO_2-layer must have a π-shifted counterpart in the adjacent CuO_2-layer, forming antiferromagnetic "bipolarons". Such objects account very well for the sine-square modulation of the 2D rods. Within this polaronic scenario, the reentrant behavior (characterized by a decrease of the ordered moment and a enhancement of the 2D contribution) can be attributed to an increase of the frustration originating from a progressive localization of holes on lowering the temperature. An order of magnitude of the hole density can be estimated from the value of the in-plane correlation length ξ_a, by using the 2D relation $n_h \approx (a / \xi_a)^2$ (which means that the in-plane correlation length is directly controlled by the average polaron-polaron distance). For $x \approx 0.37$, this relation yields a hole density $n_h \approx 0.015$ hole/Cu. Thus in YBCO, a hole density as small as 0.02 hole/Cu is sufficient to destroy completely the long range AF ordering. Furthermore, very similar q- and temperature dependencies have been found in zinc-substituted YBCO samples with almost the same oxygen contents. This confirms that the unusual behavior reported in Figure 6 is not due to impurity effects, but is rather intrinsic (for more details, see section 4 and references therein).

2.2. MAGNETIC EXCITATION SPECTRA

2.2.1. *Spin waves in the AF-phase*
The experimental determination of the spin wave spectra in the antiferromagnetic state is essential because it allows a precise characterization of the exchange and anisotropy parameters describing the interacting spin system. In YBCO the situation is slightly complicated by the fact that the crystallographic structure exhibits two magnetic Bravais sublattices of Cu2 atoms (see section 1). NIS experiments [22,23,29] have shown that the spin dynamics of bilayers in antiferromagnetic YBCO is well accounted for by the following Hamiltonian:

$$H = J \sum_{n=1,2} \sum_{<i,j>} \left(S_{in}^x S_{jn}^x + S_{in}^y S_{jn}^y + \alpha_z S_{in}^z S_{jn}^z \right) + J_b \sum_i S_{i1} \cdot S_{i2} + J' \sum_k S_{k1} \cdot S_{k2} \quad (1)$$

where the index n labels the Bravais-sublattice number, J is the in-plane nearest-neighbor superexchange coupling constant, α_z characterizes the small out-of-plane (OP) anisotropy, J_b and J' being respectively the intrabilayer and interbilayer coupling constants. Figure 4 gives a scheme of the most relevant superexchange couplings linking the Cu2 atoms.

Within the framework of the simple spin wave theory, the diagonalization of the Hamiltonian (1) yields the magnon dispersion relations summarized in Figure 7 [30,33,4]. In agreement with the existence of the two Bravais sublattices, one obtains two different modes allowing to define acoustic and optic branches. In the special case where one neglects both the anisotropy terms and the interbilayer couplings, the magnon dispersion relations are given by the simple relations:

$$\hbar\omega^{\nu}(\mathbf{q}) \approx \sqrt{(2J + J_b)^2 - (J\gamma(\mathbf{q}) \pm \frac{J_b}{2})^2} \tag{2a}$$

where ν is the branch index (ν=ac,opt) and $\gamma(\mathbf{q}) = \cos(q_x a) + \cos(q_y a)$, a being the in-plane lattice constant.

From Figure 7, one easily see that the relevant exchange and anisotropy parameters can be derived from the knowledge of the magnon dispersions for some special directions of the reciprocal space.

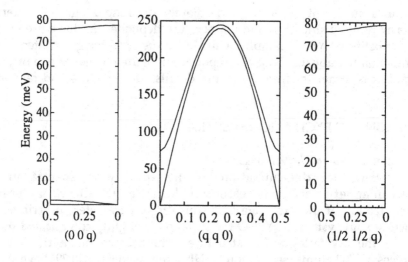

Figure 7. Calculated magnon dispersion curves for some characteristic directions in YBCO. One will note the presence of two branches (acoustic and optic).

Quantum corrections are assumed to only slightly modify the dispersion relation (2a) by a multiplicative factor $Z_c \approx 1.18$ for the S=1/2 anti-ferromagnetic square-lattice [34-36].

At a given scattering vector $\mathbf{Q} = \tau + \mathbf{q}$ (where τ is a reciprocal-lattice vector and $\mathbf{q}$ is the wave vector of the excitation) and at an energy transfer $\hbar\omega$, the dynamical structure factor is the sum of two terms (neglecting the weak in plane anisotropy):

$$S(\mathbf{Q}, \omega) = \left(1 - \frac{Q_z^2}{|\mathbf{Q}|^2}\right) S^{zz}(\mathbf{Q}, \omega) + \left(1 + \frac{Q_z^2}{|\mathbf{Q}|^2}\right) S^{xx}(\mathbf{Q}, \omega) \tag{2b}$$

Here $S^{zz}(\mathbf{Q}, \omega)$ and $S^{xx}(\mathbf{Q}, \omega)$ are the components of the dynamical structure factors associated with the OP and IP spin fluctuations, respectively. For spin wave excitations in a two-Bravais-sublattices system, one has:

$$S_{v}^{zz}(Q,\omega) = \frac{A_{v}(Q)}{\hbar\omega_{v}(q)}\delta(\omega - \omega(q)) \qquad (3a)$$

$$S_{v}^{xx}(Q,\omega) = \frac{A_{v}(Q+k_{AF})}{\hbar\omega_{v}(q+k_{AF})}\delta(\omega - \omega(q+k_{AF})) \qquad (3b)$$

expressions in which v labels the mode index. In the vicinity of the antiferromagnetic points $Q_{AF} = k_{AF} + \tau$, the coefficients $A_{v}(Q)$ are simply given by the well known expressions: $A_{ac}(Q) \approx \sin^{2}(\pi z Q_{1})$ and $A_{opt}(Q) \approx \cos^{2}(\pi z Q_{1})$. The maxima of the acoustic structure factors are at $Q_{1}=1.75+3.5n$ (n integer) and the zeros at $Q_{1}=3.5n$. For the optic structure factor, the maxima and zeros are reversed. In practice, the different polarizations can be determined by varying Q_{1}.

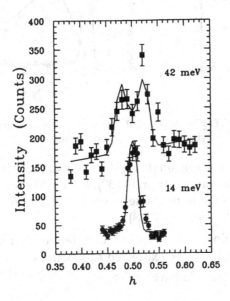

Figure 8. Acoustic magnetic excitations in $YBa_2Cu_3O_{6.2}$. Q-scans across the AF rod at low and high energy showing the stiffness of the acoustic dispersion relation (from [28]).

Figure 8 shows the wave vector dependence of the acoustic neutron-scattering cross section for two different energies, 14 and 42 meV, in the vicinity of the antiferromagnetic zone center $Q_{AF}=(1/2,1/2,Q_1)$ with $Q_1=-2$ and -4.9, respectively [28]. A two-peaks structure emerges in increasing energy leading to a very steep quasi linear acoustic dispersion curve. The spin-wave velocity which is deduced, $c \approx 650$ meV.Å, shows quantitatively the strength of the IP exchange couplings in YBCO. On Figure 9-a is depicted two energy-scans performed at two scattering vectors Q=(1/2,1/2,1.6) and (1/2,1/2,4.6) differing only by their component along c and therefore probing different polarizations.

Following the analysis of Rossat-Mignod et al [29,30,37,4], these two peaks have been unambiguously identified as corresponding to the IP and OP components, with respective energies $\Delta_{ZB} \approx 3$ meV and $\Delta_z \approx 8$ meV.

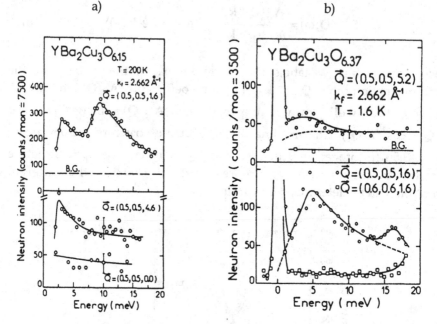

Figure 9. Energy scans in a) YBa$_2$Cu$_3$O$_{6.15}$ and b) YBa$_2$Cu$_3$O$_{6.37}$ establishing the existence of IP and OP spin components and showing their evolution upon doping.

The recent availability of large single crystals allowed to determine the high energy part of the excitation spectrum [38,39]. Figure 10 shows an example of neutron data relative to the optic mode, revealing clearly the existence of an optic gap at an energy $\Delta_{opt} \approx 67$ meV [38].

Within the simple spin wave theory, the different characteristic energy scales c, Δ_{ZB}, Δ_z and Δ_{opt} can be expressed as functions of parameters J, J$_b$, J' and α_z, according to the relations:

$$c \approx 2\sqrt{2}SZ_C Ja \qquad (4a)$$

$$\Delta_{ZB} \approx 4JS\sqrt{J'/J} \qquad (4b)$$

$$\Delta_z \approx 4\sqrt{2}JS\sqrt{\alpha_z} \qquad (4c)$$

$$\Delta_{opt} \approx 4JS\sqrt{J_b/J} \qquad (4d)$$

From the above expressions, one determines the following set of parameters for antiferromagnetic YBCO (including quantum corrections):

$$J \approx 101 \text{ meV}, \ J_b \approx 11 \text{ meV}, \ \alpha_z \approx 8\text{x}10^{-4} \text{ and } J'/J \approx 2\text{-}2.5\text{x}10^{-4}$$

which confirm quantitatively the rather good 2D-Heisenberg character of this system. Recent neutron-diffraction measurements under an applied magnetic field [26] have allowed an accurate estimation of the IP anisotropy constant, $\alpha_{xy} = \frac{J_x - J_y}{J} 5 \times 10^{-7}$, which would correspond to an IP gap $\Delta_{xy} \approx 2J\sqrt{\alpha_{xy}} \approx 0.2\text{-}0.3$ meV, not yet detected in the high-resolution NIS experiments.

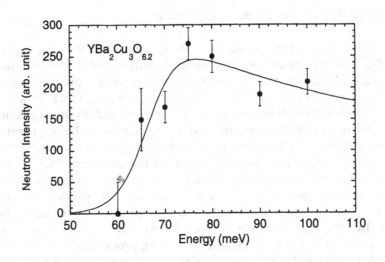

Figure 10. Optical magnon mode in YBa$_2$Cu$_3$O$_{6.2}$. The position of the optic gap corresponds to the inflexion point at about 67 meV.

2.2.2. Effects of doping

We have previously seen that the transfer of only about 1.5% of hole/Cu in the CuO$_2$ planes was strong enough to disturb the static IP spin-ordering in a YBCO$_{6.37}$ sample. Concomitantly, the spin dynamics undergo significant changes at the same doping rates. The most striking effects are seen on the low-q and low-energy part of the spin susceptibility. As an example, Figure 9-b shows the evolution upon doping of similar scans as those previously performed on the YBCO$_{6.15}$ sample, however at a much lower temperature. These constant-Q scans, combined with constant-energy scans [4], revealed that the acoustic IP and OP components behave differently in the doped sample. If the OP-fluctuations at $Q \approx Q_{AF}$ appear almost unchanged (apart from a renormalization by a factor 2 of the peak position and a small broadening), the excitations associated with the IP spin components appear strongly broadened in wave vector ($\Gamma_q \approx 0.035\text{Å}^{-1}$ at $\omega \approx 6$ meV) and damped in energy (characteristic energy scale $\Gamma_\omega \approx 15$ meV [29,30,4]). The parameters Γ_q and Γ_ω verify roughly the relation: $\Gamma_\omega / \Gamma_q \sim c$, typical of non-propagating (overdamped) magnetic excitations. A similar result has been obtained by Sato et al [40] in $YBa_2Cu_3O_{6.4}$. At larger wave vector, the magnetic excitation spectrum recovers a more and more spin-wave-like character, with a velocity weakly affected by the hole doping [43]. In fact, such a behavior is not very different to that predicted for a diluted 2D magnetic system. It is quite understandable by assuming again a disorder of the IP components induced by the creation of magnetic polarons around the holes, whereas the OP components remain better ordered [41,42,31,3244].

3.   Magnetic fluctuations in the metallic state

As discussed in the sections 1 and 2, for an oxygen content x larger than 0.4 $YBa_2Cu_3O_{6+x}$ undergoes an insulator-metal transition and a superconducting ground state builds up at low temperature, whereas simultaneously the antiferromagnetic long-range order vanishes. Clearly, there is a strong interplay between magnetism and superconductivity, and all the problem is to know if the magnetic fluctuations play a direct role in the appearance of superconductivity or if they are only undergoing its influence. Thus, the knowledge of the evolution upon doping of the magnetic excitation spectra in both the normal and superconducting states of metallic compositions is very crucial, because it might help to settle between the various (phononic, polaronic or magnetic) approaches for high-T_C superconductivity. Conceptually, the metallic state of high-T_C cuprates is very interesting to study in view of its quite unusual and fascinating properties (for reviews, see for example [5,2,45] and references therein). It is now recognized that the low-energy physics of high-T_C superconductors and most of their properties can be at least qualitatively tackled by assuming the existence of strong short range Coulomb repulsion between the charge carriers in the CuO_2 planes [46-49]. Among the numerous problems raised by the studies on YBCO, the understanding of the quite puzzling deviations from the conventional Landau's Fermi liquid theory which are observed in the normal-state of underdoped (x<0.93) compounds has particularly attracted the attention of the community. Very early, this "strange metal" behavior was attributed to the presence of relevant spin correlations

As pointed in the first chapter of this book, neutron scattering is a very powerful method for probing the condensed-matter magnetism. On the contrary to NMR which probes the magnetic fluctuations locally and at very small energy ($\omega \approx 0$), the magnetic neutron scattering allows to obtain invaluable information on the space and time spin-spin correlation functions $\langle S^\alpha(\mathbf{R}_i,0).S^\beta(\mathbf{R}_j,t)\rangle$ or equivalently on the imaginary part of the generalized susceptibility $\chi''_{\alpha\beta}(\mathbf{Q},\omega)$, related to the former through the fluctuation-dissipation theorem. In YBCO, quite accurate results have been obtained concerning the local static magnetic susceptibilities of CuO_2 planes, CuO chains and apex O1 oxygen from polarized-neutron-diffraction measurements of the magnetization densities induced by an applied magnetic field. As we will see, such measurements can be directly compared to the NMR Knight-shift measurements. More interestingly, $\chi''_{\alpha\beta}(Q,\omega)$ has been obtained in the full doping range between x_C and 1 as a function of temperature from very comprehensive unpolarized and (in the most favorable cases) polarized NIS studies.

3.1. MAGNETIZATION DENSITIES

As an illustration of the changes occurring when the underdoped-overdoped border line is crossed (see Figure 1), we will first consider the neutron-scattering determination of the local (q=0) static ($\omega \approx 0$) susceptibility on the in-plane Cu2 sites [50,51,52] in two samples with characteristic oxygen contents $x \approx 0.53$ ($T_C \approx 48$ K) and $x \approx 1$ ($T_C \approx 90$ K). Due to the weakness of the expected signals, such measurements require using the powerful polarized-neutron-diffraction technique, which allows to obtain small magnetic scattering amplitudes through the magnetic-nuclear interference term [53].

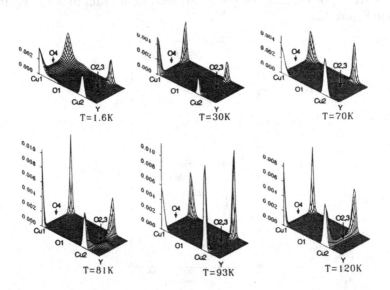

Figure 11. Magnetization densities measured at different temperature on $YBa_2Cu_3O_7$. The magnetization densities are projected along [1 -1 0] and expressed in $\mu_B / Å^2$ (from [51]).

Following the principle of the method, the polarization of the incident beam is alternately set parallel (+) or antiparallel (-) to the direction of the magnetic field applied to the sample. Depending on their initial spin state, the neutrons are scattered in a different way, giving rise to two different Bragg-reflection intensities $I^+(Q)$ and $I^-(Q)$ related to $F_N(Q)$, the nuclear structure factor, and $F_M^\perp(Q)$, the component perpendicular to Q of the magnetic structure factor. For a perfectly polarized beam, one has

$I^{\pm}(\mathbf{Q}) \propto \left| F_N(\mathbf{Q}) \pm F_M^{\perp}(\mathbf{Q}) \right|^2$. $F_M^{\perp}(\mathbf{Q})$ is directly related to the Fourier transform of the magnetization density $m(\mathbf{r})$. One experimentally accedes to $F_M^{\perp}(\mathbf{Q})$ knowing $F_N(\mathbf{Q})$ through measurements of the flipping ratios $R(\mathbf{Q})$ defined by (assuming $F_M^{\perp}(\mathbf{Q}) << F_N(\mathbf{Q})$ as it is for a high-T_C superconducting cuprate):

$R(\mathbf{Q}) = \dfrac{I^{+}(\mathbf{Q})}{I^{-}(\mathbf{Q})} \approx 1 + 4 \dfrac{F_M}{F_N} \cdot p \cdot \sin^2 \alpha$, where p is the polarization of the incident beam and α the angle between the direction of the applied magnetic field and the scattering vector $\mathbf{Q}$. The magnetization density is then deduced either by a classical inverse Fourier transform (which requires a model to be performed) or by a maximum-entropy method (which requires no a priori model [54]).

Figure 11 shows the temperature dependence of typical magnetization-density maps obtained on the overdoped compound $YBa_2Cu_3O_7$ in a field of 10 T. Similar results have been obtained for the underdoped $YBa_2Cu_3O_{6.83}$ [52] and $YBa_2Cu_3O_{6.53}$ [51] compounds. The thermal variations of the resulting local static susceptibilities on Cu2 sites of CuO_2 planes for the two extreme oxygen contents are reported and compared to equivalent "Knight-shift" NMR data [55,56] in Figure 12.

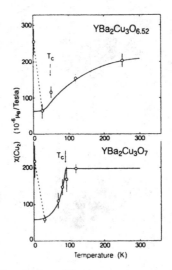

Figure 12.   Temperature dependence of the local susceptibility on Cu2 sites in the CuO_2 planes. The solid lines correspond to calculated Knight-shift thermal variations (see text). The dashed lines are guides for the eye (from [51]).

In agreement with NMR, the polarized-neutron-diffraction data in the normal state give evidence for what is called in the literature the "spin pseudo-gap" effect, which is characterized by a strong decrease of the local susceptibility well above T_C for an underdoped material, while the local susceptibility drops at T_C for an overdoped one. Another interesting result (not detected at all in the Knight-shift measurements) concerns the increase of the local susceptibility of Cu2 sites which is observed well below T_C at any doping. In addition to this contribution, the polarized-neutron-diffraction measurements show the existence of a small paramagnetic moment on the Cu1 sites for oxygen contents x>0.4 (see Figure 11), coming quite obviously from the presence of a small amount of (S=1/2) Cu^{2+} ions on the Cu1 sites. As seen in section 1, this has a direct link with the ordering mechanism of the ...Cu-O-Cu-O... chain fragments. As proposed in Reference [51], the low temperature extra contribution observed on the Cu2-sites has the same origin and should be a manifestation of the coupling existing between the CuO-chain and CuO_2-plane subsystems through the O1 apex oxygen. A comprehensive discussion of this problem has been given by G. Uimin [14,15], showing among other the relationship between the number of paramagnetic S=1/2 spins per Cu1 sites and the oxygen content x, as a function of the "quality" of the oxygen ordering (the latter being mainly controlled by the quenching-temperature value). Again, we would like to stress that n_h, the amount of holes transferred from the chain to the planes, depends also strongly on the perfection of the chain ordering.

In the following, we will show that the quite different behaviors which are observed from both sides of the optimal concentration x_{opt} are closely related to the existence of strongly doping-dependent dynamical antiferromagnetic correlations.

3.2. MAGNETIC EXCITATION SPECTRA

NMR measurements of the ^{63}Cu nuclear inverse spin-lattice relaxation rate $1/^{63}T_1$ were the first to point out unambiguously the existence of low energy antiferromagnetic correlations in the metallic state of 123-compounds, at least above T_C [56-60] (for a recent review on NMR in cuprates, see [3]). Quite complementary to NMR, NIS measurements of the magnetic excitation spectra have provided master pieces of information concerning the temperature and doping dependencies of the various scales involved both in the superconducting and normal states. As it is often the case with the neutron-scattering technique, the main difficulty is to extract the purely magnetic signal from all the other contributions. This is evidently the case with spin-1/2 high-T_C cuprates, for which one expects relatively weak magnetic neutron cross sections and obviously true for YBCO, which exhibits a very complex phonon spectrum (for more details, see the chapter by L. Pintschovius and W. Reichardt) and for which most of large size single crystals contain a large quantity of powdered parasitic "green phase" Y_2BaCuO_5 (more than 20% in some samples !). In most of cases, this task could be achieved using unpolarized NIS and medium-size samples (of typical volume V$\approx$0.4-1 cc), from a self-consistent analysis of both the momentum (Q) and temperature (T) dependencies of the neutron cross sections [4,40,61-66]. Polarized NIS, which in principle is the best suited method for separating the magnetic and nuclear scattering (see chapter 1), has been applied only in some very peculiar cases [67,65,68], however not always without some "ambiguity". Besides its intrinsic complexity, the latter technique requires also having large-size single crystals (V ≈ 2-10 cc), fortunately available for YBCO. From the analysis of all the data which have been collected so far by the various groups, a consensus about the evolution of magnetic fluctuations upon doping is beginning to emerge. In the following, we will show and discuss the most striking results which have been brought by NIS, in both the normal and superconducting states.

3.2.1. Magnetic fluctuations in the normal state

One of the key point for the understanding of high-T_C superconductivity concerns certainly the precise characterization of magnetic fluctuations spectra in the normal state.

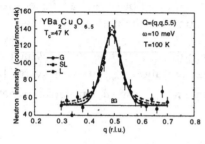

Figure 13.   Q-scan across the AF-rod at the scattering vector Q=(1/2,1/2,5.5) and an energy transfer of 10 meV for YBa$_2$Cu$_3$O$_{6.5}$.

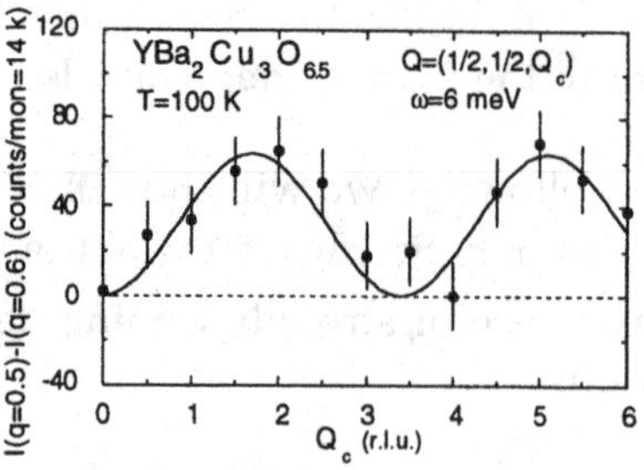

Figure 14.   Q-scan along the AF-rod at an energy transfer of 6 meV for YBa$_2$Cu$_3$O$_{6.5}$, showing the intrabilayer modulation.

Magnetic correlations.   Above T_C, all the NIS results show unambiguously the existence in a large part of the (Q,ω) space of a magnetic response peaked at the in plane wave vector $Q_{AF}=(\pi,\pi)$, suggesting quite obviously the presence of *commensurate antiferromagnetic* correlations. Figure 13 shows a typical scan along the [1,1,0] direction carried out on an underdoped YBa$_2$Cu$_3$O$_{6.5}$ sample. Quite similar results have been obtained whatever the doping rate, at least up to the optimal doping. At the reverse of LSCO, no trace of incommensurability could be detected *above* T_C in YCBO. In general, the line shape in q of the magnetic response cannot be unambiguously determined, a gaussian, a lorentzian or even a square-lorentzian function fitting as well the data (see Figure 13). Assuming for sake of simplicity a lorentzian response function given by the expression:

$$S(q) \approx S_0 \cdot \frac{1}{1+(q\xi_a)^2} \tag{5}$$

where $q = (q,q) = Q - Q_{AF}$ and S_0 is a normalization factor, orders of magnitude for the corresponding parameter ξ_a can be deduced from such measurements. In the present case ξ_a is expected to reflect closely the in-plane correlation length, although the correct determination of this parameter would require knowing the q-dependence of the energy-integrated structure factor $\int S(q,\omega)d\omega$. This is mainly due to the fact that the q-width Δq of the magnetic response is found little depending on ω for not too large energies [69,62,70,63], making the integration procedure unnecessary. In YBCO, the observed in-plane antiferromagnetic correlations involve only a small number of spins. Typically, $\xi_a / a \approx 2.5$ for $x \approx 0.5$ ($T_C \approx 47$ K) and $\xi_a / a \approx 1.1$ for $x \approx 0.92$ ($T_C \approx 91$ K), with standard error bars of the order of 10%. Table 1 summarizes the evolution of ξ_a / a for some typical oxygen contents x.

TABLE 1. In plane correlation length versus oxygen content.

oxygen content	0.45	0.50	0.60	0.83	0.92	0.97
ξ_a /a	3.5	2.5	2.0	1.6	1.1	-

At the accuracy of the neutron measurements, ξ_a appears only weakly temperature dependent [69,62,63]. For commensurate correlations (as it is the case in the normal state), this result appears in strong contradiction with the simple phenomenological model proposed by Millis, Monien and Pines [71,72] and originally intended to explain the NMR results in YBCO$_7$ (the so called "Nearly Antiferromagnetic Fermi Liquid" model). Within this approach, mainly based on the assumption of a strong coupling between the low frequency AF-paramagnon modes and the quasi-particles, the correlation length is predicted to vary with T according to $\xi(T) = \xi(0)\sqrt{\dfrac{T_x}{T + T_x}}$, where T_x is a characteristic temperature of the order of 100 K and $\xi(0)$ is at least a factor 2 larger than the neutron-scattering determination. Recently, it was shown that the ^{89}Y, ^{63}Cu and ^{17}O NMR measurements in YBa$_2$Cu$_3$O$_{6.92}$ could be reanalysed self-consistently starting from a spin susceptibility composed of a contribution arising from AF spin fluctuations delineated by the NIS data and a contribution arising from itinerant quasi-particles with a characteristic frequency increasing at low temperature [73,74]. Similarly, NMR measurements on Ni-doped YBCO [75] were found not inconsistent with the assumption $\xi(T) \approx$ constant. Here, we would like to mention that such a relation is predicted in more realistic models like the three-bands Hubbard model (see, e.g., Ref. [76] and references therein) or the t-J model (see, e.g., [77,78] and references therein). Indeed, it is a byproduct of the weak temperature dependence of J(q), the Fourier transform of the electron-electron couplings. However, from a purely experimental point of view, the in-plane correlation length could appear as well controlled by the hole density, since $\xi_a / a \approx (0.5 - 0.7) / \sqrt{n_h}$. Thus, the picture of holes inducing (overlapping) magnetic polarons might remain valid in the metallic state.

An important result obtained by NIS concerns the stability of the intrabilayer couplings, at least at low-energies. Whatever x, one observes a spin susceptibility reflecting closely the $\sin^2(\pi z Q_l)$ momentum dependence expected for the structure factor associated with a bilayer (see the section 2.2.). Figure 14 shows a typical result relative to a $YB_aCu_3O_{6.5}$ sample, which illustrates well this fact. Quite similar results have been obtained at larger doping. Thus, the intrabilayer antiferromagnetic coupling is preserved in the normal state, at least up to room temperature. Later, we will see that this is also true in the superconducting state. However, it is now well accepted that this coupling plays no fundamental role in the spin pseudo gap problem (despite some tentative in this direction [79]) since the discovery of the pseudo gap effect in a monolayer mercury-based compound [80].

Magnetic excitation spectra versus doping and temperature. As previously pointed out, the knowledge of the doping dependence of the normal-state spin susceptibility is crucial for the understanding of the intriguing changes in the electronic properties which have been observed from both sides of x_{opt}. By comparing NIS measurements at various oxygen contents performed either on the same re-oxidized single crystal [4,62,63,81] or on different single crystals (after proper phonon normalization) [66,81], it has been possible to obtain an accurate overview of the doping dependence of the imaginary part of the normal-state spin susceptibility $\chi''(Q,\omega)$ at $Q=Q_{AF}=(\pi,\pi)$. We have reported in Figure 15 the results obtained by the CEA-Grenoble/Saclay group for some typical oxygen contents $x \approx 0.5$, 0.83, 0.92 (underdoped regime) and $x \approx 0.97$ (overdoped regime), at a same temperature of about 100 K.

Three important features emerge from these data. Firstly, the low-energy part of the magnetic response is progressively depressed as the doping increases. Secondly, the spin susceptibility displays a maximum in amplitude around a well defined energy scale slightly increasing with x (typically at about 20 meV for $x \approx 0.5$ and 30 meV for $x \approx 0.8$-0.9), which is ascribed to the existence of a *spin pseudo gap* in the normal-state excitation spectrum. Thirdly, the maximum amplitude of $\chi''(Q,\omega)$, which is weakly doping dependent between x=0.5 and x=0.92, vanishes very rapidly above x_{opt}. This fact is well illustrated by the strong amplitude difference which is observed at oxygen contents 0.92 (slightly underdoped) and 0.97 (slightly overdoped). Actually, highest the energy is and more affected appears the magnetic signal. More quantitatively, between x=0.92 and x=0.97, $\chi''(Q_{AF},\omega_0)/\omega_0$ decreases by a factor 2.5 at ω_0=10 meV, while this ratio amounts to about 5 at ω_0=30 meV. This trend is consistent with NMR measurements performed on samples with very similar oxygen contents, which show almost no change in the maximum of $1/^{63}T_1T$ versus T [82], indicating that the slope of $\chi''(\omega_0)$ at very small ω_0 is only weakly doping dependent around optimal doping. As previously mentioned, all these facts have again to be ascribed to the presence of a spin pseudo gap in the normal state of underdoped materials. This pseudo gap gives rise to the slightly non linear character of the spin susceptibility at medium energies, while the spin susceptibility should remain quasi linear over a much larger energy range for the gapless overdoped $YBa_2Cu_3O_{6.97}$. A possible explanation for the surprisingly abrupt change between x=0.9 and x=1 could originate from the existence of a quantum critical point at x_{opt} [83,84].

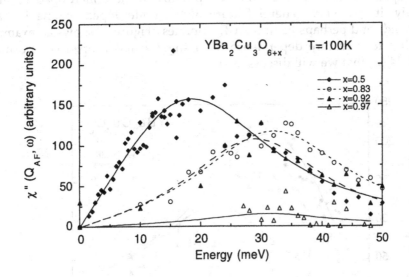

Figure 15.   Imaginary part of the normalized spin susceptibility in the normal state for four different oxygen contents (100 in vertical scale correspond to about $350\ \mu_B^2/eV$ in absolute unit). Note the rapid changes of the spin-susceptibility amplitude around x_C.

Thus, the NIS results (in agreement with NMR) demonstrate very convincingly that the disappearance of the "spin pseudo gap effect" on entering the overdoped regime results mainly from the very fast disappearance of the antiferomagnetic correlations for x larger than x_{opt}, leading the electronic system to recover a more conventional metallic character. Conversely, the presence of large AF correlations in the normal state of underdoped 123-compounds is responsible for their well pronounced non-Fermi-liquid behavior.

In the underdoped regime, the spin excitation spectra exhibit relatively weak temperature dependencies, at least in what concerns the energy scales. Figure 16 shows for example the thermal variations between 53 K and about 300 K of the spin susceptibility at $Q \approx (\pi,\pi)$ for a $YBa_2Cu_3O_{6.5}$ ($T_C \approx 47$ K) sample [4,63]. In increasing temperature, the low-energy part (slope and amplitude) of $\chi''(Q,\omega)$ decreases rapidly, whereas the maximum energy is slightly shifted to higher values (from 17 meV at 53 K to 26 meV at 287 K).

Not surprisingly, the high energy response is weakly temperature dependent, in the range of investigated T. On the low-energy side (typically below 2 meV), the spin susceptibility at $Q=(\pi,\pi)$ exhibits a maximum around a temperature $T^{max} \approx 70\text{-}80$ K, well above T_C [4]. A very similar behavior has been obtained by the "BNL" group on samples with oxygen contents x=0.5 [85,86] and x=0.6 [87,61,88]. As previously discussed, the existence of this maximum in the temperature dependence of χ'' (which is the exact analogous of that observed in the $1/^{63}T_1T$ NMR measurements [3]), reflects the *non-Fermi-liquid* character of the spin susceptibility in the underdoped regime. Actually, it is a very general feature of the underdoped regime in 123-compounds and perhaps in all high-T_C cuprates. Figure 17 shows an example of such a temperature dependence for a slightly underdoped composition x=0.92 [4,63] that we will discuss later.

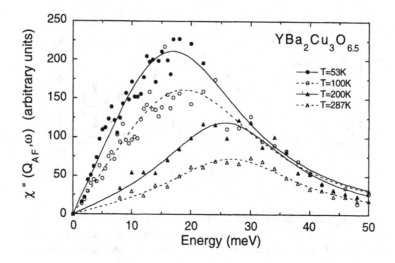

Figure 16. Imaginary part of the spin susceptibility in $YBa_2Cu_3O_{6.5}$ as a function of the energy at different temperatures. The lines are calculated as described in the text.

By performing a ω/T-scaling analysis of the NIS data obtained in the strongly underdoped regime [86,70], it appears that in a limited range of T and ω (roughly for $T > T^{max}$ and ω smaller than the maximum energy (≈ 20 meV)), the temperature dependence of the spin susceptibility converges toward a generic function of ω/T, well reproduced by the simple functional form $\tanh(\omega/\alpha kT).L(\omega)$, where $L(\omega)$ is a function weakly dependent on ω and α is a constant of the order of 1. Such a "tanh" term has been predicted for some very peculiar models [89-91]. It is also consistent with the "Itinerant-localized duality picture" of the spin fluctuations developed by Narikiyo and Miyake in order to remove the apparent inconsistency between the NMR and NIS determinations of the in-plane correlation length [92]. However, the ω/T-

scaling seems to work only in the vicinity of x_C and has never been demonstrated near x_{opt}.

For a more quantitative analysis of the NIS data near the maximum energy, one can use the following "phenomenological" spectral function, which reproduces the main features of the measurements:

$$\chi''(q,\omega) \approx \frac{\chi_0(T)}{1+(q\xi)^2}\left(\frac{\Gamma_{pg}}{\Gamma_{pg}^2+(\omega-\omega_{pg})^2} - \frac{\Gamma_{pg}}{\Gamma_{pg}^2+(\omega+\omega_{pg})^2}\right) \qquad (6)$$

where ω_{pg} and Γ_{pg} represent two energy scales characteristic of the dynamical antiferromagnetic fluctuations. This expression is typical of a damped inelastic response similar to that predicted at finite T in some 1D *gapped* quantum spin systems (like, e.g., the S=1 Haldane-gap chain [93] or the two-legs S=1/2 spin-ladder system [94],...). It is certainly not valid at very low energy, where the "than (ω/T)-scaling" should apply. In Table 2 are listed the values of ω_{pg} and Γ_{pg} for typical oxygen contents.

TABLE 2.   Doping dependence of parameters χ_0 ω_{pg} and Γ_{pg} at 100 K.

oxygen content	0.50	0.52	0.60	0.83	0.92
χ_0	199	-	-	124	113
ω_{pg} (meV)	17	25	27	32	29
Γ_{pg} (meV)	18	15	15	15	14

Thus, such a simple analysis reveals quantitatively the strong *inelastic character* of the magnetic response in the normal state. The parameters χ_0, ω_{pg} and Γ_{pg}, which in fact characterize precisely the spin pseudo gap, display relatively weak doping dependencies, at least up to x_{opt}. Above x_{opt}, nothing accurate can be said concerning ω_{pg} and Γ_{pg}. One only knows that the amplitude of the magnetic response drops rapidly to a small value and that the spin susceptibility is probably featureless, as expected for a conventional metal.

The same kind of analysis shows that ω_{pg} and Γ_{pg} depend also weakly on T. The largest effect is seen for $x \approx 0.5$ (between 53 K and 300 K, ω_{pg} increases from 16 to 27 meV, while Γ_{pg} remains roughly constant at about 13-16 meV and χ_0 decreases roughly as $1/(T+40)$). For $x \approx 0.9$ [4], apart from a decrease of χ_0, almost no temperature dependence exists.

To summarize this discussion, NIS experiments in the normal state have unambiguously established the existence of a *spin pseudo gap* in the magnetic excitation spectrum at any doping smaller than x_{opt}, with a typical energy scale ω_{pg} of the order of (0.2-0.3)J (slightly doping- and temperature-dependent) and a "quantum width" Γ_{pg} of the order of (0.1-0.15)J (very weakly doping dependent). Following the analogy with low dimensional quantum magnetism, the spin pseudo gap feature could be directly associated with the two-spinon spectrum [77,78].

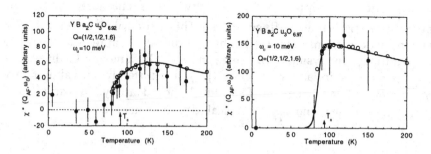

Figure 17. Imaginary part of the spin susceptibility at (π,π) and energy 10 meV as a function of temperature in $YBa_2Cu_3O_{6.92}$ and $YBa_2Cu_3O_{6.97}$. Comparison between NMR (open symbols) and NIS data [82].

At low energy, the temperature dependence of the spin susceptibility exhibits again the so called "pseudo gap effect", first reported from NMR measurements [3]. Figure 17 shows two examples of such thermal evolution at an energy of 10 meV for two $YBCO_{6+x}$ samples with oxygen contents $x \approx 0.92$ (slightly underdoped with $T_C \approx 91$ K) and $x \approx 0.97$ (slightly overdoped with $T_C \approx 92.5$ K). For comparison are also reported in Figure 17 rescaled values of $1/^{63}T_1T$ versus T obtained by NMR [82]. As previously mentioned, the low-energy part of the spin susceptibility displays a clear maximum around a temperature $T^{max} \approx 120$-130 K larger than T_C for the former, whereas the maximum occurs just above T_C for the latter. A quite similar result has been obtained in an underdoped sample with $x \approx 0.83$ [95]. In agreement with NMR data at equivalent oxygen contents [3], T^{max} is found weakly doping-dependent ($T^{max} \approx 75$ K at x=0.5, $T^{max} \approx 120$ K at x=0.83 and $T^{max} \approx 120$ K at x=0.92). Quite interestingly, T^{max} reflects closely the doping dependence of ω_{pg} ($T^{max} \sim \omega_{pg}/2k$). This unconventional behavior of χ'' has been widely interpreted in the literature as the signature that a spin pseudo gap opens below some temperature $T^* \approx 150$-160 K slightly larger than T^{max} [58,59,56,3]. The NIS results demonstrate that the situation is not so "simple". Obviously, the thermal behaviors of the low-energy and high-energy parts of χ'' start to differ below some characteristic temperature value T^* (much larger than T_C)

possibly related to ω_{pg} or Γ_{pg}. Whereas χ'' at $\omega \approx \omega_{pg}$ increases continuously when the temperature decreases down to T_C, the low-energy part of the spectrum departs from this tendency before to reach a maximum at T^{max} and then to drop at T_C, indeed showing a quite opposite temperature dependence. Because this effect is absent in the overdoped regime, one is led to ascribed it to the AF correlations. Very likely, the maximum of $\chi''(T)$ at low-energy should result from the subtle interplay between the temperature dependencies of the various parameters (amplitude, width, energy scales,...) describing the *real* magnetic response. The characteristic temperature scale $T^{max} - T^*$ deduced from NMR and NIS measurements of the "low energy" part of $\chi''(\mathbf{q},\omega)$ appears in fact quite different from the characteristic temperature scale T_{pg} deduced from macroscopic measurements of , e.g., the susceptibility or the resistivity. This is shown in particular by their completely different magnitudes and doping-dependencies [3,5]. However, it is probable that both have a common origin which should be related to the presence of the AF correlations, since both disappear in the overdoped regime, as the AF correlations do.

As it can be seen from Figure 17, the low-energy spin susceptibility vanishes completely below T_C, implying clearly the *opening of a true spin-gap* in the superconducting state. In the next sections we will discussed the main effects of superconductivity on the magnetic response.

3.2.2. *Magnetic fluctuations in the superconducting state*
A central issue in the debate over the pairing mechanism in the high-T_C cuprates concerns the extend to which the magnetic correlations are involved. NIS investigations of both the doping and temperature dependence of the superconducting-state spin susceptibility have considerably clarified the situation by showing a strong interplay between superconductivity and spin correlations.

Magnetic excitation spectra versus doping.   From the analysis of the different NIS results, a relatively consistent picture for the evolution upon doping of the magnetic excitation spectra in the superconducting state has emerged for $YBCO_{6+x}$. Figure 18 shows the doping dependence of the normalized spin susceptibility at $Q_{AF} = (\pi, \pi)$ for six different oxygen contents covering the full oxidization range [4,63,64,81]. For the record, these data have been obtained on the same single crystal, using very similar experimental conditions. Complementary studies have been reported by the BNL/Nagoya group (x=0.4 [40], x=0.6 [61], x=0.75 [40] and x=0.9 [40]), the Princeton group (x=0.5 [96], x=0.7 [96] and x $\approx$ 1 [65]) and the ORNL group (x=0.6 [68] and x $\approx$ 1 [67]). The results in the range 0.6-0.7 (covering the "60 K plateau") are particularly interesting. Apart from the fact that they nicely complement the data depicted in Figure 18, they also confirm the sensitivity of the spin susceptibility to the quality of the oxygen ordering in the Cu1 planes for x~0.6: in this range, two samples with the same x can display appreciably different magnetic properties. For a given x, higher T_C is, better is the sample and narrower is the excitation spectrum. Thus, the $YBCO_{6.6}$ sample studied at BNL ($T_C \approx 53$ K [87,61]) looks better like a $YBCO_{6.5}$ sample, whereas the $YBCO_{6.6}$ sample studied at ORNL ($T_C \approx 63$ K [68]) looks better like the Princeton $YBCO_{6.7}$ [96] or the CEA-Grenoble/Saclay $YBCO_{6.8}$ [95] samples.

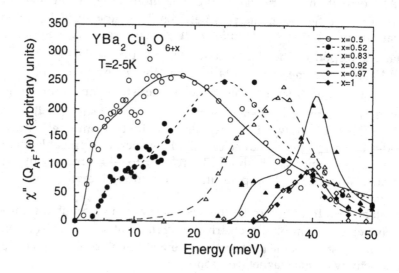

Figure 18.   Imaginary part of the spin susceptibility in the superconducting state at (π,π) as a function of energy for several oxygen contents, showing the rapid disappearance of low energy magnetic excitations in increasing doping.

Some general features emerge from the comparison of all these different excitation spectra. First, strong AF-correlations persist in the superconducting state. In increasing doping, the energy range is reduced on both the low-energy side (by the opening of a spin energy-gap E_g increasing with x) and the high energy side (by a reduction of the spectral width). Quite unexpected, the magnetic excitation spectrum condenses into a narrow contribution peaked at a well defined energy $E_r \approx 40$ meV when $x \approx 1$, defining the so called "resonance peak" [69]. The maximum energy of the spin susceptibility, E_M, evolves from about 5-7 meV for x=0.4 [40] up to 41 meV for $x \approx x_{opt}$ [69,67,65,95], whereas simultaneously the characteristic spectral width Γ_M decreases from about 15 meV in $YBCO_{6.5}$ down to about 1 meV in $YBCO_7$ (after deconvolution from the instrumental resolution, assuming a simple lorentzian function of half width Γ_M). Figure 19 summarizes the dependence on T_C of the parameters E_g, E_M,(E_r) and Γ_M for a large number of oxygen contents covering the range 0.4-1 (note that for x<0.5, the distinction between E_r and E_M appears "subtle"). From these data one can see that lower the transition temperature is, lower the energy range around which the spectral weight is localized and broader the energy range over which it appears.

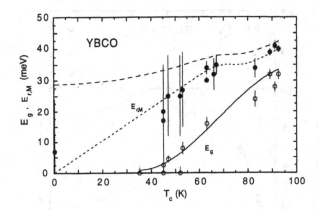

Figure 19. E_g, E_M (E_r) and Γ_M as a function of T_C. The "error bars" on E_M (E_r) represent Γ_M, the half width of the spectral response. The lower curve (E_g) and the upper curve (corresponding to the energy at half maximum) indicate roughly the extension in energy of the spectral response at very low temperature.

An interesting question concerns the possible existence of two distinct contributions in the superconducting state of underdoped samples, first reported in $YBa_2Cu_3O_{6.92}$ by the CEA-Grenoble/Saclay group [4,63,66,95]. This feature is revealed by the presence of a shoulder between E_g and E_r, only visible when the energy resolution is good enough (as low as 4.5 meV in the present case). Unfortunately, it has not been confirmed by the other groups so far, although recent measurements on $YBa_2Cu_3O_{6.6}$ by the ORNL group strongly suggest the presence of an incommensurate signal below the 35 meV resonance peak [97]. An example of such behavior is shown in Figure 18 for a slightly underdoped $YBCO_{6.92}$ sample, for which the shoulder is observed around 30-36 meV [4,63,95]. This "out-of-resonance" signal is generally very reminiscent of that observed in the normal state, in particular in what concerns the order of magnitude of the intensity and of the q-width Δq (in the present example one has typically at 5K $\Delta q \approx 0.2$ r.l.u. at energy 33 meV and $\Delta q \approx 0.15$ r.l.u. at 40 meV, while above T_C, $\Delta q \approx 0.2$ r.l.u. in the same energy range (resolution-convoluted values)). As for the normal-state component, this second contribution is also strongly doping-dependent between $x \approx 0.9$ and 1 and finally becomes hardly visible in the overdoped regime (see the evolution of the magnetic response given in Figure 18). This is probably the reason why it has not been detected by the other groups in $YBCO_7$ [67,65], while it could be seen in $YBCO_{6.92}$.

Until now, we have discussed the evolution upon doping of the spin dynamics only for energies smaller than 50 meV. Recent NIS results on $YBCO_{6.5}$ have allowed to obtain new pieces of information on the evolution of the high-energy part of the magnetic excitation spectrum in the metallic range [98,100].

For the antiferromagnetic parent compound $YBCO_{6.2}$, we have previously seen that the magnetic excitation spectrum displays two distinct branches

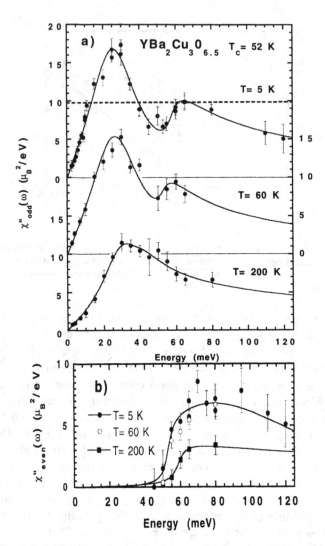

Figure 20. Acoustic (odd) and optic (even) q-integrated spin susceptibilities in $YBa_2Cu_3O_{6.5}$ as a function of energy for three temperatures, showing the presence of a double-gap structure for the acoustic modes and a single-gap structure in the optic mode (from [98]).

associated with quasi linear acoustic and gapped optic spin waves. At small wave vector , the dispersion curves of these two modes are given by the following relations (see section 2.2):

$$\omega_{ac}(q) \approx c.q \tag{7a}$$

and

$$\omega_{opt}(q) \approx \sqrt{\Delta_{opt}^2 + (c.q)^2} \tag{7b}$$

with a spin wave velocity c~650 meV.Å and a gap energy Δ_{opt}~67 meV. Figure 20 shows the energy dependence of the q-integrated susceptibilities associated with the acoustic (odd) and optic (even) modes for a $YBa_2Cu_3O_{6.5}$ sample (T_C=52 K) at three temperatures: 5 , 60 and 200 K. Interestingly, high-energy magnetic excitations persist in the metallic regime of $YBCO_{6.5}$ [98,100] and $YBCO_{6.7}$ [99]. As one can see, two important features characterize the high-energy spin dynamics for these oxygen contents. The first one concerns the persistence of an energy gap in the optic mode at an energy $\Delta_{opt} \approx 53$ meV slightly renormalized with respect to the undoped sample. The second one concerns the observation of a double-peak structure for the acoustic mode, with a dip around 50 meV which is very reminiscent of the optic gap [98,100]. There is presently no clear explanation for this effect.

The q-dependence of the magnetic response above 80 meV is consistent with the existence of damped spin wave-like excitations, exhibiting a velocity reduced by about 35% with respect to the zero-doping value (for the energy range which is involved in the present experiment). All these results can be quantitatively accounted for by introducing an effective in-plane coupling constant $J^{eff} \approx 0.65 J$ (taking into account that $\Delta_{opt} \propto \sqrt{J^{eff}J_b}$ and $c \propto J^{eff}$), which should be less and less renormalized as the energy range increases. As interesting it may be, such a behavior is typical of any strongly quantum magnetic system having a finite correlation length and, probably, it has no relation with superconductivity. At higher doping, one expects a larger renormalization of J^{eff} implying an increase of both the softening and the damping of the high-energy magnetic modes [99]. This effect might explain the doping dependence of the characteristic temperature scale T_{pg} (see Figure 1), which indeed would correspond to the crossover temperature at which the AF correlations start to develop ($T_{pg} \sim J^{eff}/k$).

AF-correlations and spin gap.   Whatever the doping rate, dynamical magnetic correlations persist in the superconducting state of $YBa_2Cu_3O_{6+x}$. As in the normal state, the magnetic response which is observed remains peaked around (π, π), suggesting again the presence of antiferromagnetic correlations, commensurate [4,40,67,65,66] or at least very weakly incommensurate [61,97]. The figure 21 shows two series of neutron-scattering data collected on a $O_{6.51}$ sample ($T_C \approx 47$ K) at 12K and on a $O_{6.97}$ sample ($T_C \approx 92.5$ K) at 5 K, which give a clear evidence for the existence of such correlations. However, if the peak positions in wave vector are almost the same in both the normal and superconducting states, the evolution upon doping of the corresponding q-widths turns out to be different. The q-width, first energy independent at small doping, becomes energy dependent near optimal doping (typically $\Delta q \approx 0.12$ r.l.u., whatever $\hbar\omega$ for x=0.51; $\Delta q \approx 0.22$ r.l.u. at $\hbar\omega \approx 33$ meV and $\Delta q \approx 0.14$ r.l.u. at $\hbar\omega \approx 40$ meV for x=0.97 [66]). As previously seen, this effect has to be ascribed to the existence of the resonance peak, which corresponds in fact to an enhancement of the spin susceptibility associated with a narrowing of the q-width at the energy E_r. We will come back to this point later.

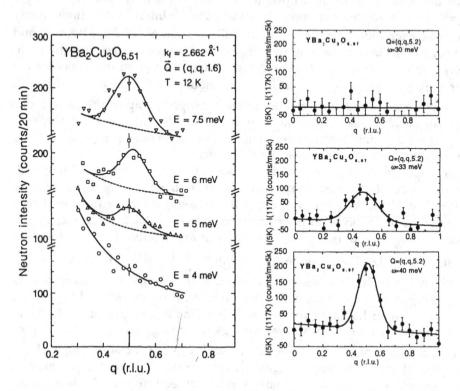

Figure 21. Q-scans across the magnetic rods in YBa$_2$Cu$_3$O$_{6.51}$ and YBa$_2$Cu$_3$O$_{6.97}$ at different energies, showing the existence of the AF correlations and the opening of a spin gap in the superconducting state.

The energy dependence of Δq for the slightly underdoped/overdoped samples has been interpreted by the CEA-Grenoble/Saclay group as arising from the existence of two different contributions [4,63,66]. In the slightly doped samples ($x \approx 0.5$), such a distinction becomes hard to do because all the contributions exhibit the same physical width, after resolution correction.

A very topical and open question (moreover not necessarily disconnected from the latter problem) concerns the possible existence in the 15-25 meV energy range of "incommensurate" correlations in the superconducting state of YBCO$_{6.6}$. Reported first by Tranquada et al [61], such a feature has been recently observed by Mook et al [97]. Quite differently to the case of La$_{0.85}$Sr$_{0.15}$CuO$_4$ [101,102], the incommensuration in YBCO$_{6.6}$ would appear at T$_C$ [97] around positions shifted from (π, π) by a rather small quantity $\delta \approx 0.06$ r.l.u.. Its existence seems constrained in a narrow doping and energy range (the "60 K plateau" between E$_g$ and E$_r$) and depends strongly on the "quality" of the sample oxygenation. By analogy with LSCO, one is tempted to ascribed the existence of such incommensurate correlations in underdoped YBCO to an underlying charge segregation, following the ideas developed by Emery et al [20,103]. However, the quite different temperature and energy dependencies which are observed for both systems strongly suggest a different origin. Although the incommensurate nature of these correlations need further experimental confirmation, one is now certain of at least one thing: it exists an additional magnetic signal below the resonance energy for x=0.6 [68,103], in agreement with other NIS measurements [4,63,66].

The Q-scans depicted in Figure 21 reveal another very interesting feature of the superconducting state: the disappearance of low-energy magnetic excitations below some doping-dependent "threshold" energy E_g. As previously seen from the energy-scans, such a behavior is typical of a gap opening in the spin-susceptibility. At low doping, the spin-gap energy is very sample dependent, with $E_g \approx 4.5$ meV for the best x$\approx$0.5 samples ($E_g / kT_C \approx 1.1$), while it amounts to 32 meV for x$\approx$0.97 ($E_g / kT_C \approx 4.1$). Figure 19 summarizes the evolution on T_C of the spin-gap energy E_g [4,61,40,62,63,65,95,96]. Thus, E_g and the corresponding E_g / kT_C ratio appears both monotonously increasing as T_C increases, implying clearly a connection with the occurrence of superconductivity.

Resonance peak.   Rossat-Mignod and coworkers [69] were the first to point out the existence of a very peculiar enhancement of the spin susceptibility in the slightly underdoped compound YBCO$_{6.92}$ around a characteristic energy $E_r \approx 41$ meV, since then referred as the "resonance peak". An example of such a behavior is shown in Figure 21 for a YBCO$_{6.97}$ sample. Since its first discovery in 91, the resonance peak has been abundantly studied, both from unpolarized [69,4,67,105,106,66] and polarized [67,65,68] neutron experiments.   All the results agree on the fact that the resonance peak is a general feature of the superconducting state in 123-compounds, clearly present both in the overdoped [69,4,67,65,66] and in the underdoped samples [4,107,68,96,95]. It corresponds to a purely magnetic signal, as established from polarized neutron experiments [67,65], although such a result was clear since the beginning from the unpolarized neutron experiments [69]. Quite remarkable properties have been underlined, especially near optimal doping, which clearly suggest a direct link with superconductivity. The first striking peculiarity concerns the extension in the (Q,ω) space of the scattering associated with the resonance peak. Quite unexpected, the spectral weight is concentrated only around a well defined commensurate point (Q_{AF},E_r) of this space. The life-time of the excitation is finite at a temperature as low as 5 K (typically one determines $\Gamma_\omega \approx 1$-2 meV, assuming a lorentzian line shape in energy), as well as the in-plane Q-width which corresponds roughly to a length scale $\xi_a \approx 10$-15 Å ($\xi_a / a \approx 2.5$-3) after resolution corrections [63,66]. Perhaps fortuitously, ξ_a corresponds roughly to the in-plane superconducting coherence length. As briefly mentioned above, the resonance peak is associated with a sizable narrowing of the in plane Q-width of the superconducting-state spin susceptibility. Strictly speaking, the in-plane correlation length is slightly anisotropic with $\xi_{100} \approx 1.4\xi_{110}$ [105]. No dispersion in wave vector could be detected for this feature, a result which eliminate any interpretation in term of conventional spin-wave excitation. This is also documented by the fact that the symmetry of the scattering is fourfold instead of being circular as expected for a system governed by a simple isotropic exchange Hamiltonian [105]. Along the c direction, the spin susceptibility obeys the sine-square modulation typical of bilayers (as in the normal state), a result indicating that the intrabilayer antiferromagnetic couplings are retained in the superconducting state.

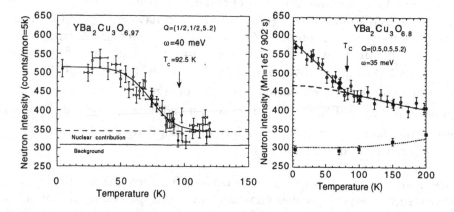

Figure 22. Temperature dependence of the resonance peak in YBa$_2$Cu$_3$O$_{6.97}$ at 40 meV (from [66]) and in YBa$_2$Cu$_3$O$_{6.8}$ at 35 meV (from [95]). The resonance amplitude represents a fraction of the normal-state intensity which decreases as T$_C$ decreases.

In order to underline the interplay between the resonance and the superconductivity, it is instructive to examine the thermal evolution of the resonance. Two examples of such temperature dependencies are reported in Figure 22 for an overdoped O$_{6.97}$ sample (T$_C \approx 92.5$ K) at 40 meV [66] and for a underdoped O$_{6.8}$ sample (T$_C \approx 83$ K) at 35 meV [95]. Similar data have been obtained for other oxygen contents (x=0.92 [4], 1 [4,106], 0.60 [68], 0.70 [96,108] and 0.5 [96,108]). In all cases, the resonance peak is associated with an enhancement of the normal-state spin susceptibility starting exactly at T$_C$, less and less important as x decreases from 1 to x$_C$. This result established the close connection with the opening of the superconducting gap. However, the relationship between these two features is not so trivial in YBCO, because only a very weak temperature dependence of the resonance energy is observed (typically, a reduction of less than 10% is observed at 0.8T$_C$ in YBa$_2$Cu$_3$O$_7$ [4,65,66]). In a conventional BCS-type superconductor, the gap energy Δ (the order parameter associated with the transition) is expected to vary as $\sqrt{1 - T/T_c}$, a behavior which actually corresponds more closely to that of the resonance-peak intensity.

Following the above discussion, one can extract the energy dependence of the resonance   peak   by   considering   the   difference

$\chi''(\mathbf{Q}_{AF}, \omega, T \approx 0) - \chi''(\mathbf{Q}_{AF}, \omega, T \approx T_C)$. In the doping range 0.6-1, this procedure allows an accurate and consistent determination of the resonance-peak characteristics (energy and intrinsic width) as a function of x (see Figure 19). At lower doping (x=0.4-0.5), the determination is not so obvious. However, for x~0.5, an enhancement of the spin susceptibility has been unambiguously observed, which may be considered as a precursor of the resonance peak [4,63,96,108]. The peak position in energy corresponds roughly to the maximum of the spin susceptibility, whereas a substantial broadening occurs as x → 0.4. Thus, the resonance peak turns out to be a general feature of 123-compounds, and perhaps of all high-T_C cuprates   since the discovery of a small enhancement of the spin susceptibility below T_C in $La_{0.86}Sr_{0.14}CuO_4$ [109] (for more details on this compound, see the chapter by S. Hayden and references therein). Unfortunately, there is so far no equivalent data concerning the mercury-, thallium- or bismuth-based compounds , essentially by lack of large single crystals suitable for NIS. Table 3 gives a compilation of some values of the resonance energy E_r and ratio E_r/kT_C taken from the literature.

TABLE 3.   Dependencies on T_C of the resonance energy E_r and the ratio E_r/kT_C.

T_C (K)	47	52	53	63	66	67	83	89	91	92.5	92.7
x	0.52	0.50	0.60	0.60	0.75	0.70	0.80	1.0	0.92	0.97	0.99
E_r (meV)	24	25	27	35	32	32	35	39	41	40	41
E_r/kT_C	5.8	5.6	5.8	6.4	5.6	6.1	5.1	5.1	5.2	5.0	5.1
Ref.	[4]	[65]	[61]	[68]	[40]	[96]	[95]	[62]	[4]	[66]	[67]

In Figure 19 is shown the corresponding dependence of E_r as a function of T_C. Without any explanation, the resonance energy does not follow a simple monotonous relation of T_C. One can be convinced of that by considering the evolution of the ratio E_r/kT_C, which is on average larger for T_C-values located in the "60 K plateau" ($E_r/kT_C \approx 6$ with a maximum at 6.4 [68]) than for T_C-values located in the "90 K plateau" ($E_r/kT_C \approx 5.1$ for $T_C > 80$ K).

How to explain the resonance peak? There is presently no definitive explanation for this unconventional feature. In most of theories advocating the role of magnetic correlations [110-114,77], the resonance peak is attributed to the creation of a quasi electron-hole pair across the superconducting energy gap. However, accounting for a sharp peak in $\chi''(\mathbf{Q}, \omega)$ at $\mathbf{Q}_{AF} = (\pi, \pi)$ requires at least two others fundamental ingredients: a d-wave gap function and the presence of relevant electron-electron interactions characterized by a Q-dependent coupling term $J(Q)$ peaked at (π, π), generally treated within the RPA formalism. In such approaches, the resonance energy E_r is expected to be proportional to $2\Delta_{max}$, where Δ_{max} is the maximum of the gap function. So, NIS measurements of the resonance energy at low temperature would allow a direct and accurate determination of the superconducting-gap energy, if the proportionality factor was known. From the values listed in Table 3, one got a ratio E_r/kT_C of the order of 5-6, while from recent ARPES [115,7] and Raman scattering [116] measurements the ratio $2\Delta_{max}/kT_C$ amounts to 7-9, roughly a factor 1.5 larger. However, this apparently good agreement hardly survives at finite temperature, since experimentally E_r remains constant whereas the superconducting order parameter $\Delta_{max}(T)$ should vanish at T_C, as predicted in conventional BCS theory. At best, this would suggest for Δ_{max} a step-like temperature dependence. By analogy with low-dimensional magnetism, such a behavior could be understood if some precursor of the superconducting gap was existing above T_C.

In order to solve this difficulty, an alternative explanation has been recently proposed by Demler and Zhang [117,118]. Following these authors, within the framework of the positive-U Hubbard model, the resonance specificities may be attributed to the existence of a new collective mode (the so called "π-resonance"), viewed as a particle-particle pair located on nearest neighbor sites in a triplet state (the so called "π-particle"). Quite interesting results have been derived for that model, which might well explain some of the most important feature of the resonance. Among other, the magnetic neutron cross-section in the superconducting state should vary as $\Delta_{max}^2(T)$ and vanish at T_C, whereas the resonance energy E_r should reflect directly the hole density n_h and therefore should be relatively temperature independent. However, the order of magnitude of the π-particle energy determined from this approach (which in principle should be related to E_r) seems to have been underestimated and is still under debate [119,120].

4. Impurity effects: $YBa_2(Cu_{1-y}Zn_y)_3 O_{6+x}$.

Among all the cationic substitutions which have an influence on the superconducting properties of YBCO, the most remarkable is without any doubt the zinc substitution of copper [121-123]. As early recognized [121], a small amount of zinc is sufficient to destroy completely the superconductivity ($T_C=0$ for only y=7%). Zn^{2+} having a non-magnetic closed-shell $3d^{10}$ configuration, this rather strong effect appears quite puzzling because, in terms of conventional superconductivity, the role of such a non-magnetic impurity was thought to be negligible [124]. Thus, zinc might affect the mechanism of the high-T_C superconductivity itself. Consequently, studies of the physical properties induced by zinc-substitution could provide valuable information on the pairing mechanism.

This section is organized as follows: First, the phase diagram of Zn-substituted YBCO will be presented. We next will discuss the effects of zinc on the normal-state spin dynamics, focusing on the occurrence of new AF excitations induced by zinc filling the spin pseudo-gap. We will then describe the effects of zinc-substitution on the superconducting-state spin dynamics. In particular, we will show how zinc suppress the resonance peak and induces in-gap spin excitations. These features will be discussed in relation with the occurrence of possible "gapless" superconductivity.

4.1. PHASE DIAGRAM

It is well established that zinc substitutes the Cu2 sites in the CuO_2 planes, at least up to 4 % [125,126]. The lattice parameters and the oxygen content at which occurs the structural Tetragonal-Orthorhombic phase transition are left unchanged by zinc substitution [125,126]. Accordingly, the oxygen ordering in the CuO plane is practically not affected. Only a negligible variation of the charge transfer has been reported from NMR-Knight-shift measurements [127]. In contrast, the electronic properties are considerably altered by zinc substitution. Figure 23 shows a comparison between the phase diagrams of pure and 2% Zn-substituted YBCO samples. From the undoped insulating state to the overdoped metallic superconducting regime, one can distinguish in the Zn-substituted system three different regimes. In between the long range AF state and the metallic states already seen in zinc-free system, an intermediate "short range" AF state without superconductivity is now present.

In the metallic phases, the superconducting transition temperature is reduced by ~ 12 K/%-Zn at optimal doping. The depression of T_C is even more important in the underdoped regime where a reduction as large as 20 K/%-Zn has been reported [121], so that 2 % of zinc are enough to remove almost completely the superconductivity in the "60 K"-plateau compounds. Moreover, numerous experiments have reported that zinc substitution induces a "gapless" superconductivity and creates a finite electronic density of states (DOS) at the Fermi level. This was suggested by the temperature and the zinc-content dependencies of the ^{63}Cu Knight shift and nuclear-spin lattice relaxation rate, T_1 [128], and also from Mössbauer spectroscopy [129], specific heat [130] and Gd^{3+} electronic spin resonance [131] results. In the underdoped metallic state, ^{89}Y NMR measurements [127,132,133] have given some evidences for the formation of local moments around zinc atoms from both the broadening of the NMR lines [127] and the observation of new satellite NMR lines [132,133]. Interestingly, the effective moment per impurity decreases sharply on increasing doping (by a factor 2.4 from $x \approx 0.6$ to $x \approx 1$ [134]). Conventional magnetic pair breaking [124] arising from these induced local moments are unable to fully account for the reduction of T_C [132,133]. Therefore, the suppression of the superconductivity should be due to other mechanisms. Most probably, it might be explained by the effect of non-magnetic impurities in presence of an anisotropic superconducting order parameter (of d-wave symmetry) [135,136]. Treated in the unitary limit, this also naturally account for the residual DOS at the Fermi level [128,136].

At low oxygen contents in the AF state, the Néel temperature is slightly reduced by zinc substitution. This has been unambiguously shown from neutron-scattering experiments [137,138], whereas previous NMR measurements have misleadingly found no change of T_N. As Zn^{2+} replace Cu2, the zinc-substitution effect on the 3d-AF ordering is actually well described by a simple dilution effect of the AF interaction by a static non-magnetic impurity [138], as it is usually observed for a quasi-2d AF-Heisenberg system.

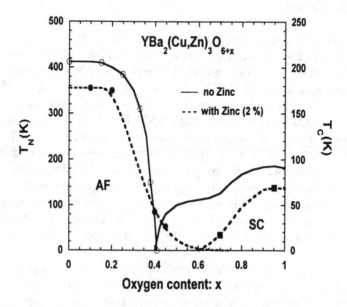

Figure 23. Comparison of the YBCO phase diagram with and without zinc: full circles and squares correspond respectively to the magnetic and superconducting transitions of some studied samples (from [138]).

In between the antiferromagnetic and superconducting phases, it exists a new region where hole doping is already large (~ 10 %) but without any trace of superconductivity. μ-SR [134] as well as neutron-scattering measurements [138] have shown that a magnetic ordering occurs in this range. From μ-SR this ordering occurs at a much lower temperature (typically ≤5 K) than that derived from neutron diffraction. Neutron-scattering measurements demonstrate that the AF ordering which is observed remains in fact short range down to the lowest temperatures [138]. This unusual temperature dependence is very reminiscent of a spin-glass behavior for which a dynamical disorder freezes at low temperature. Within the spin-glass scenario, the difference in onset ordering temperature between the neutron and the μ-SR measurements could be explained by the different time scales probed by the two techniques. Probably, this behavior already occurs in zinc-free YBCO$_{6.4}$ (see section 2), but it is sharply enhanced by zinc substitution. As we will see below, the local moments induced by zinc-substitution are a consequence of a freezing of the AF spin fluctuations from which results the AF short range ordering. In terms of transport properties, it would mean that zinc reduces the charge-carriers mobility, as evidenced by the large increase of the residual resistivity [122,123]. In this intermediate doping range, the effect is so strong that Zn might induce a metallic-insulator transition at low temperature [139].

4.2. NORMAL STATE

4.2.1. Underdoped regime (x ≈0.7)

We now emphasize the effect of zinc substitution on the AF fluctuations in the underdoped regime, especially in samples where the superconductivity is either removed [140] or repulsed to low temperature (typically about 15 K) [138,141,142].

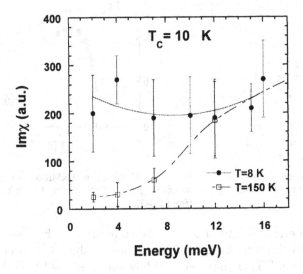

Figure 24.   Spin susceptibility at the antiferromagnetic wave vector in zinc-substituted underdoped $YBa_2Cu_{2.9}Zn_{0.1}O_{6.75}$ (from [141]).

In such cases, the low energy magnetic spectrum is strongly affected upon cooling: Figure 24 shows the low energy part of the spin susceptibility at $Q = (\pi, \pi)$ in underdoped $YBCO_{6.75}$ with about 3% of zinc. Interestingly, the magnetic response at high temperature is very similar to that of the pure system: at T=150 K, the low-energy spin susceptibility is reduced as a result of the opening of the spin pseudo gap. In contrast, the low-temperature spin susceptibility shows a large spectral weight below 12 meV down to the lowest investigated energies (≈2 meV in the present case), whereas the magnetic spectrum is left unchanged at higher energies. Similar results have been reported in underdoped YBCO with different zinc contents [138,140,142].

This particular trend is well underlined by the temperature dependence of the spin susceptibility at energies smaller than 12 meV. Figure 25 displays $\chi''(Q_{AF}, \omega = 5\,meV)$ as a function of temperature in $YBCO_{6.7}$ with 2% of zinc. From T= 200 K, the spin susceptibility first increases on cooling, then remains constant around 120 K and finally strongly increases at low temperature to reach an amplitude 4 times larger at T ≈ 20 K. The high temperature evolution is again quite similar to that of the zinc-free system. However, an enhancement now occurs at low temperature, filling the spin pseudo-gap. Undoubtedly, this corresponds to the appearance of the low energy excitations (induced by zinc) seen in Figure 24.

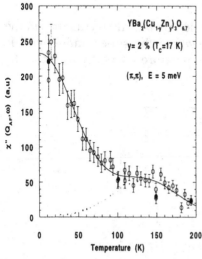

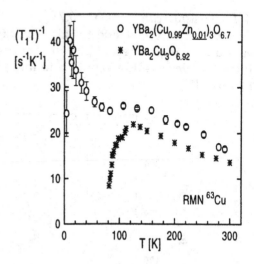

Figure 25. Temperature dependence of the spin susceptibility at energy 5 meV (from [138,142]).

Figure 26. Comparison of NMR copper relaxation in YBCO$_{6.7}$ with 1% of zinc and pure YBCO$_{6.92}$ (from [144]).

This unusual thermal behavior should be compared to that observed by NMR, especially to the temperature dependence of the ^{63}Cu nuclear spin lattice relaxation rate, $^{63}T_1$. It has been widely reported [143,144] that zinc-substitution into the CuO$_2$ planes causes an increase of $1/^{63}T_1T$ down to T_C, at the reverse of the pure system for which $1/^{63}T_1T$ start to be reduced far above T_C (see Figure 26). A recent accurate investigation in YBa$_2$ (Cu$_{2.99}$Zn$_{0.01}$)$_3$O$_{6.7}$ has even shown a non-monotonous temperature behavior in decreasing temperature characterized by a maximum of $1/^{63}T_1T$ at a temperature $T^{max} \approx 120$ K followed by a second enhancement at lower temperature [144]. This behavior is fully consistent with the neutron results, and both techniques clearly sign that zinc restores the low energy AF spin fluctuations at low temperature.

4.2.2. *Optimally doped regime (x ≈1)*

The behavior of the spin dynamics in zinc-substituted YBCO near optimal doping is quite similar to that of the underdoped regime [138,146], except that now T_C is much larger. Low-energy AF excitations are now observed in both the normal and the superconducting states, whereas they were absent from the latter in the zinc free samples, as a result of the spin-gap opening. This is illustrated in Figure 27 for an energy $\hbar\omega = 10$ meV: a Q-scan performed at a temperature much below T_C across the magnetic rod displays an intensity peaked at (π,π) in the zinc-substituted sample, while the same scan in a zinc-free sample show no sizable magnetic signal. At contrary, a quite similar response is observed in the normal state for both samples. As in the underdoped case discussed above, all these results strongly suggest that the additional low energy magnetic response is associated with spin excitations created by zinc. Quite remarkably, the new contribution is again peaked at (π,π), showing its antiferromagnetic character

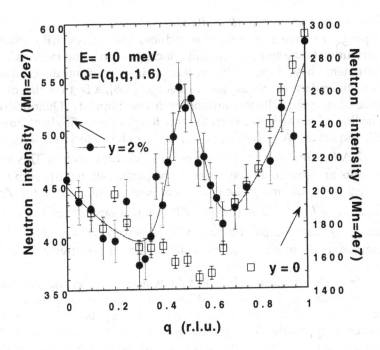

Figure 27. Q-scan across the magnetic rod at $Q_l=1.6$ and at an energy of 10 meV, in the superconducting state (T~5-10 K) for two different samples with the same oxygen content x=0.97: the zinc-free sample (open symbols, right scale) and the 2% zinc substituted sample (closed symbols, left scale), (from [66,146]).

At this stage, it is worth to notice that Ishida et al [128] have reported on the existence of two different relaxation times of the ^{63}Cu NMR lines in zinc-doped YBCO$_7$, which they have attributed to the existence of two different types of copper sites, according to the copper-zinc distance. This NMR result means that the spin susceptibility is really the superposition of two different contributions. Within this picture, the low-temperature low-energy excitations naturally arise from copper spins or charge carriers located in the near vicinity of zinc atoms. The second contribution displays characteristics which are very reminiscent of those observed in the zinc-free compositions. Thus, the temperature dependence of χ'' at $\hbar\omega \approx 10$ meV in 2%-Zn YBCO$_{6.97}$ exhibits a small enhancement above T_C [146], which reminds the maximum previously observed in underdoped YBCO (see Figure 17). By analogy with pure YBCO, the spin fluctuations present in the normal state would correspond to magnetic excitations associated with copper atoms located far from the zinc impurities and thus not disturbed by their presence. This overall temperature dependence at low energy is strongly consistent with the coexistence of two different contributions to the magnetic response, the first one seen at low temperature, almost not affected by the superconductivity and the second one occurring above T_C, displaying the same behavior as in the zinc-free system (see sections 2 and 3).

4.2.3. Zinc-induced low energy AF excitations

At any doping, we have seen that zinc induces low-energy AF excitations whose contribution to the spin susceptibility increases more and more as $T \to 0$. NIS measurements have further shown that their momentum dependence is practically independent of the oxygen content x [138], a fact consistent with the existence of magnetic clusters around each zinc impurity. Quantitatively, the intrinsic line width deduced from the low temperature NIS data shown in Figure 27, $\Delta q \approx 0.29$ Å^{-1} (FWHM), gives a characteristic length of $\xi = 2 / \Delta q \approx 7$ Å (assuming a lorentzian response in wave vector). This length scale represents an estimate of the range of the perturbation induced by zinc. At small Zn-doping, the magnetic clusters appear isolated, since the Zn-Zn average distance $a / \sqrt{y} \approx 22$ Å (for y=0.02) is much larger than the average cluster size. ξ is quite comparable to the 2D resistivity cross section of about 4-8 Å obtained from measurements of the residual resistivity in YBCO$_7$ [122,123].

However, from the comparison between the temperature dependencies of the spin-lattice relaxation times in Zn-substituted and Zn-free samples, Ishida et al [128] have attributed the "short" component of $^{63}T_1$ above T_C (associated with a smaller susceptibility) as originating from the Cu-sites near zinc. They then concluded that each zinc impurity causes, in the normal state, a collapse of the AF correlations in its neighboring. In fact, this interpretation disagrees with the available neutron data which unambiguously demonstrate that, at contrary, zinc *create* additional low-energy AF excitations. One possibility to reconcile these contradictory conclusions is to underline that the low-energy AF fluctuations observed in the neutron-scattering experiments occur mostly below ~100 K [147]. This fact is well demonstrated in the underdoped regime (see Figures 24 and 25), and is most likely also true at higher hole doping. Consequently, the effect of zinc on the spin dynamics could be simply rather weak around and above ~150 K.

Another issue, which has led to contradictory conclusions among the NMR community, concerns the effect of zinc substitution on the spin pseudo-gap. On the one hand, Knight-shift measurements, which are probing the spin susceptibility at $Q=0$, have given some evidences that the spin pseudo-gap was only slightly affected by the zinc substitution [127,128,143]. On the other hand, the thermal evolution of $1/^{63}T_1T$ has led to the conclusion that zinc removes the spin pseudo gap at $Q=(\pi,\pi)$ [143]). The neutron-scattering results show that the energy dependence of the magnetic spin susceptibility is not affected by zinc down to about 100 K (see Figure 24), and allow to define accurately T^{max} (Figure 25). Consequently, the spin pseudo-gap survives, zinc inducing rather *in-gap states*. This statement agrees with the NMR results and is also quite consistent with the picture of isolated magnetic clusters located around each zinc. Certainly, the new AF fluctuations are related to the existence of local moments. Furthermore, it is actually tempting to draw a parallel between the temperature dependence of the broadening of NMR lines [127,128] and the temperature dependence of the spin fluctuations induced by zinc at $Q=(\pi,\pi)$ seen by NIS (see again Figures 25 and 26). Both strongly increase at low temperature and remind the behavior of the macroscopic spin susceptibility [123,145]. Quite remarkably, the spectral weight associated with the low-energy excitations at low temperature is reduced by about a factor 2 at $\hbar\omega =10$ meV when the oxygen content x is increased from 0.7 to 0.97 [138], as does the effective local moment [134]. The 2D residual resistivity is also markedly larger in the underdoped samples [139]. Therefore, the NIS results appear fully consistent with the picture of CuO_2 planes in which zinc impurities create a *local phase separation*: in a range of about 7 Å, zinc atoms induce a perturbation responsible for both the residual resistivity and the additional AF fluctuations.

4.3. SUPERCONDUCTING STATE

We now discuss the modification induced by zinc on the spin dynamics in the superconducting state. The energy dependence of the spin-susceptibility at the AF wave vector is reported in Figure 28 for a zinc-free sample [66] and a 2 % zinc-substituted sample [146] at the same oxygen content x=0.97. The comparison reveals some interesting features of zinc substitution.

4.3.1. *Disappearance of the Resonance peak*
Figure 28 shows that the intensity at $\hbar\omega \approx 40$ meV (associated with the resonance peak in the zinc-free sample) is strongly reduced in the zinc-substituted sample. In the latter, the maximum amplitude of the magnetic response is now shifted to $\hbar\omega \approx 35$ meV. This might suggest a renormalization of the resonance energy by zinc. However, the temperature dependence of the neutron intensity at this energy reveals no strong change at the superconducting temperature (see Figure 29-b) meaning that this hypothesis is unlikely [148]. Most probably, the resonance peak is strongly reduced in amplitude but should remain located at the same energy. This is in agreement with the observation that the Q-width measured at $\hbar\omega \approx 38$ meV is significantly smaller than that at $\hbar\omega \approx 35$ meV [138], as also found at the resonance energy in the zinc-free system (see section 3 and [66]). The temperature dependence at $\hbar\omega \approx 39$ meV given in Figure 29-a) might support this possibility, although the change at T_C appears rather weak, in any case much weaker than in the equivalent zinc-free compound. It is worth to notice

that this scenario is compatible with the zinc-dependence of the 340 cm^{-1} B$_{1g}$ phonon mode studied by Raman scattering [149]. In the pure system, this phonon exhibits an anomaly at T$_C$ because its frequency is located close to 2Δ_{max} [150]. This effect rapidly disappears on decreasing doping [151]. Instead, upon zinc substitution, the anomaly still occurs but is reduced in amplitude on increasing the zinc content [149], as does the resonance peak.

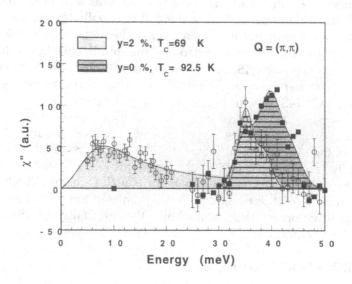

Figure 28. The complete imaginary part of the spin susceptibility at T=10.5 K in a zinc- substituted YBCO$_{6.97}$ sample (open circles). The results obtained at 4.5 K in a zinc-free sample with the same oxygen content and scaled through a longitudinal phonon measurement are reported for comparison (full squares) [66]. The lines are guides to the eye (from [146]).

Interestingly, the nearly disappearance of the resonance-peak intensity occurs also in the underdoped regime. For example, no strong feature is observed in the magnetic response of a Zn substituted YBCO$_{6.7}$ sample [138]. However, another interpretation consistent with the NIS measurements could be that zinc broadens so much the resonance peak that it becomes hardly visible. Therefore, it is difficult at present to settle definitively that question by lack of statistics of the neutron data. Investigations with smaller Zn-content are needed to clarify this point.

As emphasized in section 3, most of models consider that the magnetic resonance energy E$_r$ is essentially proportional to 2Δ_{max} [152,78,112-114,120]. Therefore, as zinc depresses dramatically T$_C$, one should expect a renormalization of E$_r$ which is far from having been seen experimentally. However, we have here implicitly assumed that the SC gap was renormalized by zinc as in conventional BCS theory. That might not be the case. In the alternative model involving the π-particles [117], the resonance energy does not depend on the superconducting gap energy, but instead is linear in doping level. As Zn-substitution leaves the carrier concentration almost unaffected, the latter approach could qualitatively accounts for the

NIS data, although the strong reduction of the resonance-peak intensity in increasing Zn-content seems unlikely in this purely magnetic approach.

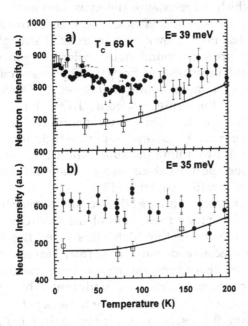

Figure 29.   Temperature dependence of the neutron intensity at a) energy 39 meV and b) energy 35 meV in $YBa_2(Cu_{0.98}Zn_{0.02})_2O_{6.97}$ [148]. Closed circles correspond to measurements at $Q=(0.5,0.5,5.2)$. Open squares represent the background measured by performing q-scans across the magnetic rod. These data have been measured on two different spectrometers.

Having no relationship with the resonance peak, one could then ask one self what is the origin of the 35 meV peak in $Zn-YBCO_{6.97}$ samples. Obviously, this peak reminds the shoulder detected in the pure system for the same oxygen content (see section 3 and [66,146]). Its observation in zinc-substituted samples seems to corroborate the existence of a *non-resonant* part to the spin susceptibility in the pure system (see the discussion in section 3.2.2.).

4.3.2.   *"Gapless" superconductivity*

Apart from the reduction of the spin-susceptibility amplitude around 40 meV, the other remarkable effect of zinc concerns the occurrence of low-energy magnetic excitations in the superconducting state. Figure 28 shows that 2% of zinc induces a broad energy band around $\hbar\omega \approx 9$ meV (extending up to 30 meV) which was totally absent in the zinc-free system. This again indicates that these additional spin excitations are created by zinc.

The magnetic response of the Zn-substituted sample exhibits an other feature already observed in the Zn-free sample, namely the existence of a "threshold" energy. In $YBCO_{6.97}$ [66], it is related to the opening of an energy gap in the spin susceptibility when entering the superconducting state (see Figure 21). In $2\%Zn-YBCO_{6.97}$, the neutron data (see Figure 28) give some evidence for a minimum of χ'' in the range 30-32 meV, which is very reminiscent of the spin-gap phenomenon [146]. Assuming a direct relationship between E_g and Δ_{max}, one is led to conclude that Δ_{max} would be weakly affected by the zinc substitution. This is consistent with the previous observation following which the resonance energy is also weakly affected by zinc. The additional low energy excitations should then be considered as in-gap states. Recently, it has been shown that an isolated strongly-scattering impurity can effectively produce bound states inside the gap in a d-wave superconductor [153]. The NIS results then bring direct pieces of information on the energy range at which these bound states occur. Obviously, these excitations are related to the finite electronic DOS at the Fermi level previously mentioned. Its observation in many experiments [128-131] has led to conclude to the existence of a "gapless" superconductivity. Indeed, Zinc reduces the coherence effect by filling the superconducting gap. This likely explains why the resonance peak weakens on increasing the zinc content, as also does, for instance, the specific heat jump [130].

The NIS results, in agreement with the NMR results, demonstrate that zinc substitution induces a local phase separation of the electronic structure of CuO_2 planes. In the superconducting state, this gives rise to a picture following which an appreciable density of normal carriers coexist with superconducting regions. In other words, zinc remove some charge carriers from the superconducting condensate, as also underlined by recent μ-SR measurements [154]. However, to fully account for the effect of Zn-substitution on the pairing mechanism, it is essential to further analyze the incidence of other impurities in 123-compounds. For instance, the substitution of Ni should provide a very interesting counterpart as Ni also creates a local magnetic moment whereas the superconductivity is much less affected [145,155].

5. Discussion and conclusion

After a decade of investigation, neutron-inelastic-scattering experiments have unambiguously established the importance of magnetic correlations in the microscopic description of high-T_C superconductors. The complete energy, wave vector and temperature dependencies of the imaginary part of the spin susceptibility, $\chi''(Q,\omega)$, have been studied for several doping. A drastic evolution is found as a function of doping, yielding a rich variety of situation when going from the undoped AF-state $YBCO_6$ (well accounted by a standard spin-wave description for a 2D spin-1/2 Heisenberg model) to the fully oxidized metallic state $YBCO_7$ (characterized by a single inelastic peak in the superconducting-state response). First, this behavior differs in many aspects from that observed in conventional superconductors, in both the normal and superconducting states. The main difference is certainly the presence in the new high-T_C materials of strong q- and energy-dependent AF correlations, which contrast sharply with the featureless and very weak magnetic response expected in standard metals.

The magnetic fluctuations in YBCO (and more generally in cuprates) are systematically found maximum around the AF wave vector (π,π). This is basically caused by a q-dependent enhancement of the spin susceptibility, for instance, related to J(q) in RPA treatments or due to band structure singularities. Of course, small shifts from (π,π) of the spin susceptibility are observed in both LSCO [101,102] and YBCO [61,97]. However, we here stress that these discommensurations have a completely different temperature and energy dependencies in both systems, a fact which strongly suggests a different origin. In particular, the incommensurability in YBCO predominantly (may be exclusively) occurs in the superconducting state.

As NMR measurements of the copper spin-lattice relaxation rate are also probing the spin susceptibility around (π,π) but in the low frequency limit, it is essential to compare the spin dynamics extracted from NMR and NIS measurements. Qualitatively, they both agree to prove the existence of strong AF fluctuations in cuprates, although a discrepancy still subsists when reaching the overdoped regime: the amplitude of the neutron spin susceptibility markedly decreases in the normal state from $YBCO_{6.92}$ to $YBCO_7$, whereas the NMR copper $1/^{63}T_1T$ remains roughly constant. Quantitatively, problems related to the NMR analysis have led to controversial debates: the standard analysis of the NMR measurements in terms of the nearly AF Fermi liquid [71] implies a clear temperature dependence of the AF correlation length whereas the corresponding NIS q-width of the magnetic response does not appreciably change. Obviously, the complete energy shape of the spin susceptibility (see Eq. (6)) should be considered to proper reproduce the NMR data. In any case, the recent attempt to deduce the spin susceptibility in absolute units in NIS experiments should help to solve this problem. In contrast to usual superconductors, the spin dynamics in YBCO is characterized by a sharp resonance peak below T_C, whose energy and relative intensity are related to the amplitude of T_C. As a matter of fact, one can generalize this observation made only in YBCO to all cuprates. Indeed, having noticed that the resonance enhancement represents only about 15 % of the magnetic intensity in $YBCO_{6.5}$ ($T_C \approx 50$ K), the absence of any report of a well marked resonance peak in LSCO ($T_C \approx 35$ K) might be simply due to the smaller value of T_C in this system. Most likely, the resonance peak is due to an electron-hole pair production across the superconducting energy gap. Interestingly, this model requires $d_{x^2-y^2}$-wave symmetry of the gap function. Actually, the resonance peak might be considered as a counterpart of the NMR Hebel-Slichter peak observed in conventional superconductors for an isotropic s-wave gap [156].

This issue of momentum dependence of the superconducting gap in high-T_C cuprates has been extensively emphasized by angle-resolved photo emission experiments (ARPES) [115,7] and electronic Raman scattering [116]. Both techniques have positively concluded to a $d_{x^2-y^2}$-wave symmetry gap with a ratio $2\Delta_{max}/kT_C \approx 7\text{-}9$, significantly larger than the ratio $E_r/kT_C \approx 5\text{-}6$ deduced from the resonance energy. This actually indicates that the resonance energy is not simply twice Δ_{max} but that it exists a proportionality factor depending on the details of the electronic band structure. In principle, such a factor could be doping dependent. Interestingly, the ratio E_r/kT_C slightly evolves with doping as shown in Figure 19. Further, the single particle spectral function, probed by ARPES in Bi-based cuprates, exhibits near the $(\pi,0)$ point of the Brillouin zone a strong sharpening only below T_C meaning than coherent quasi-particles do exist at low temperature. It has been then speculated that this peak might be related to the resonance peak through the electron-electron contribution to the self energy [157].

In the normal state, the ARPES single particle excitation spectrum exhibits also a pseudo-gap below T_{pg} (i.e. clearly above T_C) in the underdoped regime (see Figure 1). Similar observations have been made in optical conductivity [158], Raman scattering [159] or tunneling experiments [160]. It has been then widely proposed that these gap-like features correspond to a precursor of the superconducting gap, suggesting the existence of *preformed* pairs in the normal state. Similarly, one could consider that the normal-state peak observed in the NIS experiments is a precursor of the resonance peak. However, the maximum energy of the normal-state susceptibility displays a quite different doping dependence than the resonance peak. Especially, both spin susceptibilities in underdoped $YBCO_{6.83}$ and in optimally doped $YBCO_{6.92}$ exhibit a maximum at about 30 meV for $T>T_C$, whereas they are characterized by two distinct resonance energies (35 and 40 meV, respectively). Therefore, this picture of precursor effect might be too simple. For sure, the observation of unusual spin dynamics above T_C should be linked to the anomalous macroscopic properties encountered in these materials [5]. As a matter of fact, the spectral line shape of the quasi-particles, responsible for the conductivity in these materials, indicates strong many-body effects which may be accounted for by a coupling with collective excitations centered at the AF momentum [161]. Obviously, the complete scenario which would link in a self-consistent way all these experimental evidences is still missing. Finally, the existence of strong doping-dependent antiferromagnetic correlations as well as their close link with the anomalous physical properties of high-T_C superconductors naturally suggest an *electron-electron origin* to the pairing mechanism, as it has been abundantly invoked [162].

Acknowledgments

We would like to thanks our coworkers T. Fong, J.Y. Henry, B. Hennion, B. Keimer and C. Vettier. We are grateful to G. Collin and Y. Sidis for a critical reading of the manuscript as well as for very stimulating discussions. The help of E. Ressouche was greatly appreciated.

References

1. J.G. Bednorz and K.A. Müller, Z. *Phys.B Condensed Matter* **64** (1986) 189.
2. A.P. Kampf, *Physics report* **249** (1994) 219.
3. C. Berthier, M.H. Julien, M. Horvatic, and Y. Berthier, *Journal de Physique I France* **6** (1997) 2205.
4. J. Rossat-Mignod, L.P. Regnault, P. Bourges, P. Burlet, C. Vettier, and J.Y. Henry, (1993) Neutron scattering study of the high-T_C superconducting system $YBa_2Cu_3O_{6+x}$, in L.C Gupta and M.S. Multani (eds.), *Selected Topics in Superconductivity*, Frontiers in Solid State Sciences vol.1, World Sientific, Singapore, pp.265-347.

5. B. Batlogg, H.Y. Hwang, H. Takagi, R.J. Cava, H.L. Kao, and J. Kuo, *Physica C* **235-24** (1994) 130.
6. J.W. Loram, K.A. Mirza, J.M. Wade, J.R. Cooper, and W.Y. Liang, *Physica C* **235-240** (1994) 134.
7. H. Ding, T. Yokoya, J.C. Campuzano, T. Takahashi, M. Randeria, M.R. Norman, T. Mochiku, K. Kadowaki, and J. Giapintzakis, *Nature* **382** (1996) 51.
8. J. Mesot and A. Fürrer, *J. Supercon.* **10** (1997) 623.
9. M.K. Wu, J.R. Ashburn, C.J. Torng, P.H. Hor, R.L. Meng, L. Gao, Z.J. Huang, Y.Q. Wang, and C.W. Chu, *Phys. Rev. Lett.* **58** (1987) 908.
10. J.J. Capponi, C. Chaillout, A.W. Hewat, P. Lejay, M. Marezio, N. Nguyen, B. Raveau, J.L. Soubeyroux, J.L. Tholence, and R. Tournier, *Europhys. Lett.* **3** (1987) 1301.
11. J.D. Jorgensen, D.G. Hinks, P.G. Radaelli, S. Pei, P. Lightfoot, B. Dabrowski, C.U. Segre, and B.A. Hunter, *Physica C* **185-189** (1991) 184.
12. R.J. Cava, B. Battlog, M. Rabe, E.A. Rietman, P.K. Gallagher, and L.W. Rupp, *Physica C* **156** (1988) 523.
13. H. Tolentino, A. Fontaine, F. Baudelet, T. Gourieux, G. Krill, J.Y. Henry, and J. Rossat-Mignod, *Physica C* **192** (1992) 115.
14. G. Uimin and J. Rossat-Mignod, *Physica C* **199** (1992) 251.
15. G. Uimin, *Phys. Rev. B* **50** (1994) 9531.
16. P. Burlet, V.P. Plakthy, C. Marin, and J.Y. Henry, *Phys. Lett. A* **167** (1992) 401.
17. J. Fink, N. Nücker, M. Alexander, H. Rhomberg, M. Knupfer, M. Merkel, P. Aldelman, R. Claessen, G. Mante, T. Buslaps, S. Harm, R. Manzke, and M. Skibowski, *Physica C* **185-189** (1991) 45.
18. S.D. Obertelli, J.R. Cooper, and J.L. Tallon, *Phys. Rev. B* **46** (1992) 14928.
19. R. Presland, J.L. Tallon, R.G. Buckley, R.S. Liu, and N.E. Flower, *Physica C* **176** (1991) 95.
20. V.J. Emery, S.A. Kivelson, and H-Q. Lin, *Phys. Rev. Lett.* **64** (1990) 475.
21. S.A. Kivelson, V.J. Emery, and H-Q. Lin, *Phys. Rev. B* **42** (1990) 6523.
22. J.M. Tranquada, D.E.Cox, W. Kannmenn, A.H. Moudden, G. Shirane, M. Suenaga, P. Zolliker, D. Vaknin, and S.K. Shina, *Phys. Rev. Lett.* **60** (1988) 156.
23. J. Rossat-Mignod, P. Burlet, M.J. Jurgens, J.Y. Henry, and C. Vettier, *Physica C* **152** (1988) 19.
24. P. Burlet, C. Vettier, M.J. Jurgens, J.Y. Henry, J. Rossat-Mignod, H. Noel, M. Potel, P. Gougeon, and J.C. Levet, *Physica C* **153-155** (1988) 1115.
25. D. Petitgrand, and G. Collin, *Physica C* **153-155** (1988) 192.
26. P. Burlet et al, preprint (1997).
27. J. Rossat-Mignod, P. Burlet, M.J. Jurgens, C. Vettier, L.P. Regnault, J.Y. Henry, C. Ayache, L. Forro, H. Noel, M. Potel, P. Gougeon, and J.C. Levet *J. Physique* **49** (1988) C8-2119.
28. S. Shamoto, M. Sato, J.M. Tranquada, B. Sternlieb, and G. Shirane, *Phys. Rev. B* **48** (1993) 13817.
29. M.J. Jurgens, P. Burlet, C. Vettier, L.P. Regnault, J.Y. Henry, J. Rossat-Mignod, H. Noel, M. Potel, P. Gougeon, and J.C. Levet, *Physica B* **156-157** (1989) 846.
30. J. Rossat-Mignod, L.P. Regnault, M.J. Jurgens, C. Vettier, P. Burlet, J.Y. Henry, and G. Lapertot, *Physica C* **162-164** (1989) 1269.
31. P. Gawiec and D.R. Grempel, *Phys. Rev. B* **44** (1991) 2613.
32. P. Gawiec, *thesis*, Grenoble University, Grenoble(1992).
33. J.M. Tranquada, G. Shirane, B. Keimer, S. Shamoto, and M. Sato, *Phys. Rev. B* **40** (1989) 4503.
34. R.R.R.P. Singh, *Phys. Rev. B* **39** (1989) 9760.
35. E. Manousakis, *Rev. Mod. Phys.* **63** (1991) 1.
36. J. Igarashi, *Phys. Rev.* **46** (1992) 10763.
37. J. Rossat-Mignod, L.P. Regnault, M.J. Jurgens, C. Vettier, P. Burlet, J.Y. Henry, and G. Lapertot, *Physica B* **163** (1990) 4.
38. D. Reznick, P. Bourges, H.F. Fong, L.P. Regnault, J. Bossy, C. Vettier, D.L. Milius, I.A. Aksay, and B. Keimer, *Phys. Rev. B* **53** (1996) R14741.
39. S.M. Hayden, G. Aeppli, T.G. Perring, H.A. Mook, and F. Dogan, *Phys. Rev. B* **54** (1996) R6905.
40. M. Sato, S. Shamoto, T. Kiyokura, K. Kakurai, G. Shirane, B.J. Sternlieb, and J.M. Tranquada, *J. Phys. Soc. Jpn* **62** (1993) 263.
41. B.I. Shraiman and E.D. Siggia, *Phys. Rev. Lett.* **61** (1988) 467.
42. D.R. Grempel, *Phys. Rev. Lett.* **61** (1988) 1041.
43. P. Bourges, Y. Sidis, B. Hennion, R. Villeneuve, G. Collin, and J.F. Marucco, *Physica B* **213-214** (1995) 54.
44. F.P. Onufrieva, V. Kushnir, and B. Toperverg, *Physica C* **218** (1993) 463; *Phys . Rev. B* **50** (1994) 12935.
45. J.A. Wilson and A. Zahir, *Rep. Prog. Phys.* **60** (1997) 941.
46. P.W. Anderson, *Science* **235** (1987) 1196.

47. F.C. Zhang and T.M. Rice, *Phys. Rev. B* **37** (1988) 3759.
48. W. Stephen and P. Horsch, *Phys. Rev. Lett.* **66** (1991) 2258.
49. G. Dopf, A. Muramatsu, and W. Hanke, *Phys. Rev. Lett.* **68** (1992) 353.
50. B. Gillon, *Physica B* **174** (1991) 340.
51. J.X. Boucherle, J.Y. Henry, R.J. Papoular, J. Rossat-Mignod, J. Schweitzer, F. Tasset, and G. Uimin, *Physica B* **192** (1993) 25.
52. J.Y. Henry, R.J. Papoular, J. Schweitzer, F. Tasset, G. Uimin, and I. Zobkalo, *Physica C* **235-240** (1994) 1659.
53. S.W. Lovesey, (1984),*Theory of Neutron Scattering from Condensed Matter*, vol 1&2, (Clarendon, Oxford).
54. R. Papoular and B. Gillon, *Europhys. Lett.* **13** (1990) 429.
55. S.E. Barett, D.J. Durand, C.H. Pennington, C.P. Schlichter, T.A. Friedmann, J.P. Rice, and D.M. Grinsberg, *Phys. Rev. B* **41** (1990) 6283.
56. M. Takigawa, A.P. Reyes, P.C. Hammel, J.D. Thompson, R.H. Heffner, Z. Fisk, and K.C. Ott, *Phys. Rev. B* **43** (1991) 247.
57. W.W. Warren, R.E. Walstedt, G.F. Brennert, R.J. Cava, R. Tycko, R.F. Bell, and G. Dabbagh, *Phys. Rev. Lett.* **62** (19989)1193.
58. M. Horvatic, P. Segransan, C. Berthier, Y. Berthier, P. Butaud, J.Y. Henry, M. Couach, and J.P. Chaminade, *Phys.Rev. B* **39** (19989) 7332.
59. R.E. Walstedt and W.W. Waren, *Science* **248** (1990) 1082.
60. M. Horvatic, P. Butaud, P. Segransan, Y. Berthier, C. Berthier, J.Y. Henry, and M. Couach, *Physica C* **166** (1991) 151.
61. J.M. Tranquada, P.M. Ghering, G. Shirane, S. Shamoto, and M. Sato, *Phys. Rev. B* **46** (1992) 5561.
62. L.P. Regnault, P. Bourges, P. Burlet, J.Y. Henry, J. Rossat-Mignod, Y. Sidis, and C. Vettier, *Physica C* **235-240** (1994) 59.
63. L.P. Regnault, P. Bourges, P. Burlet, J.Y. Henry, J. Rossat-Mignod, Y. Sidis, and C. Vettier, *Physica B* **213&214**(1995) 48.
64. P. Bourges, L.P. Regnault, J.Y. Henry, C. Vettier, Y. Sidis, and P. Burlet, *Physica B* **215** (1995) 30.
65. H.F. Fong, B. Keimer, D. Reznick, D.L. Milius, and I.A. Aksay, *Phys. Rev. B* **54** (1996) 6708.
66. P. Bourges, L.P. Regnault, Y. Sidis, and C. Vettier, *Phys. Rev. B* **53** (1996) 876.
67. H.A. Mook, M. Yethiraj, G. Aeppli, T.E. Mason, and T. Armstrong, *Phys. Rev. Lett.* **70** (1993) 3490.
68. P. Dai, M. Yethiraj, H.A. Mook, T.B. Lindemer, and F. Dogan, *Phys. Rev. lett.* **77** (1996) 5425.
69. J. Rossat-Mignod, L.P. Regnault, C. Vettier, P. Bourges, P. Burlet, J. Bossy, J.Y. Henry, and G. Lapertot, *Physica C* **185-189** (1991) 86.
70. B.J. Sternlieb, J.M. Tranquada, G. Shirane, M. Sato, and S. Shamoto, *Phys. Rev. B* **50** (1994) 12915.
71. A. Millis, H. Monien, and D. Pines, *Phys. Rev. B* **42** (1990) 167.
72. H. Monien, D. Pines, and M. Takigawa, *Phys. Rev. B* **43** (1991) 258.
73. J.A. Gillet, T. Auler, M. Horvatic, C. Berthier, Y. Berthier, P. Segransan, and J.Y. Henry, *Physica C* **235-240** (1994) 1667.
74. T. Auler, C. Berthier, Y. Berthier, P. Carretta, J.A. Gillet, M. Horvatic, P. Segransan, and J.Y. Henry, *Physica C* **235-240** (1994) 1677.
75. J. Bobroff, H. Alloul, Y. Yoshinari, A. Keren, P. Mendel, N. Blanchard, G. Collin, and J.-F. Maruco, *Phys. Rev. Lett.* **79** (1997) 2117.
76. Q. Si, Y. Zha, K. Levin, and J.P. Lu, *Phys. Rev. B* **47** (1993) 9055.
77. F. Onufrieva and J. Rossat-Mignod, *Phys. Rev. B* **52** (1995) 7572.
78. F. Onufrieva, *Physica C* **251** (1995) 348.
79. A. Millis and H. Monien, *Phys. Rev. Lett.* **70** (1993) 2810.
80. J. Bobroff, H. Alloul, P. Mendels, V. Vialet, J.F. Maruco, and D. Colson, *Phys. Rev. Lett.* **78** (1997) 3757.
81. P. Bourges , (1998) From magnons to the resonance peak: spin dynamics in high-T_c superconducting cuprates by inelastic neutron scattering, in J. Bok, G. Deutscher and D. Pavuna (eds.), *The gap symmetry and fluctuation in high temperature superconductors*, Plenum Press, New York, under press.
82. M. Horvatic, T. Auler, C. Berthier, Y. Berthier, P. Butaud, W.G. Clark, J.A. Gillet, P. Segransan, and J.Y. Henry, *Phys. Rev. B* **47** (1993) 3461.
83. C.M. Varma, *Phys. Rev. B* **55** (1997) 1455.
84. S. Sachdev and J. Ye, *Phys. Rev. Lett.* **69** (1992) 2411.
85. P. Bourges, P.M. Gehring, B. Hennion, A.H. Moudden, J.M. Tranquada, G. Shirane,

S. Shamoto, and M. Sato, *Phys. Rev. B* **43** (1991) 8690.

86. R.J. Birgeneau, R.W. Erwin, P.M. Gehring, M.A. Kastner, B. Keimer, M. Sato, S. Shamoto, G. Shirane, and J. Tranquada, *Z. Phys. B* **87** (1992) 15.

87. P.M. Gehring, J.M. Tranquada, G. Shirane, J.R.D. Copley, R.W. Erwin, M. Sato, and S. Shamoto, *Phys. Rev. B* **44** (1991) 2811.

88. B.J. Sternlieb, J.M. Tranquada, G. Shirane, M. Sato, and S. Shamoto, *Phys. Rev. B* **47** (1993) 5320.

89. A. Virosztek and J. Ruvalds, *Phys. Rev. B* **42** (1990) 4064.

90. P.B. Littlewood and C. Varma, *J. Appl. Phys.* **69** (1991

91. V. J. Emery and S.A. Kivelson, *PhysicaC* **209** (1993) 597.

92. O. Marikyo and K. Miyake, *J. Phys. Soc. Jpn.* **63** (1993) 4169.

93. L.P. Regnault, I. Zaliznyak, J.P. Renard, and C. Vettier, *Phys. Rev. B* **53** (1994) 5579.

94. E. Dagotto and M. Rice, *Science* **271** (1996) 618.

95. P. Bourges, L.P. Regnault, Y. Sidis, J. Bossy, P. Burlet, C. Vettier, J.Y. Henry, and M. Couach, *Europhys. Lett.* **38** (1997) 313.

96. H.F. Fong, B. Keimer, D.L. Milius, and I.A. Aksay, *Phys. Rev. Lett.* **78** (1997) 713.

97 P. Dai, H.A. Mook, and F. Dogan, to appear in *Phys. Rev. Lett.* (1998).

98. P. Bourges, H.F. Fong, L.P. Regnault, J. Bossy, C. Vettier, D.L. Milius, I.A. Aksay, and B. Keimer, *Phys. Rev. B* **56** (1997) R11439.

99. T. Fong et al, to be published (1998).

100. S. Hayden et al, preprint cond-mat/9710181 (1997).

101. T.R. Thurston, R.J. Birgeneau, M.A. Kastner, N.W. Preyer, G. Shirane, Y. Fujii, K. Yamada, Y. Endoh, K. Kakurai, M. Matsuda, Y. Hidaka, and T. Murakami, *Phys. Rev. B* **40**, (1989) 4585.

102. S.W. Cheong, G. Aeppli, T.E. Mason, H.A. Mook, S.M. Hayden, P.C. Canfield, Z. Fisk, K.N. Clausen, and J.L. Martinez, *Phys. Rev. Lett.* **67** (1991) 1791.

103. V.J. Emery and S.A. Kivelson, *Physica C* **209** (1993) 597.

104. P. Bourges and L.P. Regnault, *Phys. Rev. Lett.* **80** (1998) 1793.

105. H.A. Mook, P. Dai, G. Aeppli, T.E. Mason, N.E. Hecker, J.A. Harvey, T. Armstrong, K. Salama,and D. Lee, *Physica B* **213&214** (1995) 43.

106. H.F. Fong, B. Keimer, P.W. Anderson, D. Reznick, F. Dogan, and I.A. Aksay, *Phys. Rev. Lett.* **75** (1995) 316.

107. J. Rossat-Mignod, L.P. Regnault, P. Bourges, P. Burlet, C. Vettier, and J.Y. Henry, *Physica B* **194-196** (1994) 2131.

108. B. Keimer, I.A. Aksay, J. Bossy, P. Bourges, H.F. Fong, D.L. Milius, L.P. Regnault, D. Reznick, and C. Vettier, *Physica B* **234-236** (1997) 821.

109. T.E. Mason, G. Aeppli, S.M. Hayden, and H.A. Mook, *Phys. Rev. Lett.* **77** (1996) 1604.

110. T. Tanamoto, H. Kohno, and H. Fukuyama, *J. Phys. Soc. Jpn* **61** (1992) 1886.

111. G. Stenmann, C. Pepin, and M. Lavagna, *Phys. Rev. B* **50** (1994) 4075.

112. D.Z. Liu, Y. Zha, and K. Levin, *Phys. Rev. Lett.* **75** (1995) 4130.

113. I.I. Mazin, and V.M. Yakovenko, *Phys. Rev. Lett.* **75** (1995) 4134.

114. N. Bulut and D.J. Scalapino, *Phys. Rev. B* **53** (1996) 5149.

115. Z.X. Shen, W.E. Spicer, D.M. King, D.S. Dessau, and B.O. Wells, *Science* **267** (1995) 343.

116. T.P. Devereaux, D. Einzel, B. Stadlober, R. Hackl, D.H. Leach, and J.J. Neumeier, *Phys. Rev. Lett.* **72** (1994) 396.

117. E. Demler and S.C. Zhang, *Phys. Rev. Lett.* **75** (1995) 4126.

118. S.C. Zhang, *Science* **275** (1997) 1089.

119. M. Greiter, preprint cond-mat/9705049 (1997).

120. L. Yin, S. Chakravarty, and P.W. Anderson, *Phys. Rev. Lett.* **78** (1997) 3559.

121. R. Liang, T. Nakamura, H. Kawaji, M. Itoh, and T. Nakamura, *Physica C*, **170** (1990) 307.

122. T. R. Chien, Z. Z. Wang, and N. P. Ong, *Phys. Rev. Lett.* **67** (1991) 2088.

123. S. Zagoulaev, P. Monod, and J. Jégoudez, *Phys. Rev. B* **52** (1995)10474 .

124. A. Abrikosov, (1989), *Fundamental of the Theory of Metals*, North-Holland, Amsterdam.

125. L. Raffo, F. Licci, and A. Migliori, *Phys. Rev. B* **48** (1993)1192.

126. R. Villeneuve, *thesis*, University Paris-XI, Orsay (1996).

127. H. Alloul, P. Mendels, H. Casalta, J. F. Marucco, and J. Arabski, *Phys. Rev. Lett.* **67** (1991) 3140.

128. K. Ishida, Y. Kitaoka, N. Ogata, T. Kamino, K. Asayama, J. R. Cooper and N. Athanassopoulou, *J. Phys. Soc. Jpn.* **62** (1993) 2803.

129. J. A. Hodges, P. Bonville, P. Imbert, A. Pinatel-Philippot, *Physica C* **246** (1995) 323.

130. J. W. Loram, K. A. Mirza and P. F. Freeman, *Physica C* **171** (1990) 243.

131. A. Janossy, J. R. Cooper, L. C. Brunel, and A. Carrington, *Phys. Rev. B* **50** (1994) 3445.

132. V. A. Mahajan, H. Alloul, G. Collin and J. F. Marucco, *Phys. Rev. Lett.* **72** (1994) 3100.

133. G. V. M. Williams, J. L. Tallon, and R. Meinhold, *Phys. Rev. B* **52** (1995) R7034.
134. P. Mendels, H. Alloul, J. H. Brewer, G. D. Morris, T. L. Duty, S. Johnston, E. J. Ansalto, G. Collin, J. F. Marucco, C. Niedermayer, D. R. Noakes and C. E. Stronach, *Phys. Rev. B* **49** (1994) 10035.
135. T. Hotta, *J. Phys. Soc. Jpn.* **62** (1993) 274 and references therein.
136. Y. Kitaoka, K. Ishida, and K. Asayama, *J. Phys. Soc. Jpn.* **63** (1994) 2052.
137. Y. Sidis, P. Bourges, B. Hennion, R. Villeneuve, G. Collin, and J. F. Marucco, *Physica C* **235-240** (1994) 1591.
138. Y. Sidis, *thesis*, University Paris-XI, Orsay,, (1995).
139. Y. Fukuzumi, K. Mizuhashi, K. Takenaka, and S. Uchida, *Phys. Rev. Lett.* **76** (1996) 684 .
140. K. Kakurai, S. Shamoto, T. Kiyokura, M. Sato, J. M. Tranquada, and G. Shirane, *Phys. Rev. B* **48** (1993) 3485.
141. H. Harashina, S. Shamoto, T. Kiyokura, M. Sato, K. Kakurai, and G. Shirane, *J. Phys. Soc. Jpn* **62** (1993) 4009.
142. P. Bourges, Y. Sidis, B. Hennion, R. Villeneuve, G. Collin and J. F. Marucco, *Czechoslovak Journal of Physics* **46** (1996) 1155.
143. G.-Q. Zheng, T. Odagushi, T. Mito, Y. Kitaoka, K Asayama, and Y. Kodama, *J. Phys. Soc. Jpn.* **62** (1993) 2591.
144. M. H. Julien, *thesis*, Grenoble University , Grenoble, (1997).
145. P. Mendels, J. Bobroff, G. Collin, H. Alloul, J. F. Marucco, N. Blanchard, and B. Grenier, unpublished results.
146. Y. Sidis, P. Bourges, B. Hennion, L. P. Regnault, R. Villeneuve, G. Collin, and J. F. Marucco, *Phys. Rev. B* **53** (1996) 6811.
147. H. Alloul, J. Bobroff, and P. Mendels, *Phys. Rev. Lett.* **78** (1997) 2494; K. Ishida, Y. Kitaoka, and K Asayama, *Phys. Rev. Lett.* **78** (1997) 2495.
148. P. Bourges, Y. Sidis, L. P. Regnault, B. Hennion, R. Villeneuve, G. Collin, C. Vettier, J. Y. Henry, and J. F. Marucco, *J. Phys. Chem. Solids* **56** (1995) 1937.
149. R. Gajic, S. D. Devic, M. J. Kontantibovic, and Z. W. Popovic, *Z. Phys. B* **94** (1994) 261 .
150. B. Friedl, C. Thomsen, and M. Cardona, *Phys. Rev. Lett.* **65** (1990) 915.
151. E. Altendorf, X. K. Chen, J. C. Irwin, R. Liang, and W. N. Hardy, *Phys. Rev. B* **47** (1993) 8140.
152. S. V. Maleyev, *J. Phys. I. France* **2** (1992) 181; S. Charfi-Kaddour et al, *J. Phys. I France* **2** (1992) 1853; J. P. Lu, *Phys. Rev. Lett.* **68** (1992) 125.
153. A. V. Balatsky, M. I. Salkola, and A. Rosengren, *Phys. Rev. B* **51** (1995).
154. B. Nachumi, A. Keren, K. Kojima, M. Larkin, G. M. Luke, J. Merrin, O. Tchernyshöv, Y. J. Uemura, N. Ichikawa, M. Goto, and S. Uchida, *Phys. Rev. Lett.* **77** (1996) 5421.
155. J. Bobroff, H. Alloul, Y. Yoshinari, A. Keren, P. Mendels, N. Blanchard, G. Collin, and J. F. Marucco, *Phys. Rev. Lett.* **79** (1997) 2117.
156. J.R. Schriffer, (1988), *Theory of superconductivity*, Frontiers in physics (20), Addison Wesley.
157. M. Norman, H. Ding, J.C. Campuzano, T. Takeushi, M. Randeria, T. Yokaya, T. Takahashi, T. Mochiku, and K. Kadowaki, *Phys. Rev. Lett.* **79** (1997) 3506.
158. T. Ito, K. Takenaka, and S. Uchida, *Phys. Rev. Lett.* **70** (1993) 3995.
159. R. Nemetscheck, M. Opel, C. Hoffmann, P.F. Müller, R. Hackel, H. Berger, L. Forro, A. Erb, and E. Walker, *Phys. Rev. Lett.* **78** (1997) 4837.
160. Ch. Renner, B. Revaz, J.Y. Genoud, K. Kadowaki, and O. Fisher, *Phys. Rev. Lett.* **80** (1998) 149.
161. Z.X. Shen and J.R. Schrieffer, *Phys. Rev. Lett.* **78** (1997) 1771.
162. D. Pines, *Z. Phys. B* **103** (1997) 129.

MAGNETIC EXCITATIONS IN $La_{2-x}(Ba,Sr)_x CuO_4$

S. M. HAYDEN
H.H. Wills Physics Laboratory,
University of Bristol,
Tyndall Avenue, Bristol, BS8 1TL, UK

1. Introduction

Experiments on high temperature superconductors in the decade following Bednorz and Muller's discovery have produced a tremendously rich variety of physical behaviour. As we shall see in this short review, the magnetic excitations are no exception in this respect. The characterisation of the these excitations is motivated by a number of considerations. Firstly, electrons carry spin as well as a charge, thus measurement of the motion of the spins provides an image of the electronic correlations in these complex materials. Secondly, the failure of phonon-based theory to explain the high temperature superconductor phenomenon has led to the widely-held view that magnetic excitations are involved in the pairing attraction.

In order to comprehend the excitations of a physical system, we need to know the nature of the ground state. We will therefore start by briefly reviewing what is known about the phase diagram of the $La_{2-x}(Ba,Sr)_x CuO_4$ system.

1.1. PHYSICAL PROPERTIES

1.1.1. *Structure*

The crystal structure of $La_2 CuO_4$ is shown in Fig. 1. At ambient temperature $La_2 CuO_4$ is orthorhombic, due to a rotation of the oxygen octahedra centered on the copper atoms. The electronic properties of the $La_{2-x}(Ba,Sr)_x CuO_4$ system are extremely anisotropic. Many properties are most easily understood by considering the structure as being made up of a series of square CuO_2 planes stacked along the **b** direction. There is weak overlap of the electronic orbitals in the **b** direction. The copper atoms are in the Cu^{2+} state with spin $S = \frac{1}{2}$. There is a strong exchange anisotropy, the

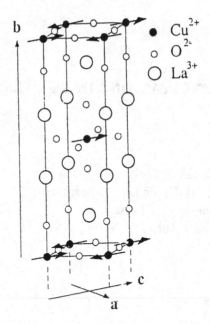

Figure 1. (a) The crystal structure of La$_2$CuO$_4$.

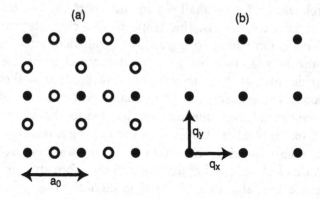

Figure 2. (a) A nearly square CuO$_2$ plane. (b) Corresponding reciprocal space lattice. The antiferromagnetic wavevector is at the centre of the squares.

coupling between copper spins in the same CuO$_2$ plane ($J_{\parallel}$) is several orders of magnitude stronger than the coupling between spins in neighbouring planes ($J_{\perp}$).

Soon after the discovery of high temperature superconductivity, neutron scattering experiments performed by Vaknin *et al.* [1] at Brookhaven National Laboratory showed that La$_2$CuO$_4$ is an antiferromagnet (see Fig. 1). There is still some controversy over the value of the ordered moment. A number of effects are known to change the value from the 1 μ_B corre-

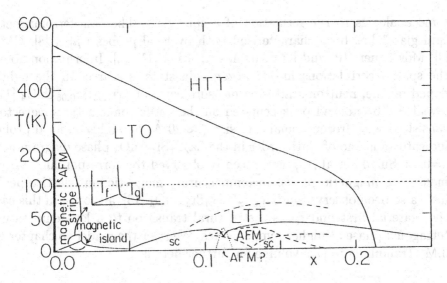

Figure 3.   The phase diagram for La$_{2-x}$Sr$_x$CuO$_4$ (solid and dotted lines) and La$_{2-x}$Ba$_x$CuO$_4$ (dashed line), after Markiewicz [6].

sponding to $S = \frac{1}{2}$: (i) the g factor of Cu^{2+} [1]; (ii) covalency effects due to hybridisation of the copper and oxygen atoms in the planes [2, 3]; (iii) oxygen stoichiometry [4]; (iv) the quantum fluctuations expected in a low-dimensional antiferromagnet with small spin (see Sec. 2.1). Experiments [1, 4, 5] have yielded ordered moments in the range 0.3-0.6 μ_B.

1.2. PHASE DIAGRAM

LDA-bandstructure calculations [7] predict that La$_2$CuO$_4$ is a metal. In fact, correlation effects mean that it is an antiferromagnetic Mott insulator. The addition of barium or strontium creates a plethora of behaviour illustrated in Fig. 3. The effect of the Sr or Ba is to add holes, moving the system away from half filling and resulting in a transition to a metallic and superconducting state for $x \approx 0.05$. It is well known that both the normal metal and the superconductor are anomalous in many ways. Although the normal metal is characterized by an increasing resistivity with temperature, it is several orders of magnitude larger than for good metals. Further, it shows a linear variation with temperature signalling an extremely temperature dependent scattering mechanism.

Doping not only affects the electrical properties, but also has a dramatic effect on the magnetic ground state. The long-range antiferromagnetic order is rapidly destroyed with increasing x. For $x > 0.02$, the antiferromagnetism is replaced by short-range antiferromagnetic correlations at low temperatures. This state appears to be formed by 'spin freezing' and has some analogies to the formation of a spin glass. The "antiferromagnetic spin glass" has been characterized both by local probes : μSR [8], NMR [9], Mössbauer [10] and by neutron scattering [11, 12]. Information about the spatial correlations in the frozen spin state is scarce. In the lightly doped regime, neutron scattering measurements on $La_{1.95}Ba_{0.05}CuO_4$ [11], revealed a broadened peak centered on the antiferromagnetic wavevector, consistent with frozen domains of size $\xi \approx 20$ Å. At higher x, local probes indicate an island of static order in the $La_{2-x}Sr_xCuO_4$ phase diagram near x=0.12. Suzuki et al [12] have recently observed the corresponding incommensurate magnetic peaks. Incommensurate magnetic superlattice peaks have also been observed in $La_{1.6-x}Nd_{0.4}Sr_xCuO_4$ [13], although in this case the material first undergoes a structural transition to a low-temperature tetragonal phase at higher temperatures. We refer the reader to chapter by J.M. Tranquada in this volume for further details.

2. The Insulating Antiferromagnet, La_2CuO_4

The crystal and electronic structure of La_2CuO_4 make it a good physical realization of a 2-D $S = \frac{1}{2}$ Heisenberg antiferromagnet. We will review the nature of the magnetic excitations for such a model system in Sec. 2.1. In the following sections, we will review the additional complications which nature has added in La_2CuO_4.

2.1. THE 2-D $S = \frac{1}{2}$ ANTIFERROMAGNET AT ZERO TEMPERATURE

The properties of Heisenberg antiferromagnets when quantum mechanics is taken into account have been investigated for much of this century. It is well known that in 1-D, quantum fluctuations destabilize the Néel state completely. In higher dimensions, quantum fluctuations lead to significant corrections to the ground state energy [14, 15] and the spin-wave excitations [16]. However, the 2-D $S = \frac{1}{2}$ square-lattice antiferromagnet is thought to be ordered at zero temperature. In practice, interplanar coupling leads to ordering at finite temperatures in real materials such as La_2CuO_4. The high anisotropy and large exchange constant make La_2CuO_4 a good realization of a 2-D $S = \frac{1}{2}$ antiferromagnet, particularly when we consider the excitations over a wide energy scale comparable with $2J$. As we shall see below, anisotropies and interplanar coupling mean that La_2CuO_4 deviates from the ideal behaviour at low frequencies.

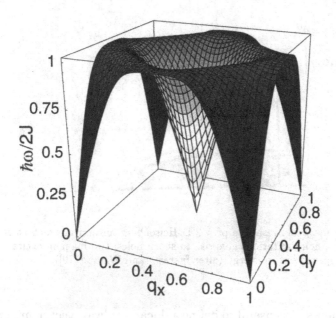

Figure 4. Spin wave dispersion in a square-lattice antiferromagnet such as La$_2$CuO$_4$. The shading shows the intensity of the spin waves, lighter shading represents higher magnetic response. Wavevectors are expressed in units of 2π.

2.1.1. *One-Magnon Excitations*

The coupling between neighbouring spins in a square-lattice antiferromagnet is described by the Heisenberg Hamiltonian,

$$H = \sum_{ij} J\,\mathbf{S}_i \cdot \mathbf{S}_j. \tag{1}$$

Conventional spin-wave theory of the Holstein-Primakoff type, in the classical large-S limit, yields the transverse dynamic susceptibility

$$\chi''_\perp(\mathbf{Q}, \hbar\omega) = Z_d(\omega)\frac{\pi}{2}g^2\mu_B^2 S\left(\frac{1-\gamma(\mathbf{Q})}{1+\gamma(\mathbf{Q})}\right)^{1/2}\delta\left(\hbar\omega \pm \hbar\omega(\mathbf{Q})\right), \tag{2}$$

where,

$$\hbar\omega(\mathbf{Q}) = 2Z_cJ\left[1-\gamma^2(\mathbf{Q})\right]^{1/2}, \tag{3}$$

and $\gamma(\mathbf{Q}) = \cos(\pi h)\cos(\pi l) = \frac{1}{2}\left[\cos(q_x) + \cos(q_y)\right]$. We have included a "quantum renormalization" of the overall spin-wave amplitude $Z_d(\omega)$. In the conventional linear spin-wave model applicable for large S, $Z_d(\omega) = 1$. In order to make comparisons with neutron scattering and nuclear reso-

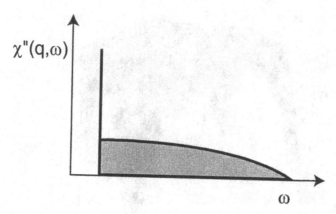

$\chi''(q,\omega)$

ω

Figure 5.   The magnetic response of a 2-D Heisenberg antiferromagnet at zero temperature. One-magnon excitations give rise to sharp poles; two-magnon excitations to a response widely spread out in energy (after Igarashi and Watabe [19])

nance experiments, it is useful to define a 'local-' or 'wavevector-integrated-' energy-dependent susceptibility $\chi''(\omega)$,

$$\chi''(\omega) \;=\; \frac{\int_{BZ} \chi''(\mathbf{Q},\omega)\,d^3 Q}{\int_{BZ} d^3 Q}. \tag{4}$$

In the low frequency limit Eq. 2 yields,

$$\chi''_\perp(\omega) \;=\; \frac{Z_d(\omega)g^2\mu_B^2 S}{2Z_c J} \tag{5}$$

2.1.2. *Multi-Magnon Excitations*

For small S, quantum corrections [17, 18] become important. The Néel state, in which the moments on each site do not fluctuate with time, is not the ground state of Eq. 1. Quantum fluctuations lead to a number of effects : (i) a reduction of the ordered moment; (ii) renormalization of the spin-wave energies and intensities; (iii) the existence of multi-magnon excitations. Igarashi [18] has recently performed a $1/S$ expansion to obtain the full magnetic response. The basic finding is that the spin-wave energies and the weight in the spin-wave pole are renormalized. This renormalization is energy dependent. In addition to the single-magnon modes, two-magnon excitations give rise to a continuum, as a function of energy, in the longitudinal response $\chi''_{zz}(\mathbf{Q},\omega)$, and three-magnon excitations result in side bands (to the one-magnon excitations) in the transverse response $\chi''_\perp(\mathbf{Q},\omega)$. The multi-magnon response is illustrated schematically in Fig. 5

For the purpose of comparing theory and experiment in the present review, we include the renormalization of the overall energy scale in the

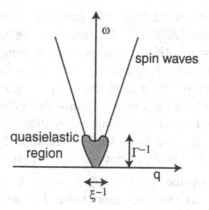

Figure 6. Schematic representation of the magnetic response in a 2D antiferromagnet at finite temperature

exchange constant, $J^* = Z_c J$, where J is the exchange constant occurring in Eq. 1. We have previously defined the renormalization of the amplitude of the magnetic response in Eq. 2. We note that, in principle, these factors are energy dependent [19]. The value of $Z_d(\omega)$ *can* be obtained by neutron scattering, if measurements are placed on an absolute intensity scale. However, Z_c *cannot* be measured directly from inelastic neutron scattering and must be estimated from theory. In the case of the $S = \frac{1}{2}$ square-lattice antiferromagnet, Singh [17] and Igarashi [18] have estimated $Z_c = 1.18$ and $Z_d(\omega = 0) = 0.61$.

2.2. THE 2-D $S = \frac{1}{2}$ ANTIFERROMAGNET AT FINITE TEMPERATURES

Since the 1950s, it has been widely believed that the 2D $S = \frac{1}{2}$ square-lattice antiferromagnet has long range order only at $T = 0$. Anderson [14], argued that even though the staggered magnetization may be reduced, the Néel state is still stable against quantum fluctuations at $T = 0$. Anderson's assertion means that this model system is an example of a 'quantum critical' phase transition i.e. one which occurs at zero temperature and is therefore dominated by quantum, rather than classical, fluctuations in the order parameter. It was not surprising that the discovery of the antiferromagnetism in La$_2$CuO$_4$ and related compounds rejuvenated theoretical [20, 21, 22, 23] and experimental [24] interest in this problem. The literature devoted to the $S = \frac{1}{2}$ square-lattice antiferromagnet is now vast, indeed there are several reviews [22] on this subject.

One of the novel features of low-dimensional antiferromagnets, in particular 2D antiferromagnets, is the strong build up of 'critical' magnetic fluctuations at temperatures considerably above the ordering temperature.

This behaviour is not confined to $S = \frac{1}{2}$ (see, for example, the review by De Jongh and Miedema [25]). Fig. 6 illustrates the magnetic excitation spectrum for a 2D $S = \frac{1}{2}$ antiferromagnet at finite temperature. For small frequencies and wavevectors, propagating excitations [26] are believed not to exist, instead the response is dominated by a 'quasielastic peak' (i.e. $\chi''(\mathbf{Q}, \omega)/\omega$ is peaked at zero frequency). Since this peak dominates the response for wavevectors near the ordering wavevector, its width in wavevector is usually characterized by the inverse correlation length ξ^{-1} obtained from the width of $S(Q)$, the Fourier transform of the equal-time correlation function. In the regime $\hbar\omega \stackrel{\sim}{<} \hbar v_s \xi^{-1}$ (v_s is the spin-wave velocity) spin-wave interactions are the main mechanism for damping. Grempel [27] estimated the order-parameter relaxation rate for $q = 0$ using a coupled-mode calculation and found that $\hbar\Gamma_2 \simeq \hbar v_s \xi^{-1}(T/2\pi\rho_s)^{\frac{1}{2}}$.

Chakravarty *et al.* [20, 21] have obtained definitive results for the low-frequency response. They showed that the long-wavelength, low-energy properties were well described by a mapping to a *classical* two-dimensional Heisenberg magnet. The effects of quantum fluctuations are absorbed into the coupling constants. Using this approach and Monte Carlo simulations they found that the temperature dependence of the correlation length ξ is,

$$\xi = C_\xi a \exp\left(2\pi\rho_s/k_B T\right). \tag{6}$$

At energies larger than a few $\hbar v_s \xi^{-1}$, propagating spin waves should exist and the response is believed to be essentially as for $T = 0$ [28], except that we must average over domain orientation. The high-frequency magnetic excitations see the moment in the 'quasielastic peak' as frozen domains of size ξ.

2.3. NEUTRON SCATTERING STUDIES OF La_2CuO_4

2.3.1. *Spin Waves and The Determination of J*

The existence of high-temperature superconductivity in $La_{1.86}Sr_{0.14}CuO_4$ and antiferromagnetism in La_2CuO_4 provided a compelling motivation to investigate the magnetic excitations in La_2CuO_4. Early measurements were performed on a conventional reactor source using an 'energy integrating' technique [24, 29] in which the two-dimensionality of the the magnetic interactions was exploited and the magnetic response was integrated in energy up to about 30 meV. This method allowed the weak signal to be seen in relatively small single crystals. The spin fluctuations were shown to have a wide energy scale (> 10 meV). Subsequent experiments were performed using epithermal neutrons produced by the "hot" source at the Institute Laue-Langevin. The use of the hot source allowed the spin-wave dispersion to be followed up to 140 meV and the two counter-propagating

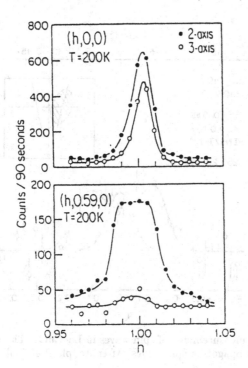

Figure 7.   Spin wave excitations in La$_2$CuO$_4$. Open circles are scans with an energy window of about 1 meV around zero. Closed circles correspond to an energy integration of about 10 meV. The lower figure demonstrates the existence of magnetic excitations with an energy scale greater than the integration window. After Shirane *et al.* [29].

branches to be resolved [30, 31]. At $T=5$ K, the spin-wave velocity was found to be $v_s = 0.85 \pm 0.03$ eV Å. Using the low-frequency limit of Eq. 3, $\hbar v_s = \sqrt{8} S J a$ and assuming $Z_c = 1.18$ [17, 21, 18] yields a Heisenberg coupling constant $J = 136 \pm 5$ meV.

In spite of the use of the hot source and 0.1 kg of aligned single crystal, the highest energy which could be achieved by reactor-based three-axis methods was about 140 meV. Spallation sources offer low background and high incident energies. The use of chopper spectrometers and the ISIS spallation source has allowed the whole single-magnon dispersion to be measured [11, 32] (see Fig. 7 in chapter 1 of this volume). The value of J determined from the ISIS experiments was $J = 132 \pm 4$ meV.

The exchange coupling constant J has been discussed widely in the literature (see, for example Ref. [33]). Our understanding is based on Anderson's theory of superexchange [34] in which $J \approx 4t^2/U$. The overlap of the copper d orbitals and the oxygen p orbitals is characterized by the hopping parameter t and U represents the on-site interaction (in the oxygen orbital).

A useful way to display the strength of the spin-wave scattering is

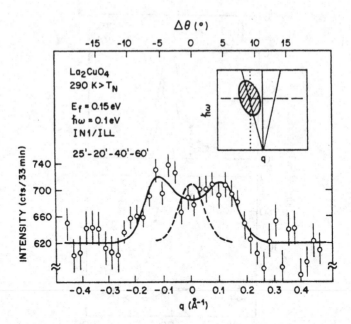

Figure 8. Hot source measurements of spin waves in La₂CuO₄. The upper panels show two resolved counter-propagating spin waves. After Aeppli *et al.* [30].

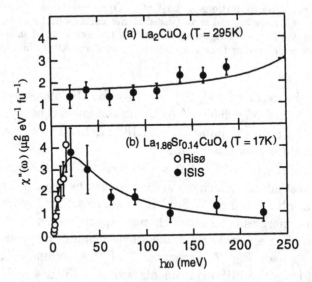

Figure 9. The local susceptibility derived from spallation source measurements (closed circles) and reactor-based measurements (open circles). After Hayden *et al.* [35].

through the local- or wavevector-integrated- susceptibility $\chi''(\omega)$. Fig. 9 shows this quantity determined in absolute units from spallation source measurements. The low-frequency limiting value of the local susceptibility for a 2D antiferromagnet is given by Eq. 5. In order to compare with experiment, we must compute the susceptibility averaged over the ordered moment direction, $\chi''_{ave} = \frac{2}{3}\chi''_{\perp}$. Thus, conventional spin-wave theory ($Z_d = 1$) predicts $\chi''_{ave}(\omega = 0) = g^2\mu_B^2 S/3J = 5.1\,\mu_B^2$ eV^{-1}. Inspection of Fig. 9 shows that the conventional spin-wave theory overestimates the spin-wave intensity as expected. Values for Z_d quoted in the literature are $Z_d = 0.34$ [32] and $Z_d = 0.39 \pm 0.1$ [35], these should be compared to Igarashi's prediction $Z_d = Z_c Z_\chi = 0.61$.

2.3.2. Interplanar Coupling and Exchange Anisotropy

In the previous section we treated La$_2$CuO$_4$ as an ideal 2D square-lattice Heisenberg antiferromagnet with nearest neighbour interactions. This provides a good description of the magnetic excitations on a large energy scale. However, at low energies and temperatures, we must consider interactions other than those in Eq. 1. The orthorhombic distortion allows a net exchange between the CuO$_2$ planes and an antisymmetric Dzyaloshinsky-Moriya (DM) exchange term. The DM exchange term introduces an in-plane anisotropy gap in the spin-wave spectrum and the interplanar coupling leads to 3D Néel order.

Early experiments [1] demonstrated that La$_2$CuO$_4$ shows a transition to 3D Néel order at about 290 K. The transition to the Néel state should occur [21] when $J_\perp(\langle S_z\rangle/S)^2(\xi/a)^2 \approx k_B T_N$. Using $\langle S_z\rangle/S = 0.6$ and $\xi(T_N) = 200a_0$, gives $J_\perp = 1.5\,\mu$eV.

The in-plane and out-of-plane zone-centre spin-wave modes which result from the presence of the DM exchange term have been studied both by infrared measurements [36] and neutron scattering [37]. The energies of the modes are 1 and 2.5 meV respectively. Peters et al. conclude that these energies are consistent with an antisymmetric exchange $J^{bc}=0.55$ meV.

2.3.3. Excitations in the Paramagnetic State

In Sec. 2.2 we noted that, from the theoretical point of view, the 2D $S = \frac{1}{2}$ antiferromagnet is an interesting system because it undergoes at $T=0$ phase transition. Endoh et al., [24] measured the temperature dependence of the magnetic correlation length and demonstrated that it could be described by the form (Eq. 6) proposed by Chakravarty et al.(CHN) [21]. Keimer et al. [38] have made a more recent study of the temperature dependences of ξ using 'carrier-free La$_2$CuO$_4$' i.e. material annealed [39] to yield the full ordered moment. They found that their data could be fairly well described by an extension to the CHN model which gives an exact analytic expression for the exponential prefactor in Eq. 6.

Yamada et al. [39] studied the frequency dependence of the response in the paramagnetic state at several temperatures and concluded that the response deviated from the CHN model in the quasielastic region. Hay-

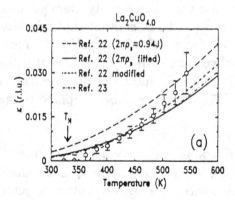

Figure 10. The inverse magnetic correlation length of a carrier-free sample of La$_2$CuO$_4$. After Keimer *et al.* Ref. [38].

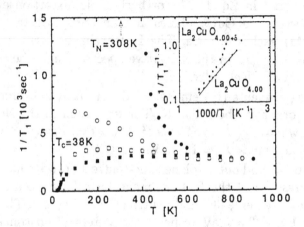

Figure 11. The temperature dependence of $^{63}T_1^{-1}$ measured by NQR for La$_2$CuO$_4$ ($x=0$, filled circles; $x = 0.04$, circles; $x = 0.075$, squares; $x = 0.15$, filled squares). After Imai *et al.* [41]

den *et al.* [31] performed higher resolution measurement and found that at T=320 K= T_N + 30 K, the response below 5 meV could not be described by damped spin waves (as used by CHN to parameterize their results). An additional quasielastic component of width $\hbar\Gamma = 1.5 \pm 0.4$ meV had to be introduced. The width of this peak is in agreement with estimates of Grempel [27] and Tyč *et al.* [40].

Nuclear resonance probes have also been used to measure the temperature dependence of the spin dynamics in La$_2$CuO$_4$. Imai *et al.* [41] have made measurements over an extremely wide temperature range, both on La$_2$CuO$_4$ and its superconducting siblings. The CHN theory can be couched in a form suitable for comparison with NQR measurements of T_1^{-1}. The solid line in the inset to Fig. 11 shows such a comparison.

3. The Normal Metal

Before considering the rather exotic metallic state of La$_{2-x}$(Ba,Sr)$_x$CuO$_4$, we will briefly review the nature of the magnetic excitations in simple metals. In metals with weak magnetic correlations such as sodium, the magnetic excitations are 'electron-hole pairs'. A magnetic field varying in space and time (such as that produced by a neutron moving through a sample) can excite electrons from below the Fermi energy to unoccupied states above it. This process creates an 'electron-hole' or 'Stoner' excitation. In the case of the sodium, the large band width and weak exchange interaction mean that the electron-hole excitations are spread out over several eV in energy. This makes them difficult to observe directly.

The response of a paramagnetic metal due to the electron-hole excitations is given by the so-called Lindhard function [42, 43],

$$\chi_0(\mathbf{q},\omega) = \frac{g^2\mu_B^2}{4N}\sum_k \frac{f[\epsilon(\mathbf{k})] - f[\epsilon(\mathbf{k}+\mathbf{q})]}{\epsilon(\mathbf{k}+\mathbf{q}) - \epsilon(\mathbf{k}) - \hbar\omega - i\Gamma} \tag{7}$$

where, the sum is over both spin states, $\epsilon(\mathbf{k})$ is the band energy, $f(E)$ the Fermi-Dirac function and,

$$\chi_0''(\mathbf{q},\omega) = \frac{\pi g^2\mu_B^2}{4N}\sum_k (f[\epsilon(\mathbf{k})] - f[\epsilon(\mathbf{k}+\mathbf{q})])$$
$$\times \delta[\epsilon(\mathbf{k}+\mathbf{q}) - \epsilon(\mathbf{k}) - \hbar\omega]. \tag{8}$$

The Lindhard response function describes a continuum of electron-hole excitations. For small energy transfers the response can be related to electron states near the Fermi energy (cf. Eq. 8) and may therefore reflect the Fermi surface. The response can be strong in materials with highly nested Fermi surfaces or with an exchange interaction which favours (but does not result in) magnetic order. An example of the former is the Cr$_{1-x}$V$_x$ alloy system [44, 45, 46]. Chromium is an incommensurate antiferromagnet with a strongly nested Fermi surface. The antiferromagnetism may be destroyed by alloying with V, producing a highly-nested paramagnetic metal. Fawcett et al. [45] demonstrated the existence of strong overdamped excitations at incommensurate positions in the paramagnet Cr$_{0.95}$V$_{0.05}$, which have similarities to the low-frequency excitations in La$_{2-x}$Sr$_x$CuO$_4$. These 'spin-density-wave paramagnons', as they were named, exist up to at least 400 meV [46].

In the presence of magnetic interactions, the response can be calculated within the RPA approximation,

$$\chi(\mathbf{q},\omega) = \frac{\chi_0(\mathbf{q},\omega)}{1 - \lambda\chi_0(\mathbf{q},\omega)}, \tag{9}$$

where $\chi(\mathbf{q}, \omega)$ includes the effects of magnetic correlations in this approximation and $\chi_0(\mathbf{q}, \omega)$ is the 'non-interacting' susceptibility, calculated for example using Eq. 7. The effect of the magnetic interactions are incorporated into the parameter λ. Within the strict RPA approximation, λ is determined using mean-field arguments, however the RPA form of the response is believed to have more general applicability [47]. The effect of introducing exchange interactions, for example using Eq. 9, is to lower the characteristic energy scale of the response [48]. The low-frequency magnetic fluctuation can then play an important part in determining the thermal properties [47, 48, 49].

Moriya [49] and Lonzarich and Taillefer [47] have noted that low-frequency low-q expansions of Eqs. 7–9 can provide a model for the magnetic excitations in a metal. For a nearly ferromagnetic metal,

$$\chi^{-1}(\mathbf{q}, \omega) = \chi_0^{-1} + c|\mathbf{q}|^2 + a\omega^2 - i\frac{\omega}{\gamma|\mathbf{q}|} + \cdots, \tag{10}$$

where the constants have been named as in Ref. [47]. An analogous expansion [49, 50] may be made for a nearly *antiferromagnetic* metal,

$$\chi^{-1}(\mathbf{q}, \omega) = \chi_\delta^{-1} + c\mathcal{R}^2(\mathbf{q}) + a\omega^2 - i\beta\omega + \cdots, \tag{11}$$

where we have expanded about the position $\mathbf{Q}_0$, $\mathbf{q} = \mathbf{Q} - \mathbf{Q}_0$ and $\chi_\delta = \chi'(\mathbf{Q} = \mathbf{Q}_\delta, \omega = 0)$. $\mathcal{R}(\mathbf{q})$ is a function which tends to zero as $\mathbf{q} \to \mathbf{Q}_\delta - \mathbf{Q}_0$. The imaginary part of Eq. 11 is

$$\chi''(\mathbf{q}, \omega) = \frac{\beta\omega}{\left[\chi_\delta^{-1} + c\mathcal{R}(\mathbf{q}) + a\omega^2\right]^2 + \beta^2\omega^2}. \tag{12}$$

In the case of a (nearly) commensurate antiferromagnet, $\mathcal{R}(\mathbf{q}) = q^2$ and we obtain the response function used by Millis, Monien and Pines [51] to fit nuclear resonance data in High-T_c materials. As we shall see below, the magnetic response in metallic $La_{2-x}Sr_xCuO_4$ is peaked at four positions near the antiferromagnetic wavevector of La_2CuO_4. In this case, $\mathcal{R}(\mathbf{q})$ must be suitably chosen to have four symmetry-related zeroes (to give peaks in the susceptibility) around $\mathbf{Q}_0$. Mason, Aeppli and Mook [52] proposed a simple form for $\mathcal{R}(\mathbf{q})$ appropriate for $La_{2-x}Sr_xCuO_4$ by adapting a function used by Noakes *et al.* [53] to describe paramagnetic scattering in Cr alloys:

$$\mathcal{R}(\mathbf{q}) = \frac{\left[(q_x^2 + q_y^2) - \pi^2\delta^2\right]^2 + 4q_x^2 q_y^2}{4a_0^2\pi^2\delta^2}. \tag{13}$$

Note that, in Eq. 13, a_0 is the lattice parameter i.e. the Cu-Cu separation and $\mathbf{q}$ in measured in units of $1/a_0$.

Eq. 12 has several useful limiting forms at low frequency. Firstly, if we define the peak height in $\chi''(\mathbf{Q}, \omega)$ for a constant energy trajectory, as

$$\chi_p''(\omega) = \chi''(\mathbf{Q} = \mathbf{Q}_\delta, \omega), \tag{14}$$

then

$$\lim_{\omega \to 0} \frac{\chi_p''(\omega)}{\omega} \sim \beta \chi_\delta^2. \tag{15}$$

If we are very close to a magnetic phase transition one would expect $\chi_\delta^{-1} a \ll \beta^2$. In this limit and for $\omega \to 0$, we have

$$\chi''(\mathbf{q}, \omega) = \frac{\beta \omega}{\left[\chi_\delta^{-1} + c\mathcal{R}(\mathbf{q}) \right]^2 + \beta^2 \omega^2} \tag{16}$$

$$= \frac{\chi_p''(\omega) \left[\kappa_0^4 + \kappa_1^4(\omega) \right]}{\left[\kappa_0^2 + \mathcal{R}(\mathbf{q}) \right]^2 + \kappa_1^4(\omega)}. \tag{17}$$

This limit was used is Refs. [52] and [53]. In the opposite limit $\chi_\delta^{-1} a \gg \beta^2$ and $\omega \to 0$, Eq. 12 becomes:

$$\chi''(\mathbf{q}, \omega) = \frac{\beta \omega}{\left[\chi_\delta^{-1} + c\mathcal{R}(\mathbf{q}) + a\omega^2 \right]^2} \tag{18}$$

$$= \frac{\chi_p''(\omega) \kappa^4(\omega)}{\left[\kappa^2(\omega) + \mathcal{R}(\mathbf{q}) \right]^2}. \tag{19}$$

This limit was used by Aeppli et al. [54] (see Sec. 3.1).

3.1. INCOMMENSURATE RESPONSE AT LOW FREQUENCY

We saw in Sec. 2.3.1 that the magnetic excitations in undoped La_2CuO_4 are characteristic of an insulating antiferromagnet with large exchange coupling. In this section, we will see that at higher doping levels, $x > 0.05$, the magnetic excitations of $La_{2-x}(Ba,Sr)_xCuO_4$ are characteristic of a metal and can be understood, at least qualitatively, in terms of the ideas introduced in the preceding section.

Experiments by Thurston et al. [55] on metallic $La_{1.89}Sr_{0.11}CuO_4$ samples showed the low-frequency magnetic response was peaked at incommensurate positions. Fig. 12 [55] shows scans through the antiferromagnetic zone centre in two different Brillouin zones. The weaker signal observed for larger $|\mathbf{Q}|$ suggests that the peaks are of magnetic origin. Thus, the magnetic excitations are completely different to those of the antiferromagnet

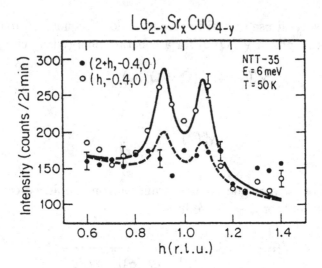

Figure 12. Constant energy scans through the antiferromagnetic zone centre in La$_{1.89}$Sr$_{0.11}$CuO$_4$. Energy transfer is 6 meV and the scan trajectory is a square diagonal in Fig. 13(d). After Thurston *et al.* Ref. [55].

La$_2$CuO$_4$, which would yield a single sharp peak in the centre of the scan under the same experimental conditions. The data shown in Fig. 12 are collected along a square diagonal in Fig. 2(b). The wavevector dependence of the excitations at low frequency was further characterized by Cheong *et al.* [56]. This experiment was conducted in a fashion such that the response could be mapped out as a function of the in-plane wavevector. Two compositions were investigated, La$_{1.925}$Sr$_{0.075}$CuO$_4$ and La$_{1.86}$Sr$_{0.14}$CuO$_4$. Fig. 13(a) shows the map of Cheong *et al.* revealing that the wavevectors of the peaks are $Q = (\pi \pm \delta\pi, \pi)$ and $Q = (\pi, \pi \pm \delta\pi)$ where δ=0.14 and 0.24 for x=0.075 and 0.14 respectively. Cheong *et al.* also found that the peaks for the more highly doped composition were quite sharp in wavevector. In particular, their width corresponded to a length $\xi = \kappa^{-1} = 17 \pm 4$ Å or about 4.5 lattice constants. Such sharp peaks are incompatible with their width being determined by the dopant separation. This observation provided strong support for models in which the conduction electrons were responsible for the incommensurate peaks and the Sr atoms are screened.

Fig. 13 demonstrates graphically how the incommensurate peaks move with doping x, in a simple picture, the band filling changes hence the area of the Fermi surface changes. Yamada *et al.* [58] have recently made a comprehensive study of the doping dependence of the incommensurate peak positions. They conclude that the onset of incommensurability, with increasing x, coincides with the appearance of metallic and superconducting behaviour near $x = 0.05$. Fig. 14 shows the measured doping dependence of the incommensurability.

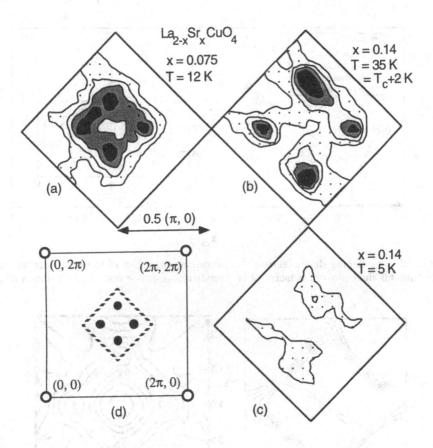

Figure 13. Low-energy magnetic scattering near (π, π) in La$_{2-x}$Sr$_x$CuO$_4$. (a) Contour plots of the intensity for non-superconducting (x=0.075, T=12 K, $\hbar\omega = 1$ meV); (b) superconducting (x=0.14, $T = T_c + 2$ K, $\hbar\omega = 2$ meV) (c) superconducting (x=0.14, $T = 5$ K, $\hbar\omega = 2$ meV) (d) The dashed region indicates the part of the Brillouin zone covered in (a)-(c). (a) is after Cheong *et al.* [56], (b),(c) after Mason *et al.* [57].

A number of groups [59, 60, 61, 62, 63, 64] have many explicit calculations of the dynamic susceptibility in a nested Fermi-liquid scenario. Incommensurate peaks arise from nesting across the Fermi surface as depicted in Fig. 15(a). In a 2-D system, the low frequency response, $\lim_{\omega\to 0} \chi''(\mathbf{q}, \omega)/\omega$, is singular for any $\mathbf{q}$ such that the Fermi surface is mutually tangential to the Fermi surface displaced by $\mathbf{q}$ (see, for example, Ref. [59]). The Fermi surface can be mapped onto curves in reciprocal space corresponding to such $\mathbf{q}$ vectors. For a circular Fermi surface, these would be simply circles with radius $2k_F$ centred on reciprocal lattice points. In La$_{2-x}$Sr$_x$CuO$_4$ the four peaked structure can be understood, at the simplest level, as the points of intersection of these curves. Fig. 15(b) shows the result of a calculation including a realistic bandstructure. The pattern is in good agreement with

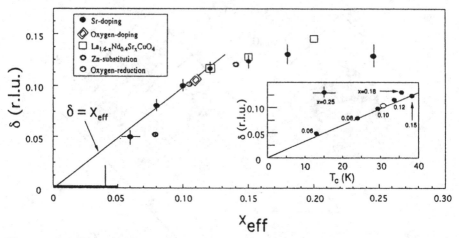

Figure 14.   Sr-doping dependence of the incommensurability of the spin fluctuations. Note that δ is defined to be a factor of two smaller than in the text. After Yamada *et al.* [58].

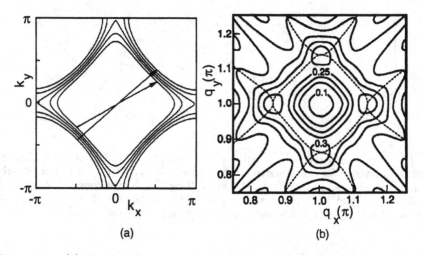

Figure 15.   (a) Fermi surface contours for different Fermi levels in $La_{2-x}Sr_xCuO_4$. Contours are the result of a rigid three-band model. The arrows correspond to transitions between mutually tangential pieces of Fermi surface yielding a large value of $\lim_{\omega \to 0} \chi''(\mathbf{q},\omega)/\omega$. (b) Contour plot of $\chi''(\mathbf{q}, \omega = 1\text{meV})$. Dashed curves indicate singularities described in the text. After Littlewood *et al.* [59].

the experimental observations.

In addition to characterizing the wavevector dependence of the magnetic response, it is interesting to also characterize its energy dependence. The expansions discussed in Sec. 3 predict that the response will broaden in wavevector with increasing frequency, and that peak amplitude will increase initially linearly with frequency (this is, in fact, simply a consequence of the antisymmetry required for the imaginary part of $\chi(\mathbf{q},\omega)$). Thurston *et*

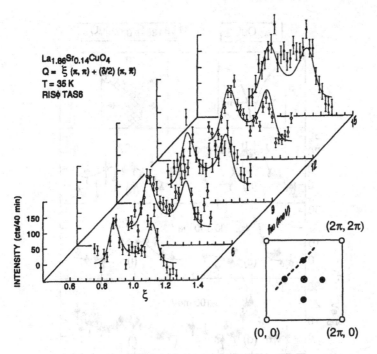

Figure 16.   Magnetic scattering in La$_{1.86}$Sr$_{0.14}$CuO$_4$ at T=35 K=T_c+2 K. A series of constant energy scans along the path in reciprocal space indicated in the inset is shown. Note how the incommensurate peaks broaden with increasing energy. After Mason *et al* [57].

al. [55] and Mason, Aeppli and Mook (MAM) [52], made energy-dependent studies in thermal neutron range. The widths of the incommensurate peaks varies rapidly with energy. MAM parameterized their data using Eq. 17.

3.2. MAGNETIC EXCITATIONS UP TO 2J

We saw earlier that the magnetic excitations of insulating La$_2$CuO$_4$ are those of a Heisenberg antiferromagnet with an exchange constant $J \approx$ 130 meV. Since the excitations derive from fluctuations in the Cu spins, the magnetic response is constrained by the moment sum rule,

$$\langle m^2 \rangle = \frac{3\hbar}{\pi} \int_{-\infty}^{\infty} \frac{\chi''(\omega)\,d\omega}{1 - \exp(-\hbar\omega/kT)}, \qquad (20)$$

where $\langle m^2 \rangle$ is the mean squared moment on the Cu sites. For $S = \frac{1}{2}$, $\langle m^2 \rangle = g^2 \mu_B^2 S(S + 1) = 3\ \mu_B^2$. We might ask, "What happens to this response when La$_{2-x}$Sr$_x$CuO$_4$ becomes metallic?" In order to address this issue, we need to study the magnetic response on the energy scale $2J$. As we saw in Sec. 2.3.1, this can be done using a chopper spectrometer on a spallation source.

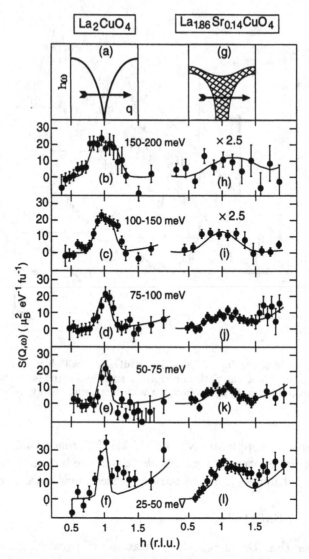

Figure 17. A comparison of the magnetic response in insulating La_2CuO_4 and metallic $La_{1.86}Sr_{0.14}CuO_4$. (a)-(f) Magnetic scattering from insulating La_2CuO_4 ($T=296$ K). (g)-(l) Magnetic scattering from metallic $La_{1.86}Sr_{0.14}CuO_4$ ($T=17$ K). The effects of entering the superconducting state on the magnetic response is small at the energies investigated. The reciprocal space trajectory is along the square diagonal passing through the antiferromagnetic zone centre in Fig. 2(b). After Hayden *et al.* [35].

Yamada *et al.* [65] and Hayden *et al.* [35] have performed spallation-source neutron scattering experiments to investigate the higher frequency excitations. Yamada *et al.* demonstrated that a magnetic response exists up to about 300 meV, as in the insulator. They also showed that the response was peaked near the antiferromagnet wavevector of the insulator up to about 120 meV.

Fig. 17 shows a comparison of the magnetic response in insulating La$_2$CuO$_4$ and superconducting La$_{1.86}$Sr$_{0.14}$CuO$_4$ from Hayden *et al.* Figs. 17(b)-(f) show constant energy scans through the antiferromagnetic position for increasing energy transfer. The peaks in the data are due to the spin waves discussed in Sec. 2.3.1. They broaden in wavevector at higher energies due to the dispersion of the spin waves. Two peaks due to spin waves propagating in opposite directions are not seen because of the poor vertical resolution used in the experiment. Figs. 17(h)-(l) shows the corresponding response for metallic La$_{1.86}$Sr$_{0.14}$CuO$_4$ collected under the same conditions and in the same units. Comparing the data collected from the two materials, we see that (i) the scattering in La$_{1.86}$Sr$_{0.14}$CuO$_4$ is considerably broader at all energies and (ii) the response decreases more rapidly with increasing frequency in La$_{1.86}$Sr$_{0.14}$CuO$_4$. A quantitative comparison of the local susceptibility in the two materials, is shown in Fig. 9. This figure basically answers the question which we posed at the beginning of this section. The formation of the metallic state in La$_{2-x}$Sr$_x$CuO$_4$ results in a tremendous redistribution of magnetic response in frequency. In particular, La$_{2-x}$Sr$_x$CuO$_4$ shows a peak in the local susceptibility near 20 meV. The more striking feature of the response is its strength in the metal. For energies in the range $\approx 5 - 60$ meV it is actually *bigger* in the paramagnetic metal than in the ordered antiferromagnet! Where has this response come from? When we compare with the antiferromagnet, we see there is a suppression of the higher frequency response. Further, the antiferromagnet has a ordered moment which accounts for $\langle m^2 \rangle \approx 0.6^2 = 0.36 \ \mu_B^2$.

In spite of the redistribution of spectral weight towards lower frequencies, a significant high-frequency magnetic response is observed [65, 35] in La$_{1.86}$Sr$_{0.14}$CuO$_4$ up to energies comparable with $2J$ in La$_2$CuO$_4$. There appears to be be a small reduction in the energy of the highest excitations. In the metal the highest excitations occur at about 260 meV [35] compared to 320 meV in the parent insulator.

3.3. NEARLY SINGULAR BEHAVIOUR OF THE LOW-FREQUENCY RESPONSE

In the this section we will review the effect of temperature on the magnetic excitations in the normal state of La$_{2-x}$Sr$_x$CuO$_4$. We defer the superconducting state to Sec. 4. It is well-know that the physical properties in the normal state of the copper-oxides superconductors are highly anomalous [66, 67]. For example, perhaps the simplest property to measure, the electrical resistivity, shows a linear variation over a very wide temperature range. What is remarkable about this, is not that it is linear: many metals have an almost linear variation of resistance with temperature, but that

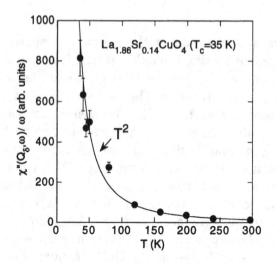

Figure 18. The temperature dependence of $\lim_{\omega\to 0} \chi''(\mathbf{Q} = \mathbf{Q}_\delta, \omega)$ for $La_{1.86}Sr_{0.14}CuO_4$. Note the diverging response at low temperatures. After Aeppli *et al.* [54].

the linear coefficient is large, suggesting that its origin is electronic rather than due to conventional electron-phonon scattering. Direct probes of the charge fluctuations such as infrared spectroscopy [68, 69] also yield dramatic temperature dependences. Since electrons carry spin as well as charge it is reasonable to ask what is the effect of temperature on the spin dynamics.

In this review we concentrate on the temperature dependence for metallic and superconducting samples. We note for completeness that that there are also several investigations [11, 70, 38] of lightly doped compositions. We saw in Sec. 3 that, for low doping levels, the antiferromagnetism in La_2CuO_4 is replaced by a 'magnetic spin glass' phase. In this regime, the low frequency response shows a rapid variation with temperature. The response has been analyzed in terms of scaling functions of ω/T.

In the more heavily doped metallic and superconducting regime, the incommensurate response increases and sharpens in wavevector as the temperature is lowered [55, 56, 52]. Such behaviour is characteristic of a system approaching a phase transition where low-frequency critical fluctuations build up at wavevectors close to the ordering wavevector. Fig. 18 shows the temperature dependence of the low-frequency response in $La_{1.86}Sr_{0.14}CuO_4$. The rapid increase of the low-frequency susceptibility with decreasing temperature is characteristic of a system close to magnetic order. Aeppli it et al. [54] have recently reported detailed measurements on $La_{1.86}Sr_{0.14}CuO_4$ of the incommensurate peak width as a function of energy and temperature. The low-frequency response was described using Eq. 19, the resulting variation of $\kappa(T, \omega)$ is shown in Fig. 19. Aeppli *et al.* [54] show that the

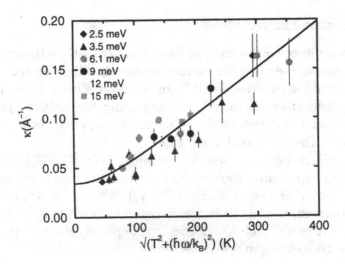

Figure 19. Temperature dependence of the inverse length scale $\kappa(\omega, T)$ obtained when the low-frequency magnetic response is described by Eq. 19. After Aeppli *et al.* [54].

energy *and* temperature dependence of κ can be described by the scaling relation,

$$\kappa^2 = \kappa_0^2 + a_0^{-2}\left[\left(\frac{k_B T}{E_T}\right)^2 + \left(\frac{\hbar\omega}{E_\omega}\right)^2\right], \qquad (21)$$

where $E_T = E_\omega$. That the response can be described in this way demonstrates the interchangeability of temperature and energy. Increasing temperature or energy causes the peak width at low frequencies to increase. This description is motivated by the idea that La$_{1.86}$Sr$_{0.14}$CuO$_4$ may be close to a 'quantum critical point' (QCP) in an appropriate parameter space. In other words, the system is near to incommensurate magnetic order (before superconductivity intervenes). If a suitable control parameter 'y' were available we could approach the QCP. In the proximity of a QCP the inverse coherence length is governed by the "distance", in $T - \omega$ space from the QCP, this then suggests Eq. 21. Similar ideas have also applied to lightly doped compositions under the label of "ω/T scaling" [11, 70, 38]. However, in this case the material does "order" into a magnetic spin glass state.

Finally, one might ask: "What is the QCP?" We might expect that metallic La$_{2-x}$Sr$_x$CuO$_4$ is close to magnetic order because it is derived from an antiferromagnet. However, this is not thought to be the appropriate QCP, since the magnetic order in La$_2$CuO$_4$ is commensurate. More likely is that the QCP corresponds to incommensurate magnetic order, such as that induced following certain lattice distortions [12, 13].

4. The Superconducting Metal

The main driving force for the study of $La_{2-x}(Ba,Sr)_xCuO_4$ is its superconductivity. Studies of the magnetic excitations can shed light on the nature and mechanism of the superconductivity in two ways. Firstly, by searching for the boson excitations responsible for the pairing. Secondly, by yielding information about the nature of the superconducting state. In particular the anisotropy of the superconducting gap.

It is now widely believed (see, for example, Refs. [71, 72]) that the superconducting gap is anisotropic in the cuprates. The case is stronger for the higher T_c materials (e.g. $YBa_2Cu_3CuO_{6+x}$), where more experiments have been performed, and the larger gap is better matched to spectroscopic probes. A number of different symmetries for the pairing state are possible. Three popular choices of gap functions are,
(i) isotropic s wave

$$\Delta(\mathbf{k}) = \Delta_0, \tag{22}$$

(ii) $d_{x^2-y^2}$ symmetry,

$$\Delta(\mathbf{k}) = \Delta_0 \left[\cos(\hat{k}_x) - \cos(\hat{k}_y)\right], \tag{23}$$

(iii) d_{xy} symmetry,

$$\Delta(\mathbf{k}) = \Delta_0 \sin(\hat{k}_x)\sin(\hat{k}_y), \tag{24}$$

4.1. COHERENCE EFFECTS IN THE SUPERCONDUCTING STATE

One of the achievements of the BCS theory of superconductivity was to demonstrate that when a metal becomes superconducting, correlations are introduced between the motion of the electrons. In a 'conventional' singlet s-wave superconductor, Cooper pairs are formed such that the states $|\mathbf{k} \uparrow>$ and $|-\mathbf{k} \downarrow>$ are simultaneously occupied. The correlations introduced can be treated more formally in the Gor'kov theory in which the order parameter is related to an 'anomalous propagator'

$$F_{\uparrow\downarrow}(\mathbf{r}, t; \mathbf{r}', t') = \langle\psi_\uparrow(\mathbf{r}, t)\psi_\downarrow(\mathbf{r}', t')\rangle, \tag{25}$$

which measures the average phase coherence of the wave function of the electrons of opposite spin at points points $\mathbf{r}$ and $\mathbf{r}'$, at times t and t'. Since the spin susceptibility also measures correlations between electrons, it is not surprising that the formation of the superconducting state can dramatically affect $\chi''(\mathbf{Q}, \omega)$. In conventional superconductors, the small value of the superconducting gap Δ precluded direct measurements of changes in $\chi''(\mathbf{Q}, \omega)$

using inelastic neutron scattering. However, measurements of the spin susceptibility in the superconducting state were made by other means. For example, nuclear magnetic resonance (NMR) measured the low-frequency local susceptibility and yielded the celebrated 'Hebel-Slichter peak' [73]. While Shull and Wedgewood [74] used the interference cross term between the nuclear and magnetic neutron scattering to measure $\chi'(\mathbf{Q} = 0, \omega = 0)$.

BCS theory makes definite predictions for the magnetic response [75] (at least in a simple metal). We must effectively evaluate a Lindhard function for the superconductor. In the superconducting state, the quasiparticle excitation spectrum is given by,

$$E^2(\mathbf{k}) = [\epsilon(\mathbf{k}) - \mu]^2 + |\Delta(\mathbf{k})|^2, \tag{26}$$

where $\epsilon(\mathbf{k})$ in the electron energy in the normal state and μ is the chemical potential. For a conventional s-wave superconductor the gap $\Delta(\mathbf{k})$ is nonzero and $\mathbf{k}$-independent. The existence of the BCS pairing means that the $|\mathbf{k}\uparrow\rangle$ and $|-\mathbf{k}\downarrow\rangle$ are coupled, this introduces 'coherence factors' into the Lindhard expression. In the superconducting state, the magnetic response is given by [75],

$$
\begin{aligned}
\chi_0(\mathbf{q}, \omega) = {} & \frac{g^2\mu_B^2}{4N} \sum_{\mathbf{k}} \frac{1}{2}\left[1 + \frac{\xi(\mathbf{k}+\mathbf{q})\xi(\mathbf{k}) + \Delta(\mathbf{k}+\mathbf{q})\Delta(\mathbf{k})}{E(\mathbf{k}+\mathbf{q})E(\mathbf{k})}\right] \\
& \times \frac{f[E(\mathbf{k}+\mathbf{q})] - f[E(\mathbf{k})]}{\hbar\omega - E(\mathbf{k}+\mathbf{q}) + E(\mathbf{k}) + i\Gamma} \\
& + \frac{1}{4}\left[1 - \frac{\xi(\mathbf{k}+\mathbf{q})\xi(\mathbf{k}) + \Delta(\mathbf{k}+\mathbf{q})\Delta(\mathbf{k})}{E(\mathbf{k}+\mathbf{q})E(\mathbf{k})}\right] \\
& \times \frac{1 - f[E(\mathbf{k}+\mathbf{q})] - f[E(\mathbf{k})]}{\hbar\omega - E(\mathbf{k}+\mathbf{q}) - E(\mathbf{k}) + i\Gamma} \\
& + \frac{1}{4}\left[1 - \frac{\xi(\mathbf{k}+\mathbf{q})\xi(\mathbf{k}) + \Delta(\mathbf{k}+\mathbf{q})\Delta(\mathbf{k})}{E(\mathbf{k}+\mathbf{q})E(\mathbf{k})}\right] \\
& \times \frac{f[E(\mathbf{k}+\mathbf{q})] - f[E(\mathbf{k})] - 1}{E(\mathbf{k}+\mathbf{q}) + E(\mathbf{k}) - \hbar\omega - i\Gamma},
\end{aligned}
\tag{27}
$$

where $\xi(\mathbf{q}) = \epsilon(\mathbf{k}) - \mu$. Unfortunately, Eq. 27 neglects electron-electron interactions which may be important in high-T_c, or even conventional, superconductors. It implicitly assumes, for example, that there is no exchange interaction between the conduction electrons. More complete models have been investigated in the literature [61, 62, 64, 76, 77, 78, 79, 80, 81, 82]. In spite of the approximations inherent in Eq. 27, it may still give a qualitative picture of how the dynamic response changes as we enter the superconducting state. Fig. 20 shows calculations of $\chi''(\mathbf{Q}, \omega)$ in the superconducting state for s-wave and $d_{x^2-y^2}$-wave symmetry. For s-wave symmetry,

(a)   (b)

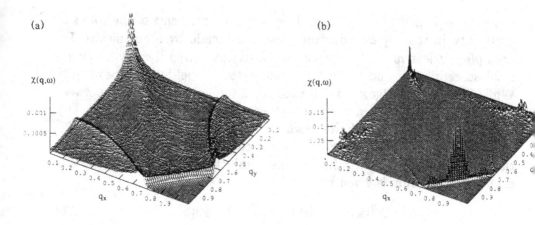

Figure 20. Calculations of $\chi''(\mathbf{Q}, \omega)$ for (a) s-wave, and (b) $d_{x^2-y^2}$ pairing. The parameters used were $\mu = -1$, $T_c = 0.02t$, $T = 0.2T_c$, $\Gamma = 0.05T_c$, $2\Delta/T_c = 3.5$ and $\omega = 0.5T_c$. After Lu [76].

the response for $\hbar\omega < 2\Delta$ is strongly suppressed. The residual pattern, present at higher temperatures for example, has the same features as the response above T_c. In the d-wave case, the response is still suppressed for $\hbar\omega < 2\Delta$, however a significant response occurs near the nesting wavevectors for $\hbar\omega > \Delta$. New "resonance peaks" may also appear below T_c.

Soon after the identification of the low-frequency incommensurate peaks, Mason *et al.* [52] found that they were strongly suppressed at low temperatures. This is strong evidence that they are derived from fermionic excitations. In view of the possibility that new features in $\chi''(\mathbf{Q}, \omega)$ might appear in the superconducting state. Later experiments [57, 84] focussed on mapping out the response near the antiferromagnetic wavevector of La_2CuO_4. Fig. 13(b) and (c) show that as superconducting $La_{2-x}Sr_xCuO_4$ is cooled through T_c, the main effect is a suppression of $\chi''(\mathbf{Q}, \omega)$ at low frequencies [57, 84]. A more recent experiment has observed an interesting coherence effect at higher energies. While the response is suppressed at low frequencies, in $La_{1.86}Sr_{0.14}CuO_4$ there is an increase in the response at $\mathbf{Q} = \mathbf{Q}_\delta$ for $\hbar\omega > 7$ meV. A more dramatic, and possibly related, effect has been observed in $YBa_2Cu_3O_{6.93}$. In this case a sharp peak appears in $\chi''(\mathbf{Q}, \omega)$ at T_c [85, 86, 87].

We still lack a detailed model for the response in the superconducting state of $La_{2-x}Sr_xCuO_4$ for $\hbar\omega < 15$ meV. Some of the available models [61, 62, 76, 77, 78, 64, 79, 80, 81] can predict the observed increase in the response for $\hbar\omega > 7$ meV, however they also predict the appearance of additional features at low frequency. The increase in scattering is a consequence of the moment sum rule (Eq. 20), what is surprising is the relatively

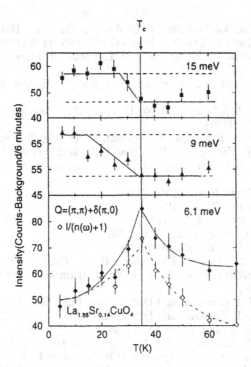

Figure 21. Temperature dependence of the magnetic response at the **Q**$_\delta$ position for various energy transfers. At low energy the response decreases below T_c, at higher energies it increases. After Mason *et al.* [83].

low frequencies at which it occurs.

Acknowledgements

I am very grateful to many colleagues, with whom I have collaborated and discussed the matters presented in this review. I would particularly like to acknowledge Gabriel Aeppli, Robert Doubble, Thom Mason, Herb Mook, Toby Perring, Piers Coleman and Gilbert Lonzarich.

References

1. D. Vaknin, S.K. Sinha, D.E. Moncton, D.C. Johnson, J.M. Newsan, C.R. Safinya and H.E. King, Phys. Rev. Lett. **58**, 2802 (1987).
2. L.J. De Jongh, Solid State Comms. **65**, 963 (1988).
3. T.A. Kaplan, S.D Mahanti, J. App. Phys. **69**, 5382 (1991).
4. K. Yamada, E. Kudo, Y. Endoh, Y. Hidaka, M. Oda, M. Suzuki, T. Murakami, Solid State Comms. **64**, 753 (1987).
5. S. Mitsuda, G. Shirane, S.K. Sinha, D.C. Johnson, M.S. Alvarez, D. Vaknin, and D.E. Moncton, Phys. Rev. B **41**, 1926 (1990).
6. R.S. Markiewicz, J. Phys. Chem. Solids. **58**, 1179 (1997).
7. J.H. Xu, T.J. Watson-Yang, J. Yu and A.J. Freeman, Phys. Lett. **12A**, 489 (1987).

8. G.M. Luke, L.P. Le, B.J. Sternlieb, W.D. Wu, Y.J. Uemura, J.H. Brewer, T.M. Rise-man, S. Ishbashi, and S. Uchida, Physica C **185-189**, 1175 (1991).

9. T. Goto, S. Kazama, K. Miyagawa, and T. Fukase, J. Phys. Soc. Jpn. **63**, 3494 (1994).

10. P. Imbert, G. Jéhanno, P. Debray, C. Garcin, and J.A. Hodges, J. Phys. (Paris) I2, 1405 (1992).

11. S.M. Hayden, G. Aeppli, H.A. Mook, D. Rytz, H.F. Hundley, and Z. Fisk, Phys. Rev. Lett. **66**, 821 (1991).

12. T. Suzuki, T. Goto, K. Chiba, T. Shinoda, T. Fukase, H. Kimura, K. Yamada, M. Ohashi, and Y. Yamaguchi, Phys. Rev. B **57**, R3229 (1998).

13. J.M. Tranquada, B.J. Sternlieb, J.D. Axe, Y. Nakamura, and S. Uchida, Nature **375**, 561 (1995).

14. P.W. Anderson, Phys. Rev. **86**, 694 (1952).

15. R. Kubo, Phys. Rev. **87**, 568 (1952).

16. T. Oguchi, Phys. Rev. **117**, 117 (1960).

17. R.R.P. Singh, Phys. Rev. B **39**, 9760 (1989).

18. J. Igarashi, Phys. Rev. B. **46**, 10763 (1992);

19. J. Igarashi and A. Watabe, Phys. Rev. B **43**, 13456 (1991); J. Igarashi and A. Watabe, Phys. Rev. B **44**, 5057 (1991); J. Igarashi, J. Phys.: Condens. Matter **4**, 10265 (1992).

20. S. Chakravarty, B.I. Halperin, and D.R. Nelson, Phys. Rev. Lett. **60**, 1057 (1988).

21. S. Chakravarty, B.I. Halperin, and D.R. Nelson. Phys. Rev. B, **38**, 2344 (1989)

22. E. Manousakis, Rev. Mod. Phys. **63**, 1 (1991).

23. A.V. Chubukov, S. Sachdev, and J. Ye, Phys. Rev. **49**, 11919 (1994).

24. Y. Endoh, K. Yamada, R.J. Birgeneau, D.R. Gabbe, H.P. Jenssen, M.A. Kastner, C.J. Peters, P.J. Picone, T.R. Thurston, J.M. Tranquada, G. Shirane, Y. Hidaka, M. Oda, Y. Enomoto, M. Suzuki, and T. Murakami, Phys. Rev. B **37**, 7443 (1988).

25. L.J. de Jongh and A.R. Miedema, Adv. Phys. **23**, 1-260 (1974).

26. For the purposes of this review we will descibe and excitation with wavevector **Q** as propaging, if it shows a peak in $\chi''(\mathbf{Q},\omega)/\omega$.

27. D.R. Grempel, Phys. Rev. Lett **61**, 1041 (1988).

28. A.V. Chubukov, private communication.

29. G. Shirane, Y. Endoh, R.J. Birgeneau, M.A. Kastner, Y. Hidaka, M. Oda, Y. Enomoto, M. Suzuki and T. Murakami, Phys. Rev. Lett. **59**, 1613 (1987).

30. G. Aeppli, S. M. Hayden, H.A. Mook, Z. Fisk, S-W. Cheong, D. Rytz, J.P. Remeika, G.P. Espinosa and A.S. Cooper, Phys. Rev. Lett. **62**, 2052 (1989).

31. S.M. Hayden, G. Aeppli, H.A. Mook, S.-W. Cheong, and Z. Fisk, Phys. Rev. B **42**, 10220 (1990).

32. S. Itoh, K. Yamada, M. Arai, Y. Endoh, Y. Hidaka, S. Hosoya J. Phys. Soc. Japan **63**, 4542 (1994)

33. H. Eskes and J.H. Jefferson, Phys. Rev. B **48**, 9788 (1993).

34. P.W. Anderson, Phys. Rev. **79**, 350 (1950).

35. S.M. Hayden, G. Aeppli, H.A. Mook, T.G. Perring, T.E Mason, S.-W. Cheong, and Z. Fisk, Phys. Rev. Lett. **76**, 1344 (1996).

36. R.T. Collins, Z. Schlesinger, M.W. Schafer, and T.J. Negran, Phys. Rev. B **37**, 2353 (1988).

37. C.J. Peters, R.J. Birgeneau, M.A. Kastner, H. Yoshizawa, Y. Endoh, J. Tranquada, G. Shirane, Y. Hidaka, M. Oda, M. Suzuki, and T. Murakami, Phys. Rev. B **37**, 9761 (1988).

38. B. Keimer, N. Belk, R.J. Birgeneau, A. Cassanho, C.Y. Chen, M. Greven, M.A. Kastner, A. Aharony, Y. Endoh, R.W. Erwin, G. Shirane, Phys. Rev. B **46**, 14035 (1992).

39. K. Yamada, K. Kakurai, Y. Endoh, T.R. Thurston, M.A. Kastner, R.J. Birgeneau, G. Shirane, Y. Hidaka, and T. Murakami, Phys. Rev. B **40**, 4557 (1989).

40. S. Tyč B. I. Halperin and S. Chakravarty, Phys. Rev. Lett. **62**, 835 (1989).

41. T. Imai, C.P. Slichter, K. Yoshimura, and K. Kosuge, Phys. Rev. Lett. **70**, 1002 (1993).
42. R.M. White, *Quantum Thoery of Magnetism* (Springer, Berlin,1983).
43. S.W. Lovesey,*Theory of Neutron Scattering from Condensed Matter* (Oxford University Press, Oxford, 1984).
44. E. Fawcett, H.L. Alberts, V.Yu. Galjin, D.R. Noakes, J.V. Yakhmi, Rev. Mod. Phys. **66**, 25(1994).
45. E. Fawcett, S.A. Werner, A. Goldman, and G. Shirane, Phys. Rev. Lett. **61**, 558 (1988).
46. S.M. Hayden, R. Doubble, G. Aeppli, T.G. Perring, E. Fawcett, J. Lowden, and P.W. Mitchell, Physica B **237-238**, 421 (1997).
47. G.G. Lonzarich and L. Taillefer, J. Phys. C:Solid State Phys. **18**, 4339 (1985).
48. S. Doniach, Proc. Phys. Soc. (London), **91**, 86 (1967).
49. T. Moriya, Phys. Rev. Lett **24**, 1433 (1970)
50. H. Sato and K. Maki, Int. J. Magnetism **6**, 183 (1974).
51. A.J. Millis, H. Monien, and D. Pines, Phys. Rev. B **42**, 167 (1990).
52. T.E. Mason, G. Aeppli, H.A. Mook, Phys. Rev. Lett. **68**, 1414 (1992).
53. D.R. Noakes, T.M. Holden, E. Fawcett, P.C. de Camargo, Phys. Rev. Lett. **65**, 369 (1990).
54. G. Aeppli, T.E. Mason, S.M. Hayden, H.A. Mook, and J. Kulda, Science **278**, 1432 (1997).
55. T.R. Thurston, R.J. Birgeneau, M.A. Kastner, N.W. Preyer, G. Shirane, Y. Fujii, K. Yamada, Y. Endoh, K. Kakurai, M. Matsuda, Y. Hidaka, T. Murakami, Phys. Rev. B **40**, 4585 (1989).
56. S.-W. Cheong, G. Aeppli, T.E. Mason, H.A. Mook, S.M. Hayden, P.C. Canfield, Z. Fisk, K.N. Clausen, and J.L. Martinez, Phys. Rev. Lett. **67**, 1791 (1991).
57. T.E. Mason, G. Aeppli, S.M. Hayden, A.P. Ramirez, and H.A. Mook Phys. Rev. Lett. **71**, 919 (1993).
58. K. Yamada, C.H. Lee, Y. Endoh, G. Shirane, R.J. Birgeneau, M.A. Kastner, Physics C **282-287**, 85 (1987).
59. P.B. Littlewood, J. Zaanen, G. Aeppli, H. Monien, Phys. Rev. B **48**, 487 (1993)
60. R.J. Jelitto, Phys. Status Solidi B **147**, 391 (1988).
61. N. Bulut, D. Hone, D.J. Scalapino, and N.E. Bickers, Phys. Rev. B **41**, 1797 (1990).
62. L. Chen, C. Bourbonnais, T. Li, A.-M.S. Tremblay, Phys. Rev. Lett. **66**, 369 (1991).
63. J. Ruvalds, C.T. Rieck, J. Zhang, and A. Virosztek, Science **256**, 1664 (1992).
64. Y. Zha, Q. Si, and K. Levin, Phys. Rev. B **47**, 9055 (1993).
65. K. Yamada, Y. Endoh, C.-H. Lee, S. Wakimoto, M. Arai, K. Ubukata, M. Fujita, S. Hosoya, and S.M. Bennington, J. Phys. Soc. Japan **64**, 2742 (1995).
66. B. Batlogg in *High Temperature Superconductivity:Proceeding.* Edited by K.S. Bedell *et al.* (Addison-Wesley, Redwood City, 1990) p37.
67. H. Takagi, T. Ido, S. Ishibashi, M. Uota, S. Uchida, Y. Tokura, Phys. Rev. B **40**, 2254 (1989).
68. Z. Schlesinger, R.T. Collins, F. Holzberg, C. Feild, S.H. Blanton, U. Welp, G.W. Crabtree, Y. Fang, Phys. Rev. Lett. **65**, 801 (1990).
69. F. Slakey, M.V. Klein, J.P. Rice, D.M. Ginsberg, Phys. Rev. B **43**, 3764 (1991).
70. B. Keimer, R.J. Birgeneau, A. Cassanho, Y. Endoh, R.W. Erwin, M.A. Kastner, G. Shirane, Phys. Rev. Lett. **67**, 1930 (1991).
71. J. F. Annett, Adv. Phys. **39** 83 (1990).
72. D.J. Scalalpino, Physics Reports **250**, 329 (1995).
73. L.C. Hebel and C.P. Slichter, Phys. Rev. **113**, 1504 (1959).
74. C.G. Shull and F.A. Wedgewood, Phys. Rev. Lett. **16**, 513 (1996).
75. J.R. Schrieffer, *Theory of Superconductivity* (Addison-Wesley, NewYork, 1983).
76. J.P. Lu, Phys. Rev. Lett. **68**, 125 (1991).
77. N. Bulut and D. Scalapino, Phys. Rev. B **47**, 3419 (1993).
78. T. Tanamoto, H. Kohno, and H. Fukuyama, J. Phys. Soc. Jpn. **62**, 1455 (1993)

79. K. Maki and H. Won, Phys. Rev. Lett. **72**, 1758 (1994).
80. F. Onufrieva and J. M. Rossat-Mognod, Physica B **49**, 4235 (1994).
81. M. Lavagna and G. Stemmann, Phys. Rev. B **49**, 4235 (1994).
82. E. Demler and S.-C. Zhang, Phys. Rev. Lett. **75** 4126 (1995)
83. T.E. Mason, A. Schröder, G. Aeppli, H.A. Mook, and S.M. Hayden Phys. Rev. Lett. **77**, 1604 (1996).
84. K. Yamada, S. Wakimoto, G. Shriane, C.H. Lee, M.A. Kastner, S. Hosoya, M. Greven, Y. Endoh, R.J. Birgeneau, Phys. Rev. Lett. **75**, 1626 (1995).
85. J. Rossat-Mignod, L.P. Regnault, P. Bourges, P. Burlet, C. Vettier, J.Y. Henry, Physica B **192**, 109 (1993).
86. H.A. Mook, M. Yethiraj, G. Aeppli, T.E. Mason, T. Armstrong Phys. Rev. Lett. **70**, 3490 (1993).
87. H.F. Fong, B. Keimer, P.W. Anderson, D. Reznik, F. Dogan, I.A. Aksay, Phys. Rev. Lett. **75**, 316 (1995).

PHONON DISPERSIONS AND PHONON DENSITY-OF-STATES IN

COPPER-OXIDE SUPERCONDUCTORS

LOTHAR PINTSCHOVIUS AND WINFRIED REICHARDT
Forschungszentrum Karlsruhe, Institut für
Nukleare Festkörperphysik, P.O.Box 3640
D-76021 Karlsruhe, Germany

1.   Introduction

The discovery of high T_c superconductivity in the cuprates by Bednorz and Müller (1986) has stimulated a large number of inelastic neutron scattering experiments to explore the phonon properties of these compounds. Most of these studies were searching for electron-phonon coupling effects and thus to clarify whether or not phonons play an important role for high T_c superconductivity. Herefore, it was indispensable to study not only the superconducting members of the cuprate family, but also the insulating parent compounds, because electron-phonon coupling effects can be identified as such only when having a firm knowledge of the normal lattice dynamical properties. Phonon studies have been performed both on single crystalline and on po-lycrystalline samples. As has been outlined in section 8.2. of the first chapter in this volume by Böni and Furrer, single crystals are indispensable for determining the dispersion curves of the phonons, $\omega = \omega_s(\mathbf{q})$, whereby the branch index s = 1...3N with N the number of atoms in the unit cell. Because all the cuprate superconductors have fairly complex structures with N = 7 or larger, the phonon dispersion contains many branches which cannot be assigned to the observed maxima in the neutron scattering cross-section by rules of thumb. Rather, assignment has to be based on predictions of lattice dynamical models, whereby the parameters of the models are properly tuned to reproduce the experimental phonon frequencies and phonon intensities. Usually, this requires an iterative procedure: starting from an educated guess of the model parameters, the observed phonon groups are tentatively assigned and their frequencies are used to tune the model parameters to achieve satisfactory agreement with experiment. Then, the model is used to assign further phonon groups and to predict where in reciprocal space the missing phonons should be looked for. The results of these further measurement will be used for a fine-tuning of the parameters. and so on.

Sometimes, an unambiguous assignment of certain phonon frequencies is only possible by making measurements for the same $\mathbf{q}$ in several different Brillouin zones, i.e. with differing momentum transfers $\mathbf{Q} = \mathbf{\tau} + \mathbf{q}$, because inelastic structure factors of the various branches will behave differently for different τ's

Single crystals are not only indispensable for establishing the phonon dispersion, but also to study the linewidth of the phonons. A large linewidth might be indicative of a strong coupling between the phonon under study and the electron system and therefore be of relevance for high-T_c superconductivity. However, there are many other sources of line broadenings than just the electron-phonon coupling, so that alternative explanations should always be considered. In Fig. 6 of the introductory chapter, an example was shown where the large linewidth has to be attributed to anharmonicity. Evidence for a large contribution of the electron-phonon coupling to the linewidth of certain phonons will be reported later.

Unfortunately, an inelastic neutron scattering study requires single crystals of rather large size. Although it is true that low energy phonons have been studied on single crystals as small as a few mm^3, samples of at least 0.1 cm^3 are necessary to aim at a complete set of phonon dispersion curves. Single crystals of such a size have been grown only for a very limited number of compounds, so that studies of other compounds had to use polycrystalline samples. As mentioned in the introductory chapter, the only quantity which can be obtained in this case is the neutron cross-section weighted phonon density of states (PDOS)

$$G(E) = \sum_i g_i(E) \, exp(-2W_i) \, \sigma_i / M_i$$

where $g_i(E)$, σ_i, M_i and W_i are the partial vibrational spectra, the neutron scattering cross-section, the mass and the Debye-Waller coefficient of the ith atom in the unit cell, respectively. In case that for a certain atom different isotopes with a markedly different neutron scattering cross-section are available, measurements of $G(E)$ on samples with different isotopic composition can be exploited to extract the partial density of states of this atom. Several results obtained in this way will be presented later.

A first summary of what has been learned from inelastic neutron scattering studies on cuprates has been published by Pintschovius and Reichardt (1994). Since then, considerable progress has been achieved and that is why it was felt appropriate to review the state of the field again. This article is intended to summarize the most important results reported so far and for this reason will include some of those presented in the previous review. We have grouped these results according to the sub-families of the cuprate family, i.e. T-phase compounds, 123-compounds, etc.

2.   T-phase compounds

2.1   UNDOPED La_2CuO_4

Undoped La_2CuO_4 undergoes a structural phase transformation at $T_s \approx$ 525 K leading from a high temperature tetragonal (HTT) to a low temperature orthorhombic (LTO) phase caused by a staggered tilting of the oxygen octahedra around the Cu atoms. Attempts to unravel the phonon dispersion curves of La_2CuO_4 have largely been restricted to the HTT phase for two reasons: firstly, the LTO phase has two times the number of atoms in the unit cell than the HTT phase (14 vs. 7) and, therefore, has two times the number of phonon branches, which makes any determination of the LTO phonon dispersion a very difficult task. Secondly, cooling of La_2CuO_4 single crystals below T_s invariably leads to twinned samples, whereby the direction of the tilt of the octahedra differs by 90° between the domains. As a consequence, neutron spectra taken on such samples are often very difficult to interpret - a difficulty which had been underestimated in the beginning, and which only recently has been partially overcome (see 2.1.3).

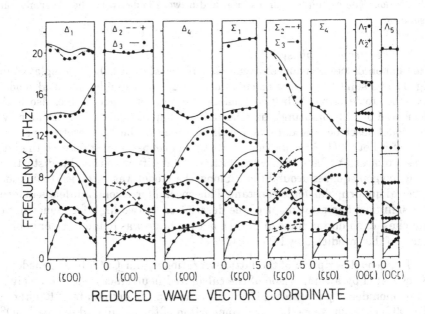

Figure 1 Phonon dispersion curves of La_2CuO_4. The solid lines were calculated from a interaction potential model developed for the lattice dynamics of several cuprates and the dash-dotted lines after inclusion of a special quadrupolar force constant (Chaplot et al., 1995). The data are from Pintschovius et al. (1991a,b) and from unpublished results of the same group.

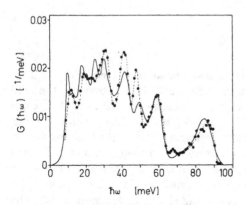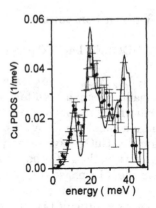

Figure 2 a) (left) Comparison of the neutron-cross-section weighted PDOS of $LaCuO_4$ calculated from a common interaction potential model (Chaplot et al., 1995) with the experimental data of Renker et al. (1992).

b) (right) Comparison of the partial PDOS of Cu calculated from the common interaction potential model with the data of Zemlyanov et al. (1993a) obtained by isotopic substitution. The calculated curves in a) and b) were folded with the experimental resolution.

2.1.1 *The tetragonal phase*

The phonon dispersion curves of the HTT phase of La_2CuO_4 are displayed in Fig. 1. The solid lines were calculated from an interatomic potential model developed to describe the lattice dynamics not only of La_2CuO_4, but also of T'-phase and of 123-compounds as reported by Chaplot et al. (1995). Evidently, the model describes the data very well. In order to check theoretical predictions of Fil et al. (1992) that the interaction of copper d orbitals with the lattice should result in a tendency to a lattice instability with rhombic distortion of the oxygen squares around the Cu ions, the model was extended to include a special quadrupolar force constant. This special force constant leads indeed to a significant improvement of the description of two high frequency optic branches (dash-dotted lines in Fig. 1), but its value was found to be only one-fourth of that predicted by Fil et al. (1992).

The good description of the lattice vibrations of La_2CuO_4 by the model is corroborated by a comparison of the calculated neutron-cross-section weighted phonon density of states (PDOS) with the experimental data of Renker et al. (1992) (Fig. 2a) as well as by a comparison of the calculated partial PDOS of Cu with that determined by Zemlyanov et al. (1993a) by means of isotopic substitution (Fig. 2b). Further, we note that the model reproduces the experimentally observed intensities quite well, which means that not only the calculated frequencies, but also the calculated eigenvectors are confirmed by experiment - which is notoriously more difficult to achieve by phenomenological models. An example is shown in Fig. 3 for the so-called scissor mode, where the in-plane oxygens vibrate perpendicular to the Cu-O bonds. By symmetry. this X-point mode is allowed to mix with another one where the axial oxygens vibrate along c. The fact that the description of the observed intensities by the model is not perfect indicates that the admixture of the axial oxygen mode is somewhat underestimated by the model.

The very good understanding of the lattice dynamics of La_2CuO_4 evident from Figs. 1-3 is the fruit of efforts of many years. Fig. 4 is meant to give an idea what sort of problems had to be overcome. This figure shows a const.-Q scan performed at the reciprocal lattice point (3,3,0), where the model predicts the observation of a longitudinal optic (LO) and of a transverse optic (TO) mode in the energy range $v = 7$ to 12 THz. In experiment, only the TO-peak can be discerned as a broad hump around 10 THz. Simulating the scan based on a lattice dynamical model on the one hand and on the resolution function of the spectrometer on the other hand reveals that the LO-peak will be rather weak and severely broadened by the finite momentum resolution in conjunction with the very steep slope of polar modes in the vicinity of the zone center. In addition, all the peaks are severely broadened by anharmonic effects because of the high temperature needed to reach the HTT phase. Last, but not least, the experiments suffer from a high multiphonon background, again a consequence of the high measuring temperature.

In the beginning, it was tried to avoid the problems associated with high temperatures by performing measurements in the LTO phase (results were presented, e.g., by Pintschovius et al. (1989)). When it gradually became clear that LTO phase data are difficult to interpret extensive measurements were carried out above T_s. Unfortunately, not all the effects related to the reduced symmetry of the LTO phase die out completely above T_s which fits to the evidence of local symmetry breaking provided by XAFS measurements (Haskel et al., 1996), and atomic pair distribution function anylasis of pulsed neutron scattering data (Egami et al., 1987). An example of a persistent peak splitting is shown in Fig. 5a. Similar, but more pronounced effects were observed in doped samples, an example of which is shown in Fig. 5b.

2.1.2 The tetragonal-to-orthorhombic phase transformation

The first inelastic neutron scattering investigations of the tetragonal-to-orthorhombic phase transformation in La_2CuO_4 have been reported by Birgeneau et al. (1987). More detailed results published later by the same group (Böni et al., 1988) are displayed in Fig. 6. The dynamic aspects of the phase transition were interpreted as classic soft-phonon behavior at the $(\frac{1}{2},\frac{1}{2},0)$ zone boundary involving rotations of CuO_6 octahedra.

On cooling below T_s, the rotation of the octahedra becomes static, whereby the tilt occurs in the (110)- or in the (1-10)-direction of the tetragonal lattice. As a consequence, these two directions are inequivalent in the LTO-phase. The tilt of the octahedra is associated with slight changes of the lattice

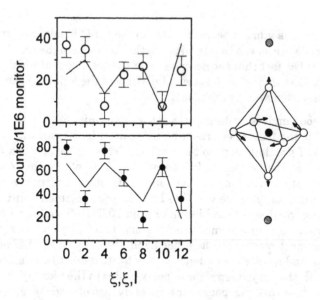

Figure 3 Calculated (lines) and observed (circles) intensities of the peaks observed at Q = (1.5,1.5,l) (top) and (2.5,2.5,l) (bottom), respectively, at an energy of $v = 15$ THz. The displacement pattern of the mode in question is indicated in the right hand part of the figure.

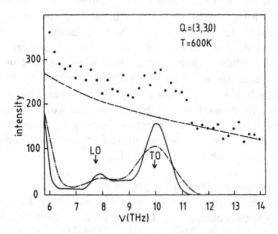

Figure 4 Constant-Q scan performed on La_2CuO_4 at T = 600 K (dots). The dashed line indicates the background level. The full line is the result of a simulation based on a lattice dynamical model and the resolution in energy and momentum transfer of the neutron spectrometer. The dash-dotted line was obtained from the full line by folding it with a Gaussian of FWHM = 1 THz to account for anharmonic broadening. The arrows denote the positions where the TO- and the LO-peak are expected in the absence of resolution effects.

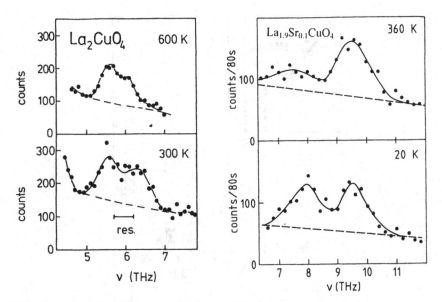

Figure 5 a) (left) Const-Q scans (Q = (0.4,0.9)) measured in La$_2$CuO$_4$ at above and below T$_s$. A peak splitting is expected only at below T$_s$.
b) (right) Const-Q scans (Q = (3.5,3.5,0)) measured in La$_{1.9}$Sr$_{0.1}$CuO$_4$ at above and below T$_s$. Only the peak at v = 10 THz is expected to persist above T$_s$.

parameters which cannot be explained on geometrical grounds, but have only a very moderate influence on the phonons. Below T$_s$, there are two tilt modes with rather different frequencies ω_L and ω_T, respectively: one in which the octahedra vibrate in the direction of the static tilt (called longitudinal mode) and one in which the octahedra vibrate perpendicular to the static tilt (called transverse mode). On cooling to T = 0, ω_L increases monotonically as expected, whereas ω_T first increases but then slightly decreases again indicative of a further incipient structural phase transition (Thurston et al., 1989). Such a further phase transition has indeed been found in the systems (La,Ba)$_2$CuO$_4$ (Axe et al., 1989) and (La,Nd,Sr)$_2$CuO$_4$ (Büchner et al., 1991). However. this transition leading to a low temperature tetragonal (LTT) phase is largely of first order and therefore does not involve a real soft mode behavior. The mode softening in orthorhombic La$_2$CuO$_4$ was successfully modelled within Ginzburg-Landau theory by Meissner and Blaschke (1992) (Fig. 7). We note that data for La$_{1.9}$Sr$_{0.1}$CuO$_4$ as published by Pintschovius (1989) have been incorporated into the figure after appropriate rescaling to account for the reduced T$_s$ = 310 K in the doped sample. The good description of all the data by the theory indicates that doping does not change the character of the transition apart from a reduction of T$_s$.

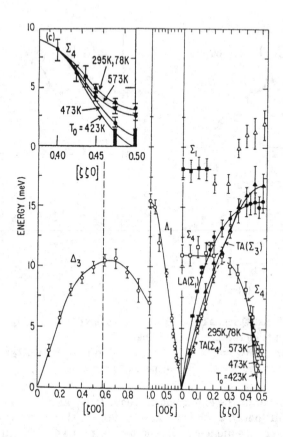

Figure 6 Summary of low-lying phonon branches in La$_2$CuO$_4$. The lines are a guide to the eye. The zone scheme and the labelling refers to the HTT phase. Only the TO phonons near the X-point (see inset) were found to exhibit a strong temperature dependence. The figure was taken from Birgeneau and Shirane (1989) showing results of Böni et al. (1988).

2.1.3   *The orthorhombic phase*

Early measurments of the PDOS of (La,Sr)$_2$CuO$_4$ samples far above and far below T$_s$ did not yield significantly different results, apart from some broadening at high temperatures attributable to anharmonic effects (Renker et al., 1987). On these grounds, it was reasonable to assume that the phase transition does not profoundly affect the lattice vibrations. In a certain sense, this conclusion has been confirmed by later findings, but the cross-sections seen in inelastic neutron scattering experiments on single crystals were found to exhibit sometimes very drastic changes on cooling below T$_s$. From various attempts to understand these changes on the basis of phenomenological models we conclude that both modifications of interatomic distances as well

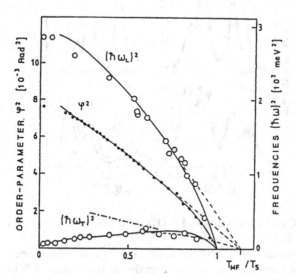

Figure 7 Temperature dependence T/T_s of the softmode frequencies ω_L and ω_T in La_2CuO_4 as well as of the order parameter ϕ. The lines are results of a calculation based on Ginzburg-Landau theory taking into account a tendency towards a LTO $\rightarrow$ LTT phase transformation (after Meissner and Blaschke, 1992).

as of interatomic force constants are of importance. The decisive step towards an appropriate model was the transition from a Born-van-Kármán (BvK) or force constant model towards a potential model, in which the distance dependence of the interatomic interactions is built in. Further important steps were the use of structural constraints and a fitting of the potential parameters to the phonon dispersions of several cuprates at a time. Results obtained in this way for the HTT phase were already given above. In Fig. 8, we show a result for the LTO phase demonstrating that rather complex spectra of a twinned sample can be very satisfactorily understood from the model and a knowledge of the resolution function of the spectrometer. We emphasize that the model is also very useful for the understanding of optical spectra observed in the LTO phase (Chaplot et al., 1996). We note that the model was not fitted to any LTO phase phonon data, but was derived from the common model solely on the basis of the known structural changes.

A close inspection of Fig. 8 reveals some systematic differences between the calculated and the observed spectra in the HTT phase: the side peaks appearing at positions of transverse acoustic phonons are larger than expected from resolution effects. Since these peaks are massively enhanced in the LTO phase, we presume that their oversize in the HTT phase is due to a local symmetry breaking as discussed at the end of section 2.1.1.

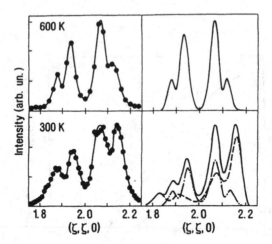

Figure 8 (left) Const-v scans at v = 1.5 THz along the [110]-direction at T above and below T_s = 520 K.
(right) Simulated scans using realistic values of the resolution in Q and v. The dashed and dotted lines correspond to the two domains of the twinned LTO sample (after Chaplot et al., 1996).

2.2 $La_{2-x}Sr_xCuO_4$

A systematic study of the phonons in $La_{2-x}Sr_xCuO_4$ has been conducted by Pintschovius et al. (1989, and unpublished results) for $x = 0.1$. Further studies of selected aspects have been reported by several groups on lightly doped and in particular on optimally doped samples. They will be referenced in the following sections.

In general, the phonons in doped samples were found to be very similar to those in undoped samples. A part of the differences is associated with the metal-insulator transition upon doping, i.e. due to screening by free carriers in doped samples (see section 2.2.1). Other differences are very likely fingerprints of a strong electron-phonon coupling in superconducting $La_{2-x}Sr_xCuO_4$ (see section 2.2.2).

2.2.1 *Screening effects*

The phonon dispersion of undoped La_2CuO_4 shows large LO-TO (or Lyddane-Sachs-Teller)-splittings typical of ionic insulators. In metallic samples, these splittings disappear as a consequence of the screening by free carriers. Examples of screening-induced changes in the phonon dispersion of $La_{1.9}Sr_{0.1}CuO_4$ are shown in Fig. 9.

The screenings effects were found to be essentially confined to small q's. This is evident from the steep dispersion of branches starting at LO-frequen-

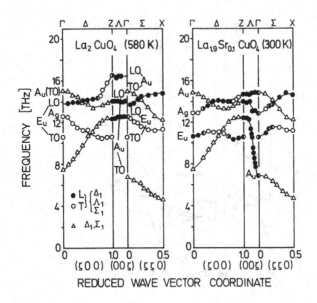

Figure 9 Dispersion of selected branches in (La,Sr)$_2$CuO$_4$ showing the effect of screening by free carriers on polar modes. Open and full symbols denote a predominant polarization in the basal plane or along c, respectively.

cies and approaching the frequencies of undoped La$_2$CuO$_4$ at q ≈ 0.2 Å^{-1}. The steep dispersion is most obvious for the Λ_1-branch starting at the A$_u$-frequency of v = 7.5 THz. For the branches polarized in the basal plane (of Δ_1- or Σ_1-symmetry) the steep dispersion is somewhat masked by an anticrossing with branches of the same symmetry, but polarized along c. The full and open symbols in Fig. 9 are meant to help the reader in following the branches from the zone center to the zone boundary. A confinement of screening effects to the small-q region means that La$_{1.9}$Sr$_{0.1}$CuO$_4$ is not a good metal or, in other words, that the bonding is largely ionic in nature as in undoped La$_2$CuO$_4$. Recent investigations of La$_{1.85}$Sr$_{0.15}$CuO$_4$ (Pintschovius et al., 1997) revealed an only slight increase of the q-range affected by screening when compared to that of La$_{1.9}$Sr$_{0.1}$CuO$_4$, which means that the bonding remains ionic in nature even in optimally doped samples.

   In view of the findings just mentioned, it was natural to attempt a description of the phonon dispersion in (La,Sr)$_2$CuO$_4$ on the basis of the model developed for undoped La$_2$CuO$_4$ suitably modified to take into account screening by free carriers. Fig. 10 demonstrates that this attempt is fairly successful. What such a model cannot reproduce is the pronounced softening in the highest Δ_1-branch around q = 0.5. As will be discussed in the next section, this softening is indicative of a strong electron-phonon coupling.

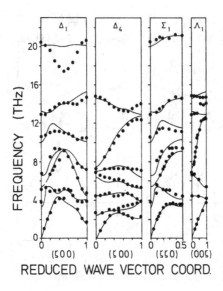

Figure 10 Dispersion of selected phonon branches of La$_{1.9}$Sr$_{0.1}$CuO$_4$. The dots denote inelastic neutron scattering results of Pintschovius et al. (1991, 1997). The lines were calculated from the common interaction potential including a term representing screening of the ionic forces by free carriers (after Chaplot et al. (1995)).

The model calculations shown in Fig. 10 were based on an ansatz for isotropic screening. Attempts were made to see whether anisotropy of the screening is needed: it turned out that anisotropic screening does not lead to substantial improvements of the fit quality. This is rather surprising in view of the well-known fact that the conductivity is much higher in the a-b plane than along c. The answer to the question whether or not one should expect screening effects in c-axis polarized modes depends on the frequency of the c-axis plasmon in comparison to the phonon frequency: as has been discussed by Falter et al. (1993), the phonon dispersion of c-axis polarized modes will look like that of an insulator or a metal, respectively, if the frequency ω_p of the c-axis plasmon ω_p is very small or very large compared to that of the phonon frequencies. In case that the two excitations have about the same frequency, a mixing will occur (Falter et al. 1994, 1995). Widely different values of ω_p have been deduced from optical data (Kim et al. (1995), Henn et al. (1996)) so that it remains unclear from these results whether c-axis polarized phonons are expected to show metallic or insulating behavior. The problem to determine ω_p by optical spectroscopy is probably related to the very strong damping of the c-axis plasmon observed in all experiments.

Recently, an attempt was made to clarify the situation by detailed inelastic neutron scattering experiments on $La_{1.85}Sr_{0.15}CuO_4$ (Pintschovius et al., 1997). In such an experiment, a severe complication arises from the finite resolution in momentum transfer: as has been pointed out by Falter et al. (1993), the plasmon energy will increase very rapidly on going away from the c-axis. As a consequence, screening will become much more effective for these off-axis phonons, and these phonons might easily dominate the neutron-scattering cross section in an experiment probing small-q phonons. Therefore,

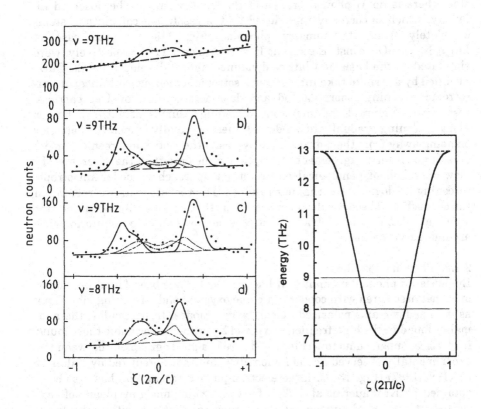

Figure 11 Constant-energy scans across the (0,0,12)-reciprocal lattice point in La_2CuO_4 (a) and $La_{1.85}Sr_{0.15}CuO_4$ (b-d). Scans a), c) and d) were performed with standard resolution in momentum transfer and scan b) with improved resolution. The lines represent results of model calculations with isotropic and anisotropic screening as is described in the text folded with the experimental resolution. The phonon dispersion in metallic $La_{1.85}Sr_{0.15}CuO_4$ and insulating La_2CuO_4 is depicted in the right hand part of the figure as thick full line and as dashed line, respectively. The scan paths are depicted by thin lines.

the c-axis phonon dispersion might look metallic in spite of insulating behavior of phonons propagating strictly along c. In order to gain information on the severity of resolution effects, measurements have been carried out with standard and with improved Q-resolution, respectively. Further, the experiments have been simulated using different functions to describe a two-dimensional, an anisotropic three-dimensional or an isotropic three-dimensional screening, respectively.

Illustrative results are depicted in Fig. 11. In the case of undoped La_2CuO_4, one would naively expect that a const-E scan at E = 9 THz across the (0,0,12)-reciprocal lattice point should give only background scattering, since there is no λ_1-phonon branch in this energy range. The observed intensity, which is correctly reproduced by the resolution calculation, stems completely from off-symmetry phonons (Fig. 11a). In the case of $La_{1.85}Sr_{0.15}CuO_4$, const.-E scans at E = 8 THz or 9 THz, respectively, were simulated on the basis of a lattice dynamical model developed for La_2CuO_4 modified by a term to take into account isotropic screening (full lines) or anisotropic screening (short dashed and dash-dotted lines) by free carriers, respectively. The model of anisotropic screening assumes insulating behavior along c, turning gradually (dash-dotted lines) or rapidly to metallic behavior on going away from the c-direction. Obviously, the isotropic screening model gives a much better agreement description of the data than the other ones. In view of the phenomenological nature of our approach to model anisotropic screening we do not claim that our results are a proof of metallic screening in the c-direction. These results show, however, that non-metallic behavior, if it occurs at all, has to be restricted to a very narrow region in reciprocal space around the c-direction.

2.2.2 Phonon anomalies
Evidence for phonon anomalies in $La_{2-x}Sr_xCuO_4$ have been found in phonon branches associated with copper - in plane oxygen bond stretching vibrations as well as for c-axis polarized apical oxygen modes. In particular, the anomalies found in the high frequency Δ_1- and Σ_1-branches resemble those found in classical superconductors (Fig. 12). There is a good agreement between the experimentally observed phonon anomalies and those predicted by Falter et al. (1993,1997) (Fig. 13). A large electron-phonon interaction has also been obtained by Krakauer et al. (1993) for the X-point mode by means of self-consistent linearized-augmented-plane wave calculations within the local-density-functional approximation (LDA). This group did not study the (0.5,0,0)-mode, but Rojeswki (1994) found on the basis of similar calculations a frequency for this mode which is in very good agreement with experiment (16.8 THz).

The anomalous nature of the phonons around (0.5,0,0) and (0.5,0.5,0) is corroborated by a pronounced line-broadening (Fig. 12, bottom). A pronounced broadening of the (0.5,0,0)-mode has also been reported by

McQueeney et al. (1996). This group claims that the line shape is not only very wide but also asymmetric, which cannot be explained by a normal electron-phonon coupling (Fig. 14). However, a recent investigation of Pintschovius et al. (unpublished) using a better energy resolution did not yield any evidence for an asymmetric line shape of the phonon in question (Fig. 16). We emphasize that the total linewidth - corrected for the instrumental resolution - obtained in this experiment is somewhat smaller than that found by McQueeney et al. Moreover, the center of mass of the phonon line is shifted to lower energies by ~1 meV. In total, the shape of the phonon line seen in the two experiments agrees on the low energy side, but not on the high energy one. This discrepancy might have a simple explanation. i.e. a macroscopic inhomogeneous Sr distribution in the McQueeney sample. As is evident from Fig. 12, underdoped regions will lead to a high energy tail of the phonon line. As will be discussed later, overdoping will not lead to a further softening of the (0.5,0,0)-phonon, and hence overdoped regions will not lead to a low energy tail of the phonon line, so that the resultant line shape will be

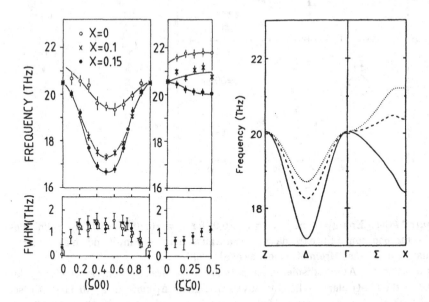

Figure 12 (left) Top: Dispersion of the topmost Δ_1- and Σ_1-branches in La$_{2-x}$Sr$_x$CuO$_4$. Lines are a guide to the eye.
Bottom: Resolution-corrected linewidths observed in La$_{1.85}$Sr$_{.15}$CuO$_4$. Full and open dots refer to measurements with different monochromators (after Pintschovius et al. (1992, 1997).
Figure 13 (right) Calculated results for the branches shown in Figure 12 using a microscopic theory (Falter et al., 1997). The broken and dotted curves represent calculations where terms representing the Cu on-site Coulomb interactions have been increased by a factor 1.5 and 2 as compared to their calculated values (full curves).

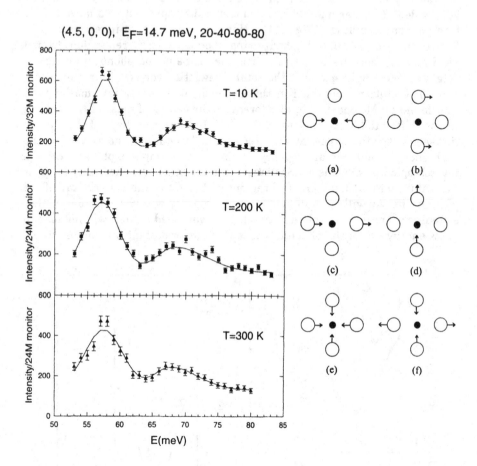

Figure 14 (left) Energy scans at Q = (4.5,0,0) for T = 10,200 and 300 K. Solid lines were obtained from a fit to the data using a Gaussian at ~ 57 meV and an asymmetric Gaussian at 69 meV (from Mc Queeney et al., 1996).

Figure 15 (right) Atomic displacement patterns for several high-energy phonons polarized in the Cu-O plane (schematically). (a) Highest Δ_1-mode at q = (0.5,0,0); (b) second highest Δ_1-mode at q = (0.5,0,0); (c) higest Δ_1-mode at q = (0,0,0) or (1,0,0); (d) highest Δ_3-mode at q = (0.5,0,0); (e) highest Σ_1-mode at q = (0.5,0.5,0) (in-plane breathing mode); (f) highest Σ_3-mode at q = (0.5,0.5,0) (quadrupolar mode). Solid circles represent copper and empty circles represent oxygen (adapted from Mc Queeney et al.. 1996).

asymmetric - in contrast to what has been discussed by McQueeney et al. (1996). In this context we would like to add that the very large linewidth of the (0.5,0,0)-phonon seen in a $La_{1.9}Sr_{0.1}CuO_4$-single crystal (Pintschovius and Reichardt, 1994) - larger than observed in $La_{1.85}Sr_{0.15}CuO_4$ - is very

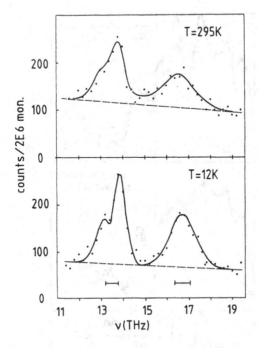

Figure 16 Sum of const.-Q scans taken at $Q=(4.5,0,0)$ and $Q=(4.5,0,1)$. The data are averages of two runs. The full line for the 12K-scan was obtained by fitting three Gaussians plus a sloping background as indicated by the broken line. The peaks at $v=16.7$ THz, 13.9 THz and 13.1 THz correspond to $\Delta_1(1)$-, $\Delta_1(2)$- and $\Delta_4(1)$-phonons, respectively, whereby the $\Delta_4(1)$-peak appears only at $Q=(4.5,0,1)$. The line for the 295K-scan was obtained by modifying the 12K-parameters in the following way: (i) the frequencies were shifted downwards by 1 %, (ii) the lines were folded with a Lorentzian of width 0.35 THz and (iii) the intensities were reduced by 17 % to account for the increase of the Debye-Waller factor.

likely not intrinsic, but partly due to an inhomogeneous Sr distribution as evident from the superconducting properties of the sample.

The phonon softening upon doping evident from Fig. 12 is reflected in the PDOS (Renker et al., 1987,1992) as a slight shift of the high-energy cut-off to lower energies (related to the softening of the breathing mode) and the appearance of a peak at E = 70 meV (related to the softening of the Δ_1 (0.5,0,0)-phonon) (Fig. 17a). Further, there is a shift of the peak at E ~ 60 meV to lower energies by ~ 2 meV related to a softening of the "scissor"-mode (for the displacement pattern of this mode, see Fig. 3). Little change has been observed in the PDOS above 50 meV after increasing the doping level to x = 0.3 (Fig. 17b), which indicates that overdoping does not lead to an enhancement of the phonon anomalies observed for x = 0.15. A further increase of the doping level to x = 0.32 even lead to a slight hardening of the PDOS (Fig.

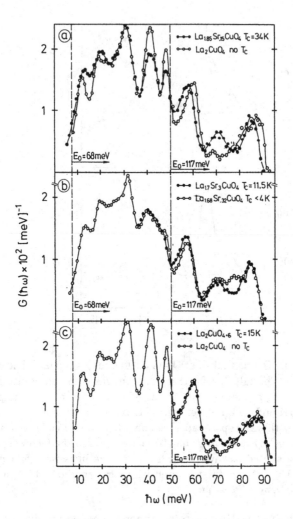

Figure 17 Neutron-cross-section weighted PDOS of $La_{2-x}Sr_xCuO_{4-\delta}$ for various doping levels as indicated in the figure (from Renker et al., 1992).

17b). In this context we note that substitution of 2% Cu by Zn in $La_{1.85}Sr_{.15}CuO_4$ - which makes the samples non-superconducting - has only a very small influence on the PDOS (Arai et al., 1991).

Falter et al. (1994,1995) predicted a strong softening of the symmetrical apical oxygen breathing mode at the Z-point (labelled O_Z^Z-mode, the displacement pattern of which is depicted in Fig. 18). Concurrently, Krakauer et al. (1993) calculated a large electron-phonon coupling for this mode. On the experimental side, it took quite some time until the O_Z^Z-mode could be pinned down because of a massive line-broadening. A strong broadening was observed even in insulating La_2CuO_4, but only at high temperatures, whereas

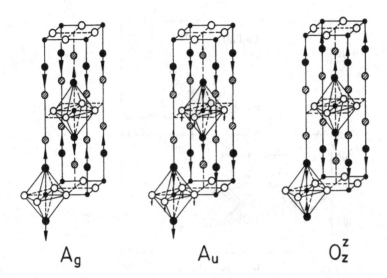

Figure 18 Displacement pattern of c-axis polarized apical oxygen modes in (La,Sr)$_2$CuO$_4$

in superconducting La$_{2-x}$Sr$_x$CuO$_4$ the broadening persists to low temperatures. For x = 0.15, the broadening is about 4 THz (FWHM) and the softening is as large as 30 % ($\sim$ 5.5 THz, see Fig. 19). Falter et al. (1995) argue that the softening is particularly strong because of a coupling of the O$_Z$^{Z}-mode to the c-axis plasmon. When studying the Λ_1-branch connecting to the O$_Z$^{Z}-mode (see the preceeding section) Pintschovius et al. (1997) searched for fingerprints of the predicted phonon-plasmon mixing: the results were inconclusive.

As was discussed in section 2.1, a special quadrupolar force constant has to be added to the common interaction potential model to arrive at a fully satisfactory description of the quadrupolar X-point mode in La$_2$CuO$_4$ (the displacement pattern is depicted in Fig. 15f). Theory (Fil et al., 1992) predicted the need for such a force constant - even a much larger one than actually required - because of a coupling of the quadrupolar mode to electronic excitations. The theory was made for insulating La$_2$CuO$_4$, so that the electron-phonon coupling dealt with in the theory has nothing to do with an electron-phonon coupling in the classical sense as related to superconductivity. Indeed, doping with Sr does not lead to a further softening of the quadrupolar mode, rather to a slight hardening.

2.2.3 *Soft-phonon behavior*

It is well known that doping La$_2$CuO$_4$ with Sr strongly reduces the HTT-to-LTO phase transition temperature T$_s$. However, as has been mentioned in Section 2.1.2, the available data for the soft-phonon behavior of doped and undoped La$_2$CuO$_4$ reveal a great similarity after allowance is made for the reduction of T$_s$. Notwithstanding this similarity, several groups spent a great

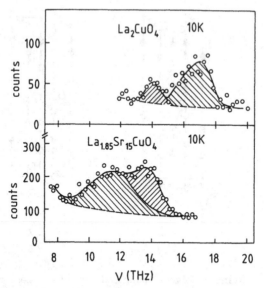

Figure 19 Const-Q scans at Q=(0,0,15) in La_2CuO_4 and $La_{1.85}Sr_{.15}CuO_4$. The differently hatched areas correspond to the contributions of the $O_z{}^Z$-mode (La_2CuO_4: $\nu \sim 17$ THz, $La_{1.85}Sr_{.15}CuO_4$: $\nu \sim 11$ THz) and another Z-point mode ($\nu \sim 14$ THz), respectively. The width of the $O_z{}^Z$-peak expected from the resolution in Q and E is $\Delta\nu = 2$ THz and 1 THz (FWHM) for La_2CuO_4 and $La_{1.85}Sr_{.15}CuO_4$, respectively.

deal of effort to investigate the soft optical phonons in $(La,Sr)_2CuO_4$ in some detail (Böni et al., 1988; Thurston et al., 1989; Pintschovius, 1990; Braden et al., 1994; Lee et al., 1996). These studies were - at least partly - motivated by widely-held ideas that the soft phonons might be important for high-T_c superconductivity.

Böni et al. (1988) took the view that the rotational nature of the soft mode leads to moderate electron-phonon coupling and hence this mode is unlikely to enhance significantly conventional phonon mediated superconductivity. In a subsequent study of the same group (Thurston et al., 1989), this view was reiterated, but the authors admitted that structural effects may still play a secondary role in the superconductivity. A theoretical investigation of the relationship between the lattice instabilities (from HTT-to-LTO and LTO-to-LTT), the electronic structure and high-T_c superconductivity by Pickett et al. (1991) conforted those who were of opinion that the lattice instability is of primary importance for the superconducting properties. From local-density-functional calculations they concluded that the effect of the LTT distortion on the electronic structure accounts for the suppression of T_c in the LTT phase, while precursor effects of the LTO-to-LTT phase transition in the LTO phase will enhance T_c related to strong anharmonicity. Such a view motivated Braden et al. (1994) to perform a rather comprehensive study of the HTT-to-LTO phase transition, from which Fig. 20 was taken. They found some

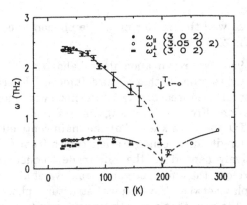

Figure 20 Soft mode frequencies in $La_{1.85}Sr_{0.13}CuO_4$ as a function of temperature. Because of a strong damping of the soft-phonon observed at $Q=(3,0,2)$, additional measurements have been performed slightly away from the superlattice peak position, i.e. at $Q=(3.05,0,2)$. The lines are a guide to the eye (from Braden et al. (1994)).

evidence that doping enhances the instability against the LTO-to-HTT phase transition not only in $(La,Ba)_2CuO_4$, but also in $(La,Sr)_2CuO_4$. The first direct evidence for an interplay between the lattice instability and superconductivity in $La_{2-x}Sr_xCuO_4$ was recently presented by Lee et al. (1996): the phonon softening related to the LTO-HTT phase transition appeared to be frozen in at

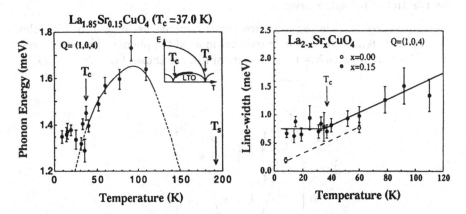

Figure 21 Left: Soft phonon energies as a function of temperature in superconducting $La_{1.85}Sr_{0.15}CuO_4$. The solid line is a guide to the eye. The dashed line is drawn by assuming a structural phase transition to the LTT phase. The temperature dependence of the soft phonon energy predicted by Landau theory is schematically drawn in the inset where the temperature region in the present experiment is indicated by a bold line.

Right: Soft phonon line-wdith as a function of temperature in superconducting $La_{1.85}Sr_{0.15}CuO_4$ and La_2CuO_4. Both solid and broken lines are guides to the eye. From Lee et al. (1996).

T_c (Fig. 21a, left) and as well the narrowing of the phonon line-width upon cooling stopped at T_c (Fig. 21b, right).

Arakawa et al. (1996) report an anomalous behavior of a transverse acoustic (TA) phonon mode similar to that of the soft mode. The scatter of the data is, however, rather large. Moreover, the TA phonons were measured in a longitudinal configuration. From an investigation of inelastic scattering cross-sections in the LTO-phase we know that the main contribution to TA-peaks measured in a longitudinal configuration comes from off-symmetry phonons which sensitively depend on the soft mode frequency (see, e.g., Figure 8 (below)). Therefore, the effects seen by Arakawa et al. (1996) might again reflect the interplay between the incipient structural phase transition and superconductivity. We note that a search for superconductivity-induced effects in TA phonons of $La_{1.85}Sr_{0.15}CuO_4$ undertaken by Chou et al. (1990) revealed only a null result.

2.3 $La_2CuO_{4+\delta}$

Since La_2CuO_4 can be made superconducting not only by replacing a part of La by Sr but also by introducing a small amount of extra-oxygen, it was natural to conduct an investigation to look for the effect of the extra-oxygen on the phonons. Serious complications arise, however, not only from the HTT to LTO phase transformation but, more importantly, from an inhomogenous distribution of the extra-oxygen.

In a series of room temperature measurements performed by Pintschovius and Braden (1996) it was found that most of the phonons in a single crystal of $La_2CuO_{4-\delta}$ with $\delta \approx 0.03$ are very similar to those of $La_{1.9}Sr_{0.1}CuO_4$. A

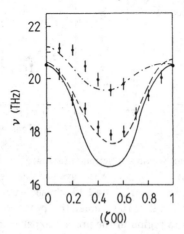

Figure 22 Peak positions observed on $La_2CuO_{4+\delta}$ at T=12 K. The dash-dotted, full and dotted lines refer to the phonon dispersion of $La_{2-x}Sr_xCuO_4$ with x=0.0.1 and 01.15, respectively. From Pintschovius and Braden (1996)

special effort was made to study the highest Δ_1-branch which shows a pronounced phonon softening upon doping with Sr (Fig. 12). Room temperature measurements yielded very broad intensity distributions, whereas in low temperature measurements a two peak structure was observed: one series of peaks corresponds to that of $La_{2-x}Sr_xCuO_4$ with $x \approx 0.1$, the other one to that of undoped La_2CuO_4 (Fig. 22). This finding is presumably associated with the well-known phase separation of $La_2CuO_{4-\delta}$ (Statt et al., 1995) into domains with $\delta \sim 0.01$ and $\delta \sim 0.06$ which are insulating and superconducting with T_c = 32 K, respectively. This indicates that doping with extra-oxygen produces very similar phonon anomalies as doping with Sr. This conclusion is somewhat corroborated by oxygen-doping induced effects in the PDOS observed by Renker et al. (1992) (see Fig. 17c). Certainly, the effects are considerably smaller than those found for $La_{1.85}Sr_{0.15}CuO_4$, but on the other hand, the low T_c of the sample indicates that the extra-oxygen content was quite small.

2.4 $La_{1.6-x}Nd_{0.4}Sr_xCuO_4$

Recently, the system $La_{1.6-x}Nd_{0.4}Sr_xCuO_4$ has attracted considerable interest because of the observation of superlattice peaks which were interpreted as evidence of stripe-phase order of holes and spins (Tranquada et al., 1996,1997). A necessary ingredient for the appearance of these superlattice peaks seems to be the occurrence of the LTO-HTT phase transition, which is the case for samples in which part of the La is replaced by Sr (Büchner et al., 1991). A widely held hypothesis is that dynamic stripe correlations are a general property of cuprate superconductors (see, e.g., Zaanen et al. (1996)),

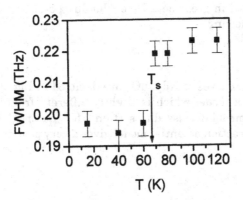

Figure 23 Temperature dependence of the linewidth of a transverse acoustic phonon in $La_{1.48}Nd_{0.4}Sr_{0.12}CuO_4$.

and that the transition to the LTT-phase is important only in providing a pinning potential to make the correlations static.

An abrupt increase of the phonon heat transport observed at the LTO $\rightarrow$ LTT transition in $La_{1.88-y}Nd_ySr_{0.12}CuO_4$ was discussed by Büchner et al. (1996) in relationship to the presence of stripe correlations of holes and spins. These results motivated a recent inelastic neutron scattering study of Braden et al. (1997) of $La_{1.48}Nd_{0.4}Sr_{0.12}CuO_4$. Preliminary results show an abrupt decrease of a phonon linewidth at the LTO $\rightarrow$ LTT transition (Fig. 23), which fits well to the increase of the phonon heat transport. However, little can be said about whether the line narrowing has anything to do with a freezing of dynamical stripe correlations. We note that the line narrowing sets in abruptly at T_s - as does the increase of the phonon heat transport - whereas the appearance of the superlattice peaks is sluggish.

3.   T'-phase compounds

The phonons of T'-phase compounds have been investigated much less intensively than those of T-phase compounds. The major reason for this state of affairs is the lack of large superconducting single crystals which strongly hampers any attempt to establish a link between the phonons and high T_c superconductivity in this class of compounds. A further complication arises from the very strong magnetic scattering contributions associated with crystalline electric field (CEF) transitions of the rare earth ions which makes it very difficult to extract information on the phonon spectrum from powder data, so that the lack of superconducting single crystals is even more serious. Therefore, nearly all phonon studies on T'-phase compounds reported so far only elucidate the lattice dynamical aspect, but do not shed any light on the coupling between electrons and phonons. The available results for the two compounds investigated in great detail, i.e. Nd_2CuO_4 and Pr_2CuO_4, will be summarized in the following.

3.1 Nd_2CuO_4

The phonon dispersion curves of Nd_2CuO_4 are depicted in Fig. 24. The lines were calculated from a model which is slightly different from the common interaction potential model discussed in section 2 for $(La,Sr)_2CuO_4$. It is not that this common interaction potential model gives a very poor descrip-

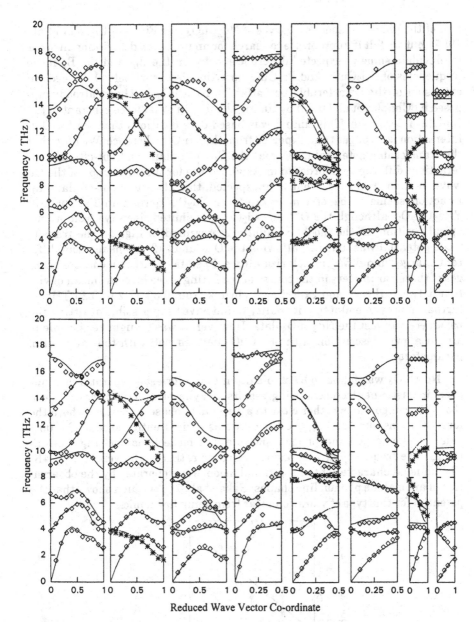

Figure 24 Phonon dispersion curves of Nd_2CuO_4 (top) and Pr_2CuO_4 (bottom) after Pintschovius et al. (1991) and Pyka et al. (1998). The arrangement of the panels is as follows (from the left to the right): Δ_1; Δ_2 (stars) and Δ_3 (diamonds); Δ_4; Σ_1; Σ_2 (stars) and Σ_3 (diamonds); Σ_4; Λ_1 (diamonds) and Λ_2 (stars); Λ_3. The lines were calculated from an interaction potential model with common parameters for Nd_2CuO_4 and Pr_2CuO_4 (Reichardt, 1997a).

tion of the phonon dispersion curves of Nd_2CuO_4 (see Fig. 4 of Chaplot et al., 1995), but we felt it appropriate to search for an improved description in order to elucidate some unexpected differences between the B_u- and the B_g-mode frequencies of Nd_2CuO_4 and Pr_2CuO_4 (see the discussion below). Indeed, an extension of the transferable potential for Nd_2CuO_4 had been already proposed by Chaplot et al. (1995) by tuning the longitudinal force constant of the Cu-O(2) interaction. This tuning was based on indications that the Cu-O(2) interaction is stronger than expected from the Cu-O(2) distance. We note that the transferable potential does treat the Cu-O in-plane and out-of-plane interactions differently in order to make allowance for the anisotropy of the Cu wavefunction. However, what is unexpected, that the Cu-O out-of-plane interaction seems to be of longer range in Nd_2CuO_4 than in La_2CuO_4 or $YBa_2Cu_3O_7$, although the O(2) in Nd_2CuO_4 is shifted sideways from a true apical position. In the model calculations presented in Fig. 24, the transferable potential was tuned by allowing the O(1) and O(2) atoms to adopt different charges and different polarizabilities. In spite of rather moderate changes of these parameters in the course of the fitting procedure, the mean deviation between theory and experiment shrunk from 0.35 THz to 0.25 THz. A further improvement of the fit quality would have been possible by fitting the parameters to just the Nd_2CuO_4 data. However, it was thought reasonable to aim at a good description of the Pr_2CuO_4 data as well with the same set of parameters.

Nd_2CuO_4 was found to be stable down to the lowest temperatures. However, inelastic-neutron-scattering results of Pyka et al. (1992) have shown that this compound is rather close to a structural phase transition where the four O(1) ions surrounding the Cu rotate as a rigid entity around the 001-axis. The corresponding phonon shows a soft mode behavior (Fig. 25), although the frequency of this phonon remains rather high even at T = 4K. The unusual character of the rotational mode is corroborated by the observation of anomalous phonon line shapes (Pyka et al., 1992) which shows that the potential for this type of vibration is far from being harmonic.

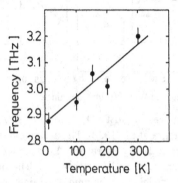

Figure 25 Temperature dependence of the frequency of the rotational mode in Nd_2CuO_4 (after Pyka et al., 1992).

From model calculations we learned that the low frequency of the rotational mode is an inherent feature of the T'-structure and that the frequency of
this mode depends very sensitively on the radius of the rare earth ion. Replacing Nd by Pr makes this mode - which is the endpoint of the lowest Σ_3-
branch - more stable (see Fig. 24). On the other hand, replacing Nd by Gd
makes this mode completely unstable , as Gd is a slightly smaller ion. As a
consequence, Gd_2CuO_4 shows a superstructure at room temperature (Braden
et al., 1994b). Unfortunately, an attempt to study the soft mode behavior in
Gd_2CuO_4 was unsuccessful because of too strong absorption in spite of using a
sample containing mostly the non-absorbing [158]Gd isotope (Braden and
Reichardt, 1996).

Attempts have been made to determine the PDOS of doped (Lynn et al.,
1991) and undoped (Renker, 1993) Nd_2CuO_4. However, as has been discussed
by Pintschovius and Reichardt (1994), little can be learned about the phonons
from these results, as the measured spectrum is dominated by magnetic contributions. The calculated phonon spectrum was presented in Fig. 14 of the
article of Pintschovius and Reichardt (1994).

3.2 Pr_2CuO_4

The phonon dispersion curves of Pr_2CuO_4 are displayed in the lower part of
Fig. 24. Evidently, they are very similar to those of Nd_2CuO_4. The most
noticeable differences concern the branches connecting to the B_{1g} and B_{1u}
zone center modes. For the sake of clarity, they are depicted in a separate
figure (Fig. 26). In addition to the unexpected frequency differences between
Nd_2CuO_4 and Pr_2CuO_4, the behavior of these modes as a function of temperature is anomalous: in Pr_2CuO_4, the B_{1g} and the B_{1u} modes (and related ones
throughout the Brillouin zone) soften appreciably on cooling, whereas in
Nd_2CuO_4, these modes harden more than expected from the usual behavior of
other modes associated with the thermal contraction.

The elongation pattern of these modes involves out-of-plane displacements of the oxygen atoms leading to a warping of the Cu-O planes or the
Nd,Pr-O planes or both (see Pintschovius and Reichardt 1994, Fig. 26). The
reason for the unusual behavior of these modes remains unclear. Since the
lattice parameters, the atomic masses and the ionic radii of Nd_2CuO_4 and
Pr_2CuO_4 are only slightly different, the common interaction potential model
of Chaplot et al. (1995) discussed previously is unable to reproduce the observed differences between the frequencies of the B_{1g}- and B_{1u}-modes in
Nd_2CuO_4 and Pr_2CuO_4. Model caluclations do not even allow us to answer
the question in which of the compounds the behavior of the modes in question
has to be considered as "normal" and in which one it has to be considered as
anomalous: as can be seen from Fig. 24, a model fitted to the wealth of data
for both Nd_2CuO_4 and Pr_2CuO_4 does not lead to a fully satisfactory descrip-

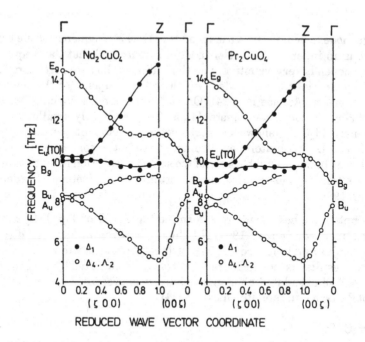

Figure 26 Selected phonon branches in Nd_2CuO_4 and Pr_2CuO_4. Lines are a guide to the eye.

tion of the phonon branches in question for either compound, but rather predicts an intermediate behavior.

It was said above that it is difficult to extract information on the phonon spectrum from powder data because of very strong magnetic contributions to the scattering. Recently, however, the isotope substitution technique was successfully used to extract the partial PDOS of Cu in Pr_2CuO_4 as well as in $Pr_{1.85}Ce_{0.15}CuO_4$ (Parshin et al., 1997). We note that determination of a partial PDOS requires very precise measurements already in the case of non-magnetic compounds, because the scattering contrast between different Cu isotopes is only moderate. In the case of Pr_2CuO_4, however, the project of determining the partial PDOS of Cu was even more ambitious, as not only the contributions of Pr- and O-vibrations, but also the magnetic contributions had to be eliminated. As can be seen from Fig. 26 (top, left), a comparison of the calculated and the experimental Cu-PDOS for the undoped compound shows quite good agreement. The data for $Pr_{1.85}Ce_{0.15}CuO_4$ indicate a softening of the Cu-spectrum for energies $E < 20$ meV (Fig. 25 top, right). As a complement, we also show the total PDOS of undoped Pr_2CuO_4 calculated from the model of Fig. 25, bottom.

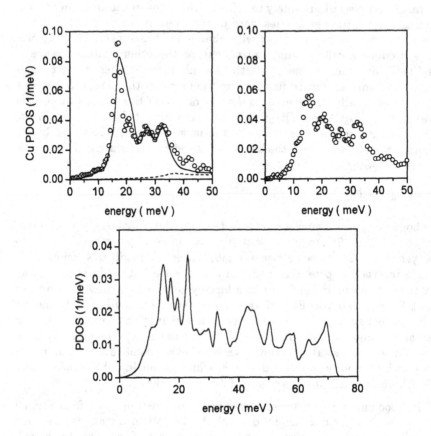

Figure 27 top: partial PDOS of Cu as determined experimentally by an isotope substitution technique by Parshin et al. (1997) for Pr_2CuO_4 (left) and $Pr_{1.85}Ce_{0.15}CuO_4$ (right), respectively. The full line is the result of a simulation based on an interaction potential model and the instrumental parameters. The dashed lines denotes the multiphonon contribution.

bottom: total PDOS of Pr_2CuO_4 as calculated from an interaction potential model and folded with a Gaussian of 1 meV (FWHM), which corresponds to the mean deviation between the calculated and the experimental phonon dispersion curves (Reichardt, 1997a).

4.   123-compounds

Soon after the discovery of high T_c-superconductivity in the system $YBa_2Cu_3O_{7-x}$ it was found that the phonon spectrum changes remarkably as a function of oxygen content (Renker et al., 1988). Starting from the idea that

the strong changes in the phonon spectrum cannot be explained solely by structural changes but are likely to reflect differences in the electron-phonon coupling, several further studies were undertaken in which changes in the phonon spectrum were correlated with changes of the superconducting properties brought about by substituting one or the other atomic species of $YBa_2Cu_3O_7$ by other atoms. Nevertheless, all these studies using powder samples lead only to limited insight into the origin of the observed changes. Progress was finally achieved by a determination of the phonon dispersion curves of $YBa_2Cu_3O_6$ and $YBa_2Cu_3O_7$, whereby even today the O_7-data are far from being complete due to the lack of large untwinned samples. In the following, we will start with the best understood representative of the series which is $YBa_2Cu_3O_6$.

4.1 $YBa_2Cu_3O_6$

The phonon dispersion curves as calculated from the transferable potential of Chaplot et al. (1995) together with the available experimental data are displayed in Fig. 28. To be precise, the model used is a modified version of the common interaction potential model in that the $Cu(1)$-$O(4)$ interaction was not treated as an ionic bond, but as a largely covalent one described by two special force constants. Fits of the model parameters to the O_6-phonon dispersion had shown that there is very little charge on the $Cu(1)$-ion, and hence any attempt to describe the $Cu(1)$-$O(4)$ bond as a predominantly ionic one would be unphysical. The view of the $Cu(1)$-$O(4)$ bond as a covalent one is supported by the very short bond length, which is considerably shorter than in O_7, where the $Cu(1)$ ion has regained its normal charge.

The good agreement between model and experiment obvious from Fig. 28 indicates that a basic understanding of the O_6-lattice dynamics has been achieved. This claim is corroborated by the good agreement between the calculated and experimental neutron-cross-section-weighted PDOS as well as the partial PDOS of Cu (Fig. 29).

Several studies of the Raman modes of $YBa_2Cu_3O_x$ as a function of oxygen content have concluded a softening of the topmost Ag mode when going from $x=7$ to $x=6$, i.e. from 502 cm^{-1} to $\sim$ 477 cm^{-1} (Hangyo et al., 1988; Thomsen et al., 1988a; Macfarlane et al., 1988). Neutron measurements were, however, unable to find a phonon peak at the predicted position in O_6. A later Raman study reported by Burns et al. (1991) came to the conclusion that the mode seen in $YBa_2Cu_3O_x$ with $x \sim 6.3$ around 480 cm^{-1} is a defect mode, and that the real Ag mode for x close to 6.0 has a rather high frequency, i.e. 600 cm^{-1}. As can be seen from Fig. 30, the neutron results are in line with the picture proposed by Burns et al. (1991). We note that models fitted to all the available data are not able to reproduce an Ag mode frequency as low as $\sim$ 480 cm^{-1} unless long range forces are included, whereas the high value of the

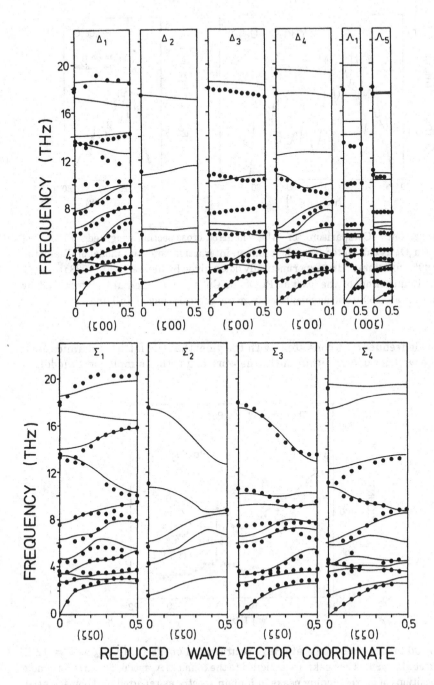

Figure 28 Phonon dispersion curves of $YBa_2Cu_3O_6$ (Data: Pyka et al., 1994; lines: calculated from the interaction potential model of Chaplot et al., 1995)

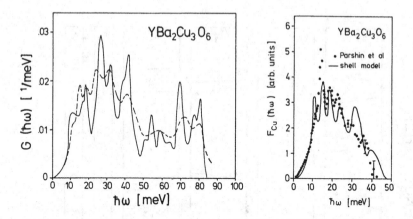

Figure 29 left: Comparison of the neutron-cross-section-weighted PDOS of $YBa_2Cu_3O_6$ as calculated from the common interaction potential model of Chaplot et al. (1995) (full line) and as experimentally determined by Renker et al. (1988a).
right: Comparison of the partial PDOS of Cu as experimentally determined by Parshin et al. (1990) and as calculated from the potential model (full line).

A_g mode frequency is consistent with the view that the lattice dynamics of O_6 can be well described within the frame work of a rather simple ionic model.

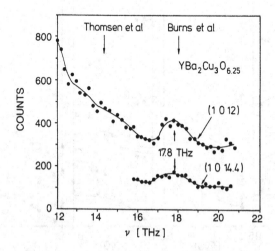

Figure 30 Inelastic neutron scattering spectra taken on $YBa_2Cu_3O_{6.25}$ at T = 12 K (Pyka et al., 1994). The peaks are assigned to the topmost A_g mode. The arrows denote the positions of corresponding peaks in Raman spectra as reported by Thomsen et al. (1988a) and Burns et al. (1991).

4.2 $YBa_2Cu_3O_7$

4.2.1   Lattice dynamics

In spite of tremendous efforts, the lattice dynamics of $YBa_2Cu_3O_7$ (O_7) is still not as well understood as that of other cuprates, e.g. La_2CuO_4. The common interaction potential model of Chaplot et al. (1995) yields a good overall description of the data, but particular phonon branches are described rather poorly. In some cases, as will be discussed in the next section, there is good reason to believe that the poor description is associated with the occurrence of phonon anomalies as was the case in $La_{1.85}Sr_{0.15}CuO_4$. In other cases, however, it remains unclear whether a large deviation between model and experiment has to be interpreted as an indication of a phonon anomaly or rather as an inadequate treatment of short range forces. For instance, the frequencies of ab-plane polarized chain oxygen vibrations are significantly underestimated, whereas the frequencies of c-axis polarized chain oxygen modes are considerably overestimated.

After tuning the parameters of the model and adding a term to soften the frequencies of breathing-type vibrations, the single crystal data are very satisfactorily described by the model (Fig. 31). However, a comparison of the neutron-cross-section weighted PDOS calculated from the model with that experimentally determined on a powder sample by Renker et al. (1988a) shows less good agreement than expected from the description of the dispersion curves (Fig. 32 left, dotted line). The agreement is markedly improved after adding a term to describe the anomalously low frequency of Δ_1-and Δ_4-branches (see section 4.2.3) (full line in Fig. 31), but serious discrepancies remain below 35 meV. We note that in this energy range, the description of the PDOS is less satisfactory than was found for earlier models which reproduce the single crystal data less well (see, e.g., Fig. 8 of Pintschovius and Reichardt, 1994).

There are two obstacles for a better understanding of the lattice vibrations in O_7: one is the enormous complexity of the phonon dispersion due to the large number of atoms in the unit cell and the other is the lack of large untwinned single crystals. The lack of untwinned samples is most crucially felt in studies of the 100/010-direction, where it is difficult to know to which of the two directions the observed phonon peaks have to be assigned. Whereas early assignments had been based on the assumption that the frequency differences between the 100- and the 010-directions are small, we now know that this assumption is not always justified. In studies of the 110-direction, twinning is less a problem rather than the loss of tetragonal symmetry which leads to many interactions between phonon branches which do not interact in O_6 (see Figs. 28/31).

Results obtained in the low frequency region on a small untwinned single crystal (12 mm^3) are depicted in Fig. 33. The most important outcome of this

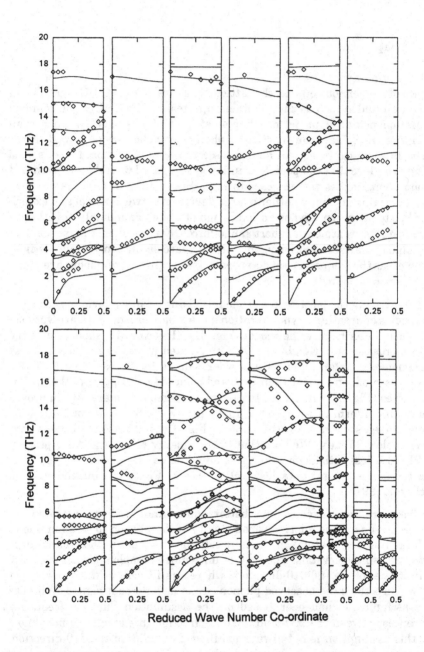

Figure 31 Phonon dispersion curves of YBa$_2$Cu$_3$O$_7$. Data: Reichardt et al. (1994b) and Reichardt (1996). Lines: model calculations of Reichardt (1997b). The arrangement of the panels is as follows (from the left to the right): (top) Δ_1, Δ_2, Δ_3, Δ_4, Δ'_1, Δ'_2: (bottom) Δ'_3, Δ'_4, Σ_1, Σ_2, Λ_1, Λ_2, Λ_3

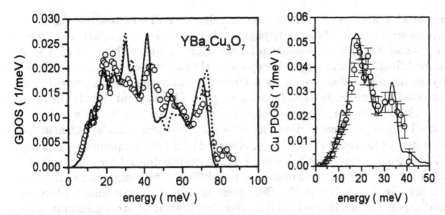

Figure 32 left: Neutron-cross section weighted PDOS of O7 as experimentally deter-
mined by Renker et al. (1988a) and as calculated (full and dotted lines, see text)
(Reichardt 1997). Right: Partial PDOS of Cu as determined by isotope substituion
(Parshin et al., 1990) and as deduced from the lattice dynamical model of Fig. 31.

study was an unambigous assignment of transverse chain oxygen vibrations.
From the rather high frequency of these modes ($v \approx 5$ THz), the moderate
phonon linewidth, and the absence of an appreciable frequency shift with
temperature, Pyka et al. (1993b) concluded that the chain-oxygen vibrations
are not strongly anharmonic. This finding contradicts theoretical predictions
of Cohen et al. (1990) that the chain-oxygen atoms are sitting in wide double-
well potential.

Extended x-ray absorption fine structure (EXAFS) results of Mustre de

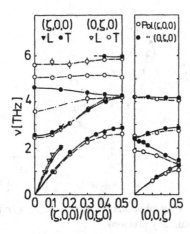

Figure 33 Low frequency branches of $YBa_2Cu_3O_7$ as obtained on a small-detwinned
single crystal (Pintschovius et al., 1991b)

Leon et al. (1990) promoted the idea of a double-well potential for the axial oxygens with two nearly equally populated O(4) sites 0.13 Å apart. The objection that such a split O(4)-position is incompatible with diffraction data lead to a modification of the original proposal in that a double-well was postulated only for the infrared active c-axis polarized apical oxygen mode, whereas the associated Raman mode should show a single-well behaviour (Mustre de Leon et al., 1992). However, even this sophistication did not lead to full agreement with all the experimental data: not only the O(4)Raman mode, but also the associated IR active mode had been observed at a high frequency (~17 THz) which is clearly incompatible with a double-well potential for such a mode. Salkola et al. (1994) tried to overcome this difficulty by assigning the IR peak traditionally attributed to the O(4)-mode instead to a combination tone of Raman and IR active excitations. The appearance of strong combination tones was explained by nonlinear and nonadiabatic effects. However, a thorough inelastic neutron scattering investigation of apical oxygen vibrations by Reichardt (1996) revealed that (i) the frequency of the A_u mode is in agreement with the traditional interpretation of IR results, (ii) that the in-

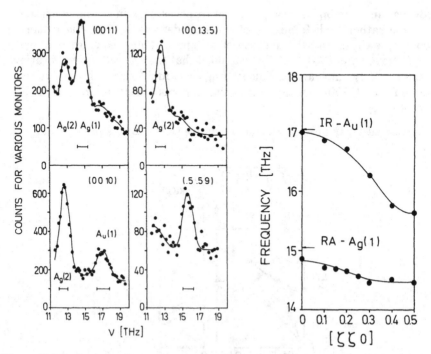

Figure 34 left: Scans at Γ showing phonons with A_u and A_g symmetry and further scans showing related zone-boundary phonons.

right: Phonon branches in the ζζ0-direction with symmetric and antisymmetry elongations parallel to c of the apical oxygen atoms (from Reichardt, 1996).

trinsic linewidth of A_u related phonons is small (see Fig. 30) and (iii) that the observed intensities are in good agreement with predictions of a harmonic phonon model. In conclusion, there is no way to reconcile the original idea of Mustre de Leon et al. (1990) about a double-well potential for the apical oxygem atoms in O7 with the experimental observations.

In addition to the PDOS data of Renker et al. (1988) further data of this kind have been published by Rhyne et al. (1987), Arai et al. (1992) and Chumachenko et al. (1996). From the present knowledge of the lattice dynamics of 123-compounds the early results reported by Rhyne et al. for O6 and O7 appear to be poor. On the other hand, the O_7-spectrum of Arai et al. (1992) is in very good agreement with the data of Renker et al. (1988), except for an isolated high frequency peak at $E = 87$ meV found by Arai et al. but not by Renker et al. A search for phonon peaks above 80 meV was unsuccessful (Pyka et al., 1994), as had to be expected from all the lattice dynamical models considered so far. Finally, the results of Chumachenko et al. (1996) are also in satisfactory agreement with those of Renker et al. (1988), not only for O_7, but for O_6 and intermediate oxygen concentrations as well. Unfortunately, these data yield little information on the precise shape of the upper end of the spectrum due to insufficient resolution.

Hung Fai Fong et al. (1995) tried to separate magnetic and phonon contributions to the scattering cross section at excitation energies $E \sim 41$ meV by a detailed analysis of the momentum dependence of the scattered intensity. They came to the conclusion that a broad peak around $q = (\pi/a, \pi/a)$ in the normal state sometimes ascribed to magnetic scattering can be entirely accounted for by a phonon which primarily involves vibrations of the in-plane oxygens. The lattice dynamical model used in this study is certainly oversimplified, but calculations based on more realistic models performed by the authors of this review essentially confirm the conclusions of Hung Fai Fang et al. (1995) in regard to the assignment of the 41 mev peak.

4.2.2 Screening effects

Screening effects in $YBa_2Cu_3O_7$ have been less extensively studied than in $La_{1.85}Sr_{0.15}CuO_4$, mainly because there is no analog of the steeply dispersive Λ_1-branch which was used as a testing ground for different screening models (see Section 2.2.2). The absence of steep dispersion in Λ_1-branches points to a more efficient screening in O_7 than in $La_{1.85}Sr_{0.15}CuO_4$, but it is difficult to assess the difference quantitatively because the dispersion of Λ_1-branches can be studied only up to $q = 0.5$ in O_7 instead of $q = 1$ in $La_{1.85}Sr_{0.15}CuO_4$. As to the ab-plane polarized branches, the determination of an inverse screening length is complicated by an anticrossing of ab-plane polarized and of c-axis polarized branches, very similar to the situation in $La_{1.85}Sr_{0.15}CuO_4$. As a consequence, determination of an inverse screening length has to be based on the observation of an exchange of eigenvectors rather than on the dispersion itself, which is, of course, less precise. Notwithstanding these difficulties, it is safe to say that screening in O_7 does very little affect large q-phonons. In other words, the bonding remains essentially ionic in nature.

4.2.3 Phonon anomalies

Similar to what has been observed in the system $(La,Sr)_2CuO_4$, evidence of pronounced phonon anomalies have been found in branches associated with longitudinal Cu-in-plane-oxygen bond stretching vibrations. A strong renormalization of phonon frequencies was found both in the 110- and the 100/010-directions, whereby the effects are especially large in the 100/010-direction. The interpretation of the data for this direction is, however, complicated by the loss of well-defined phonon peaks about half-way to the zone boundary due to strong electron-phonon coupling effects and/or effects of twinning and anticrossing with other branches. As a consequence, the interpretation of the data as presented, e.g., by Reichardt et al. (1994) and Pintschovius and Reichardt (1994), remained longtime tentatively. The issue was finally settled by complementary studies carried out by Reichardt (1996). Progress was achieved (i) by using a sample of very good quality, especially a high oxygen content, (ii) by studying not only Δ_1- and Σ_1-phonons, but also phonons of Δ_4- and Σ_2-symmetry and (iii), most importantly, by extending the measurements to directions along the zone boundary. The new results confirmed the interpretation of the data presented in earlier publications (Reichardt et al., 1994; Pintschovius and Reichardt, 1994).

The presently available data are summarized in Figs. 35. In order to emphasize the similarity between the picture emerging from the recent study with that proposed earlier, two panels shown previously are included in Fig. 30 although they should have been somewhat modified in the light of the new results. In particular, a broad peak observed in early experiments at Γ and ascribed to both the B_{2u}- and the B_{3u}-phonons has been resolved in recent experiments. The splitting associated with the a-b anisotropy has been found to be quite large, i.e. $\Delta v \sim 1.4$ THz, in agreement with Raman results of McCarty et al. (1990).

Evidently, phonons associated with Cu-O2/3 bond-stretching vibrations of Δ_1- and Δ_4-symmetry, respectively, behave very similarly in that the frequencies strongly decrease towards the zone boundary - in contrast to the behavior observed in O_6 which has to be considered as the normal one, i.e. in the absence of electron-phonon coupling effects. The similarity between branches of Δ_1- and Δ_4-symmetry is not astonishing in view of the fact that the two types of vibrations are distinguished from each other only by the relative phase between the in-plane elongations in the two planes of the $YBa_2Cu_3O_7$- unit cell: the elongations are in phase and in opposite phase for phonons of Δ_1- Δ_4-symmetry, respectively. For both types of phonons, no well defined peaks could be observed around q $\sim$ 0.25. Unfortunately, the twinning of the samples precludes any detailed analysis in how far the ill-definedness of the peaks is due to the finite Q-resolution in conjunction with an anticrossing of

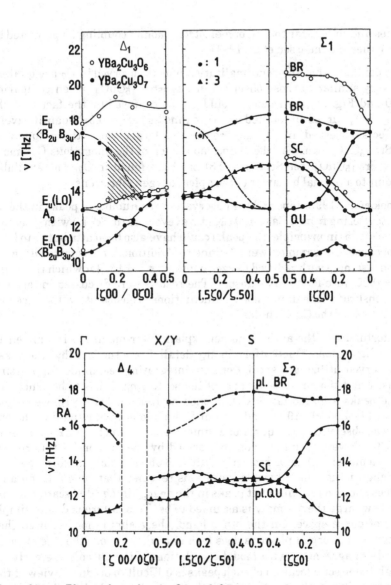

Figure 35 top: Phonon branches in O6 and O7 associated with bond-stretching and bond-angle-bending elongations of the planar oxygen atoms (the elongations are in phase between the two Cu-O layers). The left and right panels are adapted from an earlier publication (Pintschovius and Reichardt, 1994).

bottom: Phonon branches in O7 associated with bond-stretching (dots) and bond-angle-bending (triangles) elongations of the planar oxygen-atoms (the elongations are of opposite phase in the two layers). The arrows denote Raman results of Mc Carty et al. (1990). The figure was taken from Reichardt (1996).

branches and in how far to a strong electron-phonon coupling as predicted by ab-initio theory (Andersen et al, 1991).

We point out that the renormalization effects observed in O_7 are qualitatively very similar to those observed in $La_{1.85}Sr_{0.15}CuO_4$ (when comparing Fig. 20 and Fig. 30, the reader should not be confused by the fact that the $(La,Sr)_2CuO_4$-curves are plotted up to $q = 1$ instead of $q = 0.5$). Quantitatively, the effects observed in O_7 are somewhat larger than those found in $La_{1.85}Sr_{0.15}CuO_4$. On the other hand, no analog of the anomalous $O_Z Z$-mode of $La_{1.85}Sr_{0.15}CuO_4$ (see Section 2.2.3) has been found in O_7. The available data point to a normal behavior of the apical oxygen vibrations.

Mook et al. (1996) reported the observation of a number of peaks in the inelastic scattering from $YBa_2Cu_3O_{6.93}$ at wavevectors (h^*,k^*,l^*) with a component $h^* \sim 0.25$ (in principle, the peaks could have also been attributed to $k^* \sim 0.25$, because the samples were twinned). Additional measurements were made on a sample with reduced oxygen content $(x = 6.15)$, for which the peaks at $h^* \sim 0.25$ were found to be absent. The results were discussed in terms of incommensurate one-dimensional fluctuations consistent with accepted values of $2k_F$ for the Cu-O chains.

Unfortunately, the authors do not explain their picture of incommensurate one-dimensional fluctuations in any detail. From the fact that the peaks were observed utilizing a special energy integrating technique suppressing the elastic and the quasielastic part of the scattering, it has to be concluded that the peaks are associated with phonon scattering. Therefore, we conjecture that Mook et al. (1996) had a giant Kohn anomaly in mind like the one which was observed in the quasi-one-dimensional conductor KCP (Renker et al., 1973). However, simulations performed by the authors of this review based on a model with an arbitrarily built-in Kohn anomaly in the dispersion of the longitudinal chain oxygen vibrations revealed that such a Kohn anomaly does not lead to significant peaks in the energy integrated scattering, no matter how large the anomaly is assumed to be. This failure is due to simple reasons of phase space. On the other hand, these simulations showed that scattering from normal phonons does lead to peaks of about the size of those observed - in agreement with model calculations of the authors themselves - although the precise location of such peaks is difficult to predict in view of the many parameters entering the simulations (checks would have been much easier for conventional triple-axis spectrometer data).

Whereas Mook et al. (1996) admit that peaks stemming from normal phonons are somewhat a problem for the interpretation of their low resolution data, they claim that normal phonons can be excluded as origin of a very sharp peak observed in a (single) high resolution scan. Firstly, we have to say that simulations based on normal phonons do not necessarily lead to an absence of peaks in the q-range investigated. Secondly, and more importantly, a Kohn anomaly related to $2k_f$ of the Cu-O chains can never be that narrow as to be compatible with the very small q-width of the observed peak ($\Delta h \sim 0.01$ $2\Pi/b$) because of the finite length of the Cu-O chain fragments in the sample investigated: $x = 6.93$ translates into an average length of about 14b, whereas $\sim 100b$ would be needed.

4.2.4   Superconductivity-induced phonon effects

Superconductivity-induced phonon self-energy effects in $YBa_2Cu_3O_{7-\delta}$ have been first observed by Raman scattering (Macfarlane et al., 1987; Thomsen et al., 1988). It took quite some time until sufficiently large single crystals of $YBa_2Cu_3O_7$ became available to search for such effects also by inelastic neutron scattering. The first successful experiment of this kind was reported by Pyka et al. (1993). It yielded the first information on superconductivity-induced frequency shifts for $q \neq 0$ phonons in $YBa_2Cu_3O_x$ (Fig. 36). For branches starting from the Raman-active mode at 340 cm-1 (42.5 meV) - for which the largest effects had been found by Raman scattering - the shifts were found to decrease only slowly with increasing momentum transfer ($x = 7.0$) or even exhibit a maximum at finite q ($x = 6.92$) when going along the (100)/(010)-direction (Fig. 37). The shifts were found to be much smaller for the A_u mode at 307 cm-1 and the E_u mode at 343 cm-1 and for $q \neq 0$ phonons related to these modes.

A later study using improved experimenal conditions (Reichardt, 1995) revealed that the shifts for phonons related to the E_u mode at 343 cm-1 are definitely non-zero, but indeed rather small, i.e. of about the same size as previously observed for the A_u mode at 307 cm-1: $\Delta v/v = (0.4 \pm 0.1)\%$.

Further results on the temperature dependence of the peak position and linewidth of phonons related to the 340 cm-1 Raman active mode were reported by Reznik et al. (1995). In agreement with the results of Pyka et al. (1993a) it was found that superconductivity-induced frequency shifts decrease only slowly when going along the (100)/(010) direction from the zone

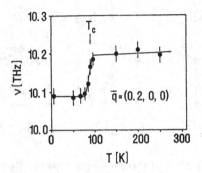

Figure 36 Frequency of a phonon with q = (0.2,0,0) in $YBa_2Cu_3O_{6.92}$ vs temperature. The line is a guide to the eye (from Pyka et al., 1993a).

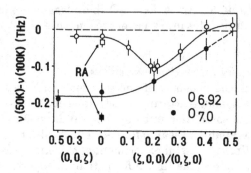

Figure 37 Frequency shifts observed in $YBa_2Cu_3O_x$ for phonons related to the A_g-mode at 240 cm^{-1} vs momentum transfer. The lines are a guide to the eye. The Raman results (Krantz et al., 1988) refer to x = 7.0 (full square) and x = 6.87 (open square), respectively. The figure was taken from Pyka et al. (1993a).

center to the zone boundary (Fig. 39, left). On the other hand, the decrease seems to be rather fast when going along the (110)-direction, although the data are too scant to assess the difference between the two directions quantitatively (we note that the first data point along (1,1,0), i.e. for q = 0.25, corresponds to q = 0.35 along (1,0,0), where the effect is not very large either).

The very large size of their sample (~10 cm^3) enabled Reznik et al. to improve the instrumental resolution to an extent as to investigate not only frequency shifts but also line broadenings. They found a significant increase in linewidth below T_c with a maximum along the (1,0,0)-direction half-way to

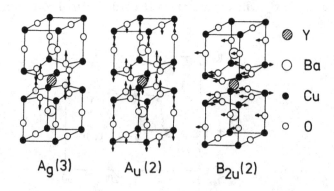

Figure 38 Displacement pattern of three q = 0 phonons in $YBa_2Cu_3O_7$. Left: A_g mode at 340 cm^{-1} (10.2 THz); middle: A_u mode at 307 cm^{-1} (9.2 THz); right: B_{2u} mode at 347 cm^{-1} (10.4 THz).

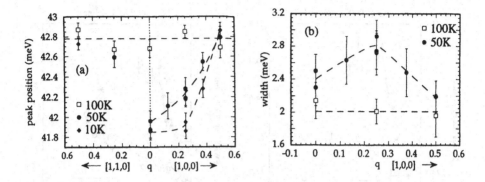

Figure 39 q-dependence of (left) peak position and (right) linewidth of phonons related to the A_g-mode at 340 cm^{-1} (42 meV) at different temperatures. Dashed lines are a guide to the eye (from Reznik et al., 1995).

the zone boundary (Fig. 39, right). Data taken at q=(0.25, 0.25, 0) and q=(0.5,0.5,0) rule out any significant broadening below T_c at these q values which supports an anisotropy between the (100)/(010)- and the (110)-directions. Reznik et al. (1995) argue that the anisotropy in the superconductivity induced phonon self-energy effects between the (100)/(010) and (110) directions is indicative of a d-wave superconducting order parameter and/or a strong anisotropy of the electronic structure close to the Fermi surface.

In a recent investigation, Harashina et al. (1996) observed superconductivity-induced phonon anomalies for a c-axis polarized B_{2u} mode (see Fig. 40). The eigenvector of this mode is very similar to that of the $A_g(3)$-mode displayed in Fig. 38, the difference lying only in the relative phase of the elongations in the two Cu-O planes: the B_{2u}-mode under discussion is associated with in-phase instead of out-of-phase elongations. A further mode of A_u symmetry at E ~18 meV did not show any appreciable anomaly. This mode has in-phase motions of the 0(2,3) ions with respect to the Cu atoms (and a large Y displacement amplitude).

For some time, the superconductivity-induced phonon self-energy effects seemed to be well explained by the theory of Zeyer and Zwicknagl (1990). In particular, Friedl, Thomsen and Cardona (1990) concluded that the anomalous broadening of the $A_g(3)$ as a function of phonon frequency by substituting Y by various rare earths can be qualitatively and quantitatively explained by the above theory. This study yielded a single, sharp superconducting gap at $2\Delta = 316$ cm^{-1} (39.5 meV). Since the frequency of the $A_g(3)$ mode in YBa$_2$Cu$_3$O$_7$ is very close to this gap value, it should show a much larger shift than other modes with frequencies a above or far below the gap. In the course of the time, more and more results have been accumulated which do not fit into this simple picture, which applies also to the neutron results of Pyka et

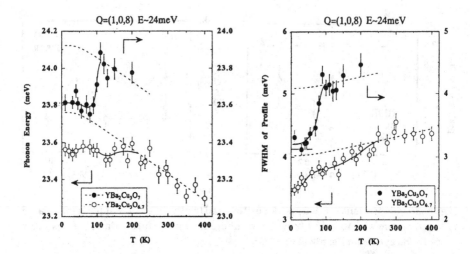

Figure 40 left: Energies of the B_{2u} mode vs. temperature for $YBa_2Cu_3O_7$ and $YBa_2Cu_3O_{6.7}$. The dotted lines show the T-dependence expected from ordinary anharmonicity. The solid lines are guides to the eye.

right: Full widths at half-maximum of the profiles of the B_{2u} mode vs. temperature for $YBa_2Cu_3O_7$ and $YBa_2Cu_3O_{6.7}$. The instrumental resolution is estimated as 3.13 meV. Dashed and solid lines are guides to the eye (from Harashina et al. 1996).

al. (1993a) and Harashina et al (1996): whereas the B_{2u} mode at 43 meV $\approx 2\Delta$ shows only a very small shift, a rather large shift was found for the B_{2u} mode at 24 meV $<< 2\Delta$ (Fig. 40). Obviously, the eigenvector plays a very important role. Harashina et al. (1996) argue that c-axis polarized B-symmetric modes should couple effectively to electrons in the Fermi surface region around the **k**-points ($\pm\pi/a,0$) and ($0, \pm\pi/a$) where a large gap opens below T_c if the symmetry of the order parameter is $d_{x^2-y^2}$-like. For A-symmetric modes, the same considerations predict no significant superconductivity-induced effect (in this context, we note that the mode labelled above as $A_g(3)$ in the tetragonal notation should be correctly labelled as B_{1g}).

As indicated by the dashed and dotted lines in Fig. 40, Harashina et al. (1996) interpret their results for the oxygen depleted sample as evidence for anomalies which they associate with the opening of a spin-gap. Similar phenomena had been observed by optical spectroscopy and interpreted in the same way (see, e.g., Litvinchuk, Thomsen and Cardona, 1994).

4.3   OTHER 123-COMPOUNDS

A variety of other 123-compounds than O_6 or O_7 have been investigated by inelastic neutron scattering, but exclusively on powder samples. The aim of

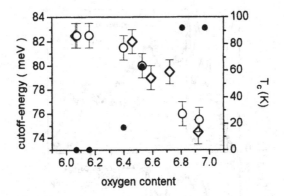

Figure 41 Cut-off energy of the phonon spectrum (open symbols) and superconducting transition temperature T_c (full dots) of $YBa_2Cu_3O_x$ vs oxygen content. Circles denote results of Chumachenko et al. (1996), diamonds and full dots results of Renker et al. (1988a). The figure was adapted from Chumachenko et al. (1996).

all these investigations was to correlate superconducting properties with special features of the phonon spectrum. To this end, the superconducting transition temperature was varied by substituting one or the other atomic species of $YBa_2Cu_3O_x$ partially or completely and/or varying the oxygen content x.

To begin with, x has been varied over nearly the entire range $6 \leq x \leq 7$ both by Renker et al. (1988a) and Chumachenko et al. (1996). It was found that the high-energy cut-off of the phonon spectrum changes continuously between x = 6 and x = 7 in close correlation with T_c (Fig. 41). This signifies that the phonon anomalies observed in high-energy branches of O_7 (see section 4.2) correlate with the superconducting properties of the samples. The close correlation between T_c and the cut-off energy is further corroborated by the observation that increasing the number of free carriers in $YBa_2Cu_3O_{6.5}$ - and thereby increasing T_c - by replacing a part of the Y atoms by Ca atoms lowers the cut-off energy considerably (Fig. 42(b)). Furthermore, an increase of the cut-off energy was observed on suppressing T_c by substituting Ba by La, although in this case, the increase is not quite as large as generated by a reduction of the oxygen content (Fig. 42(c)).

In this context we note that the strong changes in the phonon spectra of O_x below 60 meV (see 42(a)) are probably largely related to structural differences and not to phonon anomalies, so that it is not astonishing that suppressing T_c by other means than depleting the oxygen content does not lead to strong changes in the lower part of the phonon spectrum.

A similar correlation between T_c and cut-off energy as discussed above was found in the system $YBa_2Cu_{3-x}Zn_xO_7$ by Renker et al. (1989) (Fig. 43) and Parshin et al. (1992) (Fig. 44). In this system, reduction of T_c does not

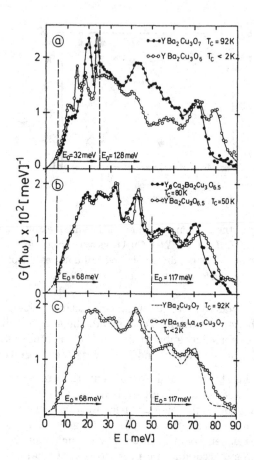

Figure 42 Comparison of the neutron-cross-section weighted PDOS of various 123-superconductors and non-superconducting reference compounds. The upper and lower parts of the spectra have been studied using different incident energies E_0 in order to optimize the resolution. The figure was taken from Renker et al. (1992).

only depend on the oxygen content and on the total Zn content, but also on the distribution of the Zn atoms on sites in the Cu-O planes or the Cu-O chains, respectively, which in turn depends on the heat treatment of the sample. The results of Parshin et al. (1992) indicate that the increase of the cut-off energy correlates - for the same oxygen and Zn content - with the decrease of T_c (Fig. 44).

As is well known - but still poorly understood - T_c can also be suppressed by replacing Y by Pr. The effect of this modification on the phonon spectrum was investigated by Renker et al. (1988b, 1989). We note that a determination of the PDOS of Pr_2CuO_4 and Nd_2CuO_4 from powder data is somewhat complicated by magnetic contributions to the scattering. However, the high

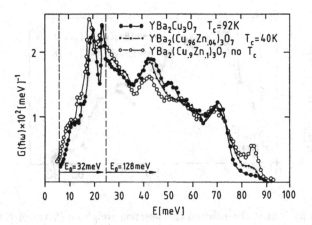

Figure 43 Comparison of the neutron-cross-section weighted PDOS of YBa$_2$(Cu$_{1-x}$n$_x$)$_3$O$_7$ for x = 0,0.04 and 0.1 respectively (after Renker et al., 1989).

energy part of the spectrum is not affected by this complication. As can be seen from Fig. 45, a comparison of the PDOS of NdBa$_2$Cu$_3$O$_7$ and of PrBa$_2$Cu$_3$O$_7$ fits into the trend found in other systems, although the increase in cut-off energy associated with the T$_c$ suppression is again not as large as in O$_6$.

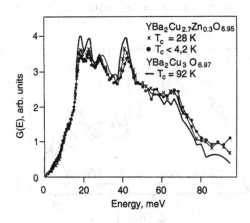

Figure 44 Neutron-cross-section weighted PDOS of YBa$_2$Cu$_3$O$_{6.97}$ and two samples of YBa$_2$Cu$_{2.7}$Zn$_{0.3}$O$_{6.95}$ prepared with different distributions of the Zn atoms on the Cu(1) and Cu(2) sites (redrawn from a figure of Parshin et al., 1992).

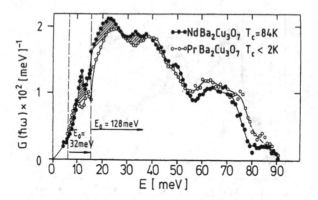

Figure 45 Comparison of the neutron-cross-section weighted PDOS of $NdBa_2Cu_3O_7$ and $PrBa_2Cu_3O_7$. The hatched areas denote magnetic scattering contributions (after Renker et al., 1988a).

5.   Bi-based compounds

The first investigation of the PDOS of Bi-based compounds was reported by Renker et al. (1989). The aim of this study was to search for effects associated with the transition from a superconducting to a semiconducting material in $BiSr_2(Ca_{1-x}Y_x)Cu_2O_8$ as a function of the Y content. As can be seen in Fig. 46, significant shifts were observed at the high energy end of the spectrum, similar to, but smaller than, those found in the system $YBa_2Cu_3O_x$. Therefore, it is obvious to assume that the lowering of the cut-off energy when going from a semiconducting to a superconducting material indicates the appearance of anomalies in phonon branches associated with Cu-O bond-stretching vi-

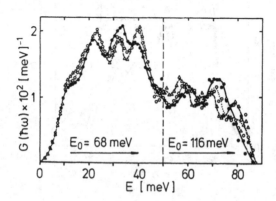

Figure 46 Comparsion of the neutron-cross-section weighted PDOS of $Bi_2Sr_2(Ca_{1-x}Y_x)cu_2O_8$ for $x = 0, 0.6$ and 1, respectively (after Renker et al., 1989).

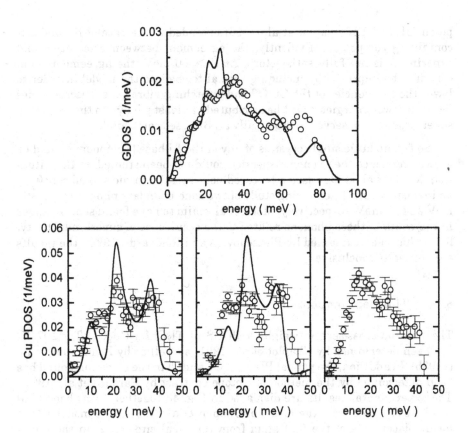

Figure 47 top: neutron-cross-section weighted PDOS of $Bi_{1.8}Pb_{0.2}Sr_2Ca_{1.1}Cu_{2.15}O_x$ as experimentally determined by Zemlyanov et al. (1993) (circles) and as calculated from an interaction potential model (full line; Reichardt, 1997c).

bottom: partial PDOS of Cu atoms in Bi-2201 (left), Bi-2212 (middle) and Bi-2223 (right) as experimentally determined by an isotope substitution technique (circles; from Parshin et al., 1996) and as calculated from an interaction potential model (full lines; Reichardt, 1997c).

brations. On the other hand, changes in the PDOS observed at E < 40 meV might be simply due to the difference in mass of Ca and Y.

Later, Zemlyanov et al. (1993) determined the PDOS of another Bi-2212 compound, i.e. $Ba_{1.8}Pb_{0.2}Sr_2Ca_{1.1}Cu_{2.15}O_x$ with a $T_c = 70$ K. The spectrum was found to be very similar to that reported by Renker et al. (1989) for $Ba_2Sr_2CaCu_2O_8$. In particular, the cut-off energy is the same for the two compounds. Subsequently, the same group (Parshin et al., 1996) studied the partial PDOS of Cu in Bi-2201, Bi-2212 and Bi-2223. The results of the measurements of Zemlyanov et al. (1993) and Parshin et al. (1996) are displayed in Fig. 47 together with the result of model calculations (Reichardt, 1997c) as far as available. The calculations were based on the common interaction potential model of Chaplot et al. (1995) extended to the case of Bi- and Ca-containing compounds. Evidently, the agreement between calculation and experiment is not fully satisfactory. At $E > 60$ meV, the agreement could certainly be improved by including special terms into the model in order to lower the frequencies of the Cu-O bond-stretching vibrations. Discrepancies found at lower energies might be attributed - at least partly - to the neglect of superstructures observed in practically all Bi-based compounds.

So far, no large single crystals of any of the Bi-based compounds could be grown and hence the phonon dispersion could not been studied. In this situation, Mook et al. (1992) used a textured polycrystal (mosaic spread ~13°) for an inelastic neutron scattering study of two zone-boundary phonons at E ~62 meV and 76 meV, respectively. Mook et al. claim to have found some changes in the width of these phonons related to the occurrence of superconductivity, but as has been explained by Pintschovius and Reichardt (1994), the results are not really conclusive.

6.   Tl-based compounds

The neutron-cross-section weighted PDOS of $Tl_2CaBa_2Cu_2O_8$ ($T_c = 107$ K) has been determined by Chaplot et al. (1991) as well as by Zemlyanov et al. (1993b). In addition, the partial PDOS of Cu and Tl in the same compound has been determined by the method of isotope contrast by Parshin et al. (1994). The experimental results are displayed in Fig. 48 together with theoretical results calculated from the model of Chaplot et al. (1991). The model allows for displacements of the O(3) atom from the ideal site ($\frac{1}{2},\frac{1}{4},z$) in the (100)-direction of about 0.5 Å and for positional disorder of the O(3)-atoms by using a supercell containing four O(3)-atoms. The potential parameters of the model were refined as to produce to a minimum in the potential energy for a structure close to that observed in diffraction experiments. As can be seen from Fig. 48, the calculated results reproduce the observed spectra rather well.

There are no data available for a non-superconducting or low-T_c reference compound, presumably because it is rather difficult to prepare such a sample in large quantities. There is, however, one indication that phonon anomalies are present also in $Tl_2CaBa_2Cu_2O_8$: the model overestimates the cut-off energy by nearly 10 meV, as do such models for O7 if no term is included to account for the anomalously low frequency of Cu-O bond-stretching vibrations.

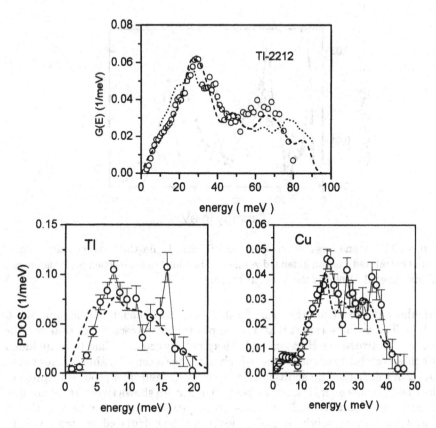

Figure 48 top: neutron-cross-section weighted PDOS of $Tl_2CaB_2Cu_2O_8$ as experimentally determined by Zemlyanov et al. (1993b) (circles) and Chaplot et al. (1991) (dotted line) and as calculated by Chaplot et al. (1991) (dashed line). The data of Chaplot et al. were not corrected for multiphonon contributions. The figure was adapted from Zemlyanov et al. (1993b).

bottom: Partial PDOS of Tl (left) and Ca (right) atoms in $Tl_2CaBa_2Cu_2O_8$ as experimentally determined by isotope substitution by Parshin et al. (1994) and as calculated from the model of Chaplot et al. (1991) (redrawn after Parshin et al. 1994).

7.   Hg-based compounds

It appears interesting to study the phonons of Hg-based superconductors because of the very high T_c's found in this family of the cuprates. However, only very small single crystals could be grown so far, and it is even very difficult to prepare large polycrystalline samples of those phases which show the highest T_c's. Therefore, only $HgBa_2CuO_4$ having a $T_c = 95$ K has been investigated by inelastic neutron scattering. Fig. 49 shows the neutron-cross-section weighted PDOS of this compound determined in the neutron-energy-gain mode at $T = 300$ K by Renker et al. (1996). The results were analyzed by model

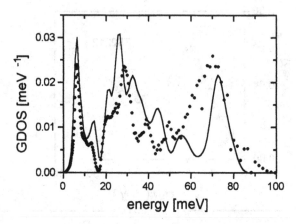

Figure 49 Neutron-cross-section weighted PDOS of $HgBa_2CuO_4$ as measured (dots) and as calculated from an extended version of the common interaction potential model of Chaplot et al. (1995) by Renker et al. (1996).

calculations based on the transferable potential model of Chaplot et al. (1995). The authors report that a large positive transverse force constant for the bond between the Hg atom and the apical oxygen atom had to be included to achieve stability of the lattice, which indicates a considerable covalent character of this bond. The additional force constant leads to a good description of the PDOS at low energies, but inspection of Fig. 49 shows that significant discrepancies remain at high energies. In particular, the experimental PDOS shows much more weight at $E > 50$ meV than the calculated spectrum which, however, seems difficult to reproduce by any reasonable model. This discrepancy was much more serious for early data of the same group (Renker et al. 1994). As was discussed by the authors, it is probably related to the presence of some unreacted precursor material which has no well defined crystallographic phase and therefore went undetected in x-ray diffraction patterns.

Additional temperature dependent measurements performed by Renker et al. (1996) in the neutron-energy-loss-mode of operation revealed some evidence for changes in the PDOS at $E \sim 60$ meV when going from $T = 100$ K to $T = 30$ K. These changes are discussed in terms of superconductivity-induced phonon effects, although the lack of measurements at intermediate temperatures does not allow one to establish a clear link of the observed changes to the occurrence of superconductivity. We emphasize that searches for such effects in the PDOS of other cuprates revealed only a null result.

8.   Conclusive remarks

Inelastic neutron scattering studies have provided a wealth of information on the phonon properties of cuprate superconductors and their non-superconducting parent compounds. Many of these studies were motivated by the hope to find evidence for a strong electron-phonon coupling. The numerous measurements of the PDOS on polycrystalline samples revealed that the high-frequency end of the phonon spectra is shifted downwards in respect to that of non-superconducting analogues. From those systems, which could be studied in single crystalline form, we know that the softening of the phonon spectrum is related to the occurrence of phonon anomalies in the dispersion of branches associated with Cu-O bond stretching vibrations. We note that these anomalies resemble very much those found in another "high-T_c" oxide superconductor, i.e. $Ba_{0.6}K_{0.4}BiO_3$ (Braden et al., 1996), whereas they are absent in the low-T_c oxide superconductors $Sr_{0.98}Nb_{0.02}TiO_3$ (Reichardt, 1994) and Sr_2RuO_4 (Braden et al., 1998). Herewith, we do not want to claim that the phonon anomalies observed in cuprate superconductors are evidence for a phonon mechanism of high-T_c superconductivity, but on the other hand, we find it hard to believe that the appearance of phonon anomalies associated with the insulator-to-superconductor transition is completely accidental. The observation of superconductivity-induced phonon self-energy effects by optical and neutron spectroscopy corroborates the idea that phonons are involved in high-T_c superconductivity at least to some extent.

A widely discussed idea was the enhancement of the electron-lattice coupling due to strong anharmonicity. Although neutron scattering has found evidence for incipient phase transitions also in structurally stable compounds like Nd_2CuO_4, strong anharmonicity is hardly essential for high-T_c superconductivity, as anharmonicity was found to be small in the prototypical high-T_c compound $YBa_2Cu_3O_7$. Here, inelastic neutron scattering played a 'destructive' role by proving that the various ideas about strong anharmonicity in O_7 do not hold.

Recently, stripe order has been proposed as a key phenomenon for the understanding of the unusual properties of the cuprates, including high-T_c superconductivity. Microscopic segregation of charges and spins has also been evoked to explain some unusual phonon properties, but so far, such explanations remain completely speculative. The situation is not likely to change until theory will tell us more about the signature of stripe order in the phonon response.

Finally, we would like to point out that the benefits of the phonon studies of the cuprates go well beyond their contribution to our understanding of high-T_c superconductivity: the lattice dynamical models developed for the cuprates have proved to be very helpful for modelling the phonon dispersion of other oxides, in particular, but not only, of oxides with perovskite related structures like the manganates.

References

Andersen, O.K., Liechtenstein, A.I., Rodriguez, O., Mazin, I.I., Jepsen, O., Antropov, V.P., Gunnarsson, O. and Gopalan, S. (1991) Electrons, phonons, and their interaction in $YBa_2Cu_3O_7$, *Physica C* **185-189**, 147-155.

Arai, M., Yamada, K., Hidaka, Y., Taylor, A.D. and Endoh, Y. (1991) Phonon density of states of super- und non-superconducting states in $La_{1.85}Sr_{0.15}Cu_{1-x}Zn_xO_4$ (x=0,0.002), *Physica C* **181**, 45-50.

Arai, M., Yamada, K., Hidaka, Y., Itoh, S., Bourden, Z.A., Taylor, A.D., and Endoh, Y. (1992) Anomaly of phonon state of superconducting $YBa_2Cu_3O_7$ studied by inelastic neutron scattering, *Phys. Rev. Lett.* **69**, 359-362.

Arakawa, T., Arai, M., N. Nishima, T., Fujita, M., Murakami, Y., Kawada, H., Yamada, K. Lee, C.H. and Endoh, Y. (1996) Anomalous structure and phonon dynamics in $La_{1.85}Sr_{0.15}CuO_4$, *Journ. Low Temp. Phys.* **105**, 819-824.

Axe, J.D., Moudden, A.H., Hohlwein, D., Cox, D.E., Mohanty, K.M., Moodenbaugh, A.R. and Youven Xu (1989) Structural phase transformation and superconductivity in $La_{2-x}Ba_xCuO_4$, *Phys. Rev. Lett.* **62**, 2751-2754.

Bednorz, G.A. and Müller, K.A. (1986) Possible high T_c superconductivity in the Ba-La-Cu-O system, *Z. Phys. B - Condensed Matter* **64**, 189-193.

Birgeneau, R.J., Chen, C.Y, Gabbe, D.R., Jenssen, H.P., Kastner, M.A., Peters, C.J., Picone, P.J., Tineke Thio, Thurston, T.R. and Tuller, H.L. (1987) Soft-phonon behavior and transport in single-crystal La_2CuO_4, *Phys. Rev. Lett.* **59**, 1329-1332.

Birgeneau, R.J. and Shirane, G. (1989) Neutron scattering studies of structural and magnetic excitations in lamellar copper oxides - a review, in D.M. Ginsberg (ed.) *Physical Properties of High Temperature Superconductors I*, World Scientific, pp. 151-212.

Böni, P., Axe, J.D., Shirane, G., Birgeneau, R.J., Gabbe, D.R., Jenssen, M.P., Kastner, M.A., Peters, C.J., Picone, P.J. and Thurston, T.R. (1988) Lattice instability and soft phonons in single-crystal $La_{2-x}Sr_xCuO_4$, *Phys. Rev. B* **38**, 185-194.

Braden, M., Schnelle, W., Schwarz, W., Pyka, N., Heger, G., Fisk, Z., Gamyunov, K., Tanaka, I., and Kojima, H. (1994a) Elastic and inelastic neutron scattering studies of the tetragonal to orthorhombic phase transition of $La_{2-x}Sr_xCuO_{4-a}$ *Z. Phys. B* **94**, 29-37.

Braden, M., Paulus, W., Cousson, A., Vigoureux, P., Heger, G., Goukassov, A., Bourges, P. and Petitgrand, D. (1994b) Structure analysis of Gd_2CuO_4: a new modification of the T'-phase, *Europhys. Lett.* **25**, 625-630.

Braden, M. and Reichardt, W. (1996) Inelastic-neutron-scattering study of Gd_2CuO_4, unpublished results.

Braden, M., Pintschovius, L, Nakamura, F. and Fujita, T. (1997) Inelastic neutron scattering study of $La_{1.48}Nd_{0.4}Sr_{0.12}CuO_4$, unpublished results.

Braden, M., Reichardt, W. and Maeno, Y., Inelastic neutron scattering study of Sr_2RuO_4, unpublished results.

Büchner, B., Braden, M., Gramm, M., Schlabitz, W., Schnelle, W., Hoffels, Q., Braunisch, W., Müller, R., Meyer, G. and Wohlleben, D. (1991) Low temperature phase transition and superconductivity in (La,Nd)-Sr-Cu-O, Physica C 185-189, 903-904.

Büchner, B., Lang, A., Baberski, O., Hücker, M. and Freimuth, A. (1996) Transport properties of rare earth doped La2-xSrxCuO4 (1996) Journ. Low Temp. Physics 105, 921-925

Burns, G., Dacol, F.H., Feild, C. and Holtzberg, F. (1991) Raman modes of YBa$_2$Cu$_3$O$_x$ with variable oxygen content, Physica C 181, 37-44.

Chaplot, S.L., Dasannacharya, B.A., Mukhopadhyay, Rao, K.R., Vijayaraghavan, P.R., Iyer, R.M., Phatak, G.M. and Yakhmi, J.V. (1991) Phonon density of states in Tl$_2$CaBa$_2$Cu$_2$O$_8$, Physica C 174, 378-381.

McCarty, K.F., Liu, J.Z., Shelton, R.N. and Radovsky, H.B. (1990) Raman active phonons of a twin-free YBa$_2$Cu$_3$O$_7$ crystal: a complete polarization analysis, Phys. Rev. B 41, 8792-8797.

Chaplot, S.L., Reichardt, W., Pintschovius, L. and Pyka, N. (1995) Common interaction model for the lattice dynamics of several cuprates, Phys. Rev. B 52, 7230-7242.

Chaplot, S.L., Reichardt, W. and Pintschovius, L. (1996) Influence of the orthorhombic distortion on the lattice vibrations in La$_2$CuO$_4$, Physica B 219 & 220, 219-221.

Chou, H., Yamada, K., Axe, J.D., Shapiro, S.M., Shirane, G., Tanaka, I., Yamane, K. and Kojima, H. (1990) Inelastic-neutron-scattering study of the electron-phonon interaction in a superconducting La$_{1.85}$Sr$_{0.15}$CuO$_4$ single crystal, Phys. Rev. B 42, 4272-4275.

Chumachenko, A.V., Zemlyanov, M.G., Parfionov, D.E., Parshin, P.P., and Shikov, A.A. (1996) Experimental study of the effect of carrier concentration on the vibrational spectrum of YBa$_2$Cu$_3$O$_{7-y}$, JETP 82 (English edition), 107-110.

Egami, T., Dmowski, W., Jorgensen, J.D., Hinks, D.G. and Capone, II, D.W. (1987) Local atomic structure of La$_{2-x}$(Sr,Ba)$_x$CuO$_4$ determined by pulsed neutron scattering, Reviews of Solid State Science 1, 247-257.

Falter, C., Klenner, M., and Chen, Q. (1993) Role of bonding, reduced screening, and structure in the high-temperature superconductors, Phys. Rev. B 48, 16690-16706.

Falter, C. and Klenner, M. (1994) Nonadiabatic and non-local electron-phonon interaction and phonon-plasmon mixing in the high temperature superconductors, Phys. Rev. B 50, 9426-9433.

Falter, C., Klenner, M. and Hoffmann, G.A. (1995) Phonon renormalization and c-axis phonon-plasmon mixing in La$_2$CuO$_4$, Phys. Rev. B 52, 3702-3710.

Falter, C., Klenner, M., Hoffmann, G.A. and Cheng, Q. (1997) Origin of phonon anomalies in La$_2$CuO$_4$, Phys. Rev. B 55, 3308-3313.

Fil, D.V., Tokar, O.I., Shelankov, A.L. and Weber, W. (1992) Lattice mediated interaction of Cu^{2+} Jahn-Teller ions in insulating cuprates, Phys. Rev. B 45, 5633-5640.

Friedl, B., Thomsen, C. and Cardona, M. (1990) Determination of the superconducting gap in RBa$_2$Cu$_3$O$_{7-\delta}$, Phys. Rev. Lett. 65, 915-918.

Haskel, D., Stern, E.A., Hinks, D.G., Mitchell, A.W., Jorgensen, J.D. and Budnick, J.I. (1996) Dopant and temperature induced structural phase transitions in La$_{2-x}$Sr$_x$CuO$_4$, Phys. Rev. Lett. 76, 439-442.

Hangyo, M., Nakashima, S., Mizoguchi, K., Fujii, A. and Mitsuishi, A. (1988) Effect of oxygen content on the phonon Raman spectra of YBa$_2$Cu$_3$O$_{7-\delta}$, Solid State Commun. 65, 835-839.

Harashina, H., Kodama, K., Shamoto, S., Sato, M., Kahurai, K. and Nishi, M. (1996) Supercon-
    ductivity-induced B_{2u} phonon anomalies of $YBa_2Cu_3O_{6+x}$ and symmetry of the order para-
    meter. Neutron inelastic scattering studies. *Physica C* **263**, 257-259.

Henn, R., Kircher, J., Cardona, M., Wittlin, A., Duijn, V.H.M. and Menovsky, A.A. (1996) Far-
    infrared response of $La_{1.87}Sr_{0.13}CuO_4$ determined by ellipsometry, *Phys. Rev. B* **53**, 9353-
    9360.

Hung Fai Fong, Keimer, B., Anderson, P.W., Reznik, D., Dogan, F. and Aksay, I.A. (1995) Pho-
    non and magnetic neutron scattering at 41 meV in $YBa_2Cu_3O_7$, *Phys. Rev. Lett.* **75**, 316-319.

Kim, J.M., Somal, H.S., Czyzyk, M.T., van der Marel, D., Wittlin, A., Gerrits, A.M., Duijn,
    V.H.M., Hien, N.T., and Menousky, A.A. (1995) Strong damping of the c-axis plasmon in
    high-T_c cuprate superconductors, *Physica C* **247**, 297-308.

Krakauer, H., Pickett, W.E., and Cohen, C.E. (1993 Large calculated electron-phonon interac-
    tions in $La_{2-x}M_xCuO_4$, *Phys. Rev. B* **47**, 1002-1015.

Krantz, M., Rosen, H.J., Macfarlane, R.M. and V.Y. Lee, Effect of oxygen stoichiometry of
    Raman active lattice modes in $YBa_2Cu_3O_x$, *Physica Rev. B* **38**, 4992-4995.

Lee, C.H., Yamada, K., Arai, M., Wakimoto, S., Hosoya, S., and Endoh, Y. (1996), Anomalous
    breaking of phonon softening in the superconducting state of $La_{1.85}Sr_{0.15}CuO_4$. *Physica C*
    257, 264-270.

Litvinchuk, A.P., Thomsen, C. and Cardona, M. (1994) Infrared-active vibrations of high-tempe-
    rature superconductors: experiment and theory. In D.M. Ginsberg (ed), *Physical Properties
    of High Temperature Superconductors IV*, Wordl Scientific, Singapore, 375-470.

Lynn, J.W., Sumarlin, I.W., Neumann, D.A., Rush, J.J., Peng, J.L. and Li, Z.Y. (1991) Phonon
    density of states and superconductivity in $Nd_{1.85}Ce_{0.15}CuO_4$, *Phys. Rev. Lett.* **66**, 919-922.

Macfarlane, R.M., Hal Rosen and Seki, M. (1987) Temperature dependence of the Raman spec-
    trum of the high T_c superconductor $YBa_2Cu_3O_7$, *Sol. State Commun.* **63**, 831-834.

Macfarlane, R.M., Rosen, H.J., Engler, E.M., Jacowitz, R.D. and Lee, V.Y.Lee (1988) Raman
    study of the effect of oxygen stoichiometry on the phonon spectrum of the high-Tc
    superconductor $YBa_2Cu_3O_x$, *Phys. Rev. B* **38**, 284-289.

Meissner, G. and Blaschke, H. (1992) Mode softening in orthorhombic La_2CuO_4, *Ferroelectrics*
    128, 11-116.

Mook, H.A., Dai, P., Salama, K., Lee, D., Dogan, F., Aeppli, G., Boothroyd, A.T. and Mostoller,
    M.E. (1996) Incommensurate one-dimensional fluctuations in $YBa_2Cu_3O_{6.93}$, *Phys. Rev.
    Lett.* **77**, 370-373.

Mustre de Leon, J., Conradson, S.D., Batistic, I. and Bishop, A.R. (1990) Evidence for an axial
    oxygen-centered lattice fluctuation associated with the superconducting transition in
    $YBa_2Cu_3O_7$, *Phys. Rev. Lett-* **65**, 1675-1678.

Mustre de Leon, Batistic, I., Bishop, A.R., Conradson, S.O. and Trugman, S.A. (1992) Polaron ori-
    gin for anharmonicity of the axial oxygen in $YBa_2Cu_3O_7$, *Phys. Rev. Lett.* **68**, 3236-3239.

Parshin, P.P., Zemlyanov, M.G., Parphenov, O.E., and Chernyshev, A.A. (1990) Experimental in-
    vestigation of the partial density of Cu atoms vibrational states for $YBa_2Cu_3O_{7-y}$, in
    S.Hunklinger, W. Ludwig and G. Weiss (eds.) *Phonons 89*, World Scientific, pp. 310-312

Parshin, P.P., Glazkov, V.P., Zemlyanov, M.G., Irodova, A.V., Partenov, O.E. and Chernyshev, A.A. (1992) Superconductivity, structure and phonon spectrum of YBa$_2$ (Cu$_{0.9}$Zn$_{0.1}$)$_3$O$_{7-y}$, *Superconductivity* 5 (English edition), 445-450.

Parshin, P.P., Zemlyanov, M.G., Irodova, A.V., Ozhogin, V.I., Tolmacheva, N.S. and Shustov, L.D. (1994) Vibrational spectra of Cu and Tl atoms in Tl$_2$Ba$_2$CaCu$_2$O$_8$, *Phys. Solid State* 36 (English edition), 628-631.

Parshin, P.P., Ivanov, A.S., Zemlyanov, M.G., Shustov, L.D. and Schober, H. (1997) Determination of the partial phonon-density-of states of Cu in Pr$_2$CuO$_4$ and Pr$_{1.85}$Ce$_{0.15}$CuO$_4$, unpublished results.

Pickett, W.E., Cohen, R.E. and Krakauer, H. (1991) Lattice instabilities, isotope effect, and high-T$_c$ superconductivity in La$_{2-x}$Ba$_x$CuO$_4$, *Phys. Rev. Lett.* 67, 228-231.

Pintschovius, L., Pyka, N., Reichardt, W., Rumiantsev, A.Yu., Ivanov, A.S. and Mitrofanov, N.L. (1989) Inelastic neutron scattering study of La$_2$CuO$_4$, in V.L. Aksenov, N.N. Bogolubov and N.M. Plakida (eds.), *Progress in High Temperature Superconductivity*, Vol. 21, World Scientific, Singapore, pp. 36-46.

Pintschovius, L. (1990) Lattice dynamics and electron-phonon coupling in high-T$_c$ superconductors, in U. Rössler (ed.), *Festkörperprobleme/Advances in Solid State Physics* 30, 183-195.

Pintschovius, L., Pyka, N., Reichardt, W., Rumiantsev, A.Yu., Mitrofanov, N.L., Ivanov, A.S., Collin, G. and Bourges, P. (1991a) Lattice dynamical studies of HTSC materials, *Physica B* 174, 323-329.

Pintschovius, L., Pyka, N., Reichardt, W., Rumiantsev, A.Yu., Mitrofanov, N.L., Ivanov, A.S., Collin, G., and Bourges, P. (1991b) Lattice dynamical studies of HTSC materials, *Physica* C185-1989, 156-159

Pintschovius, L. and Reichardt, W. (1994) Inelastic neutron scattering studies of the lattice vibrations of high T$_c$ compounds, in D.M. Ginsberg (ed.), *Physical Properties of High Temperature Superconductors IV*, World Scientific, Singapore, 295-374.

Pintschovius, L., and Braden, M. (1996) Phonon anomalies in La$_2$CuO$_{4+\delta}$, *Journ. Low Temp. Physics* 105, 813-816.

Pintschovius, L., Reichardt, W. and Braden, M. (1997)Lattice vibrations of (La,Sr)$_2$CuO$_4$ (unpublished results).

Pyka, N., Mitrofanov, N.L., Bourges, P., Pintschovius, L., Reichardt, W., Rumiantsev, A.Yu. and Ivanov, A.S. (1992) Inelastic-neutron-scattering study of a soft rotational mode in Nd$_2$CuO$_4$, *Europhys. Lett.* 18, 711-716.

Pyka, N., Reichardt, W., Pintschovius, L., Engel, G., Rossat-Mignod, J. and Henry, J.H. (1993a) Superconductivity-induced phonon softening observed by inelastic neutron scattering, *Phys. Rev. Lett.* 70, 1457-1460.

Pyka, N., Reichardt, W., Pintschovius, L., Chaplot, S.L., Schweiss, P., Erb, A., and Müller-Voigt, G. (1993b) Neutron-scattering study of chain-oxygen vibrations in YBa$_2$Cu$_3$O$_7$, *Phys. Rev. B* 48, 7746-7749.

Pyka, N., Reichardt, W., Pintschovius, L. and Collin G. (1994) Lattice dynamics of YBa$_2$Cu$_3$O$_6$ and YBa$_2$Cu$_3$O$_7$, unpublished results.

Pyka, N., Ivanov, A.S., Reichardt, W., Pintschovius, L., Mitrofanov, N.L. and Rumiantsev, A.Yu. (1998) Lattice dynamics of T'-phase compounds, in preparation.

Mc Queeney, R.J., Egami, T., Shirane, G., and Endoh, Y. (1996) Wide and asymmetric bond-stretching phonons in $La_{1.85}Sr_{0.15}CuO_4$, Phys. Rev. B **54**, 9689-9692.

Reichardt, W., Pintschovius, L., Pyka, N., Schweiß, P., Erb, A., Bourges, P., Collin, G., Rossat-Mignod, J., Henry, I.Y., Ivanov, A.S., Mitrofanov, N.L. and Rumiantsev, A.Yu. (1994) Anharmonicity and electron-phonon coupling in cuprate superconductors studied by inelastic neutron scattering, *Journ. of Superconductivity* **7**, 399-407.

Reichardt, W., Pintschovius, L., Pyka, N., Schweiss, P., Erb, A., Bourges, P., Collin, G., Rossat-Mignod, J., Henry, I.V., Ivanov, A.S., Mitrofanov, N.L. and Rumiantsev, A.Yu. (1994) Anharmonicity and electron-phonon coupling in cuprate superconductors studied by inelastic neutron scattering, *Journ. of Superconductivity* **7**, 3099-408

Reichardt, W. (1995) Inelastic neutron scattering study of superconductivity-induced phonon frequency shifts in $YBa_2Cu_3O_7$, unpublished results.

Reichardt, W. (1996) Cu-O bond stretching vibrations in $YBa_2Cu_3O_7$ studied by inelastic neutron scattering, *Journ. of Low Temp. Phys.* **105**, 807-812.

Reichardt, W. (1997a) Lattice dynamical model calculations for Nd_2CuO_4 and Pr_2CuO_4, unpublished results.

Reichardt, W. (1997b) Lattice dynamical model calculations for $YBa_2Cu_3O_7$, unpublished results.

Reichardt, W. (1997c) Lattice dynamical model calculations for Bi-based high-T_c compounds, unpublished results.

Renker, B., Rietschel, H., Pintschovius, L., Gläser, W., Brüesch, P., Kuse, D. and Rice, M.J. (1973) Observation of a giant Kohn anomaly in the one-dimensional conductor $K_2Pt(CN)_4Br_{0.3} \cdot 3H_2O$. Phys. Rev. Lett. **30**, 1144-1147.

Renker, B., Gompf, F., Gering, E., Nücker, N., Ewert, D., Reichardt, W. and Rietschel, H. (1987) Phonon density-of-states for the high T_c-superconductor $La_{1.85}Sr_{0.15}CuO_4$ and its non-superconducting reference La_2CuO_4, Z. Phys. B - Condensed Matter **67**, 15-18.

Renker, B., Gompf, F., Gering, E., Ewert, D., Rietschel, H. and Dianoux, A. (1988a) Strong changes in the phonon spectra of 123 superconductors by varying oxygen concentration, Z. Phys. B - Condensed Matter **73**, 309-312.

Renker, B., Gompf, F., Gering, E. and Ewert, D. (1989) Observation of phonon shifts in $Bi_2Sr_2Ca_{1-x}Y_xCu_2O_8$ and related high temperature superconductors, *Physica C* **162-164**, 462-463.

Renker, B., Gompf, F., Adelmann, P., Wolf, T., Mutka, H. and Dianoux, A. (1992) Electron-phonon coupling in HTC superconductors evidenced by inelastic neutron scattering, *Physica B* **180 & 181**, 450-452.

Renker, B. (1993) Measurement of the phonon density of states of Nd_2CuO_4, unpublished results.

Reznik, D., Keimer, B., Dogan, F. and Aksay, I.A. (1995) q dependence of self-energy effects of the plane oxygen vibration in $YBa_2Cu_3O_7$, Phys. Rev. Lett. **75**, 2396-2399.

Rojewski, E. (1994) Mikroskopische Berechnung von Normalleitereigenschaften des Hochtemperatursupraleiters $(La_{2-x}Sr_x)CaO_4$, Thesis, *KfK report* **5391**, Forschungszentrum Karlsruhe.

Rhyne, J.J., Neumann, D.A., Gotaas, J.A., Beech, R., Toth, L., Lawrence, S., Wolf, S., Osofsky, M. and Gubser, D.U. (1987) Phonon density of states of superconducting $YBa_2Cu_3O_7$ and the nonsuperconducting analog $YBa_2Cu_3O_6$, Phys. Rev. B **36**, 2294-2297

Salkola, M.I., Bishop, A.R., Mustre de Leon, J., and Trugman, S.A. (1994) Dynamic polaron tunneling in $YBa_2Cu_3O_7$: Optical response and inelastic neutron scattering, *Phys. Rev. B* **49**, 3671-3674.

Statt, B.W., Hammel, P.C., Fisk, Z., Cheong, S.W., Chou, F.C., Johnston, D.C., and Schirber, J.E. (1995) Oxygen ordering and phase separation in $La_2CuO_{4+\delta}$, *Phys. Rev. B* **52**, 15575-15581.

Thomsen, C., Liu, R., Bauer, M., Wittlin, A., Genzel, L., Cardona, M., Schönherr, E., Bauhofer, W. and König, W. (1988) Systematic Raman and infrared studies of the superconductor $YBa_2Cu_3O_{7-x}$ as a function of oxygen concentration, *Sol. State Commun.* **65**, 55-58.

Thomsen, C., Cardona, M., Gegenheimer, B., Liu, R. and Simon, A. (1988b) Untwinned single crystals of $YBa_2Cu_3O_{7-\delta}$: An optical investigation of the a-b anisotropy, *Phys. Rev. B* **37**, 9860-9863.

Thurston, T.R., Birgeneau, R.J., Gabbe, D.R., Jenssen, H.P., Kastner, M.A., Picone, P.J., Preyer, N.W., Axe, J.D., Böni, P., Shirane, G., Sato, M., Fukuda, K. and Shamoto, S. (1989) Neutron scattering study of soft optical phonons in $La_{2-x}Sr_xCuO_{4-y}$, *Phys. Rev. B* **39**, 4327-4333.

Tranquada, J.M., Axe, J.D., Ichikawa, N., Nakamura, Y., Uchida, S. and Nachumi, B. (1996) Neutron-scattering study of stripe-phase order of holes and spins in $La_{1.48}Nd_{0.4}Sr_{0.12}CuO_4$, *Phys. Rev. B* **54**, 7489-7499.

Tranquada, J.M., Axe, J.D., Ichikawa, N., Moodenbaugh, A.R., Nakamura, Y. and Uchida, S. (1997) Coexistence of, and competition between, superconductivity and charge-stripe order in $La_{1.6-x}Nd_{0.4}Sr_xCuO_4$, *Phys. Rev. Lett.* **78**, 338-341

Zaanen, S., Osman, D.Y., Eskes, M. and van Saarloos, W. (1996) Dynamical stripe correlations in cuprate superconductors, *Journ. Low Temp. Physics* **105**, 569-580

Zemlyanov, M.G., Krylov, I.V., Parshin, P.P. and Soldatov, P.I. (1993a) Partial oscillation spectra of Cu, La, and O atoms in La_2CuO_4, *JETP* (English edition) **77**, 148-152

Zemlyanov, M.G., Irodova, A.V., Krylov, I.V., Parshin, P.P., Tolmacheva, N.S., Shustov, L.D., Soldatov, P.I. and Suleimanov, S.Kh. (1993b) Thermal excitation spectra of Tl- and Bi-based 2212 HTSC, *Superconductivity: Physics, Chemistry, Technology* **6**, 435-440.

Zeyher, R. and Zwicknagl, G. (1990) Superconductivity-induced phonon self-energy effects in high-Tc superconductors, *Z. Phys. B - Condensed Matter* **78**, 175-190

Zeyher, R. (1991) Superconductivity-induced dynamic and static changes in $q \neq 0$ phonons of high-T_c oxides, *Phys. Rev. B* **44**, 9596-9604

PHASE SEPARATION, CHARGE SEGREGATION, AND SUPERCONDUCTIVITY IN LAYERED CUPRATES

J. M. TRANQUADA

Brookhaven National Laboratory
Upton, NY 11973, USA

1. Introduction

Experimental observations suggesting inhomogeneity in the layered cuprates have been reported since shortly after the discovery of superconductivity in these materials [1]. There are at least two forms of inhomogeneity that have, by now, been fairly well established. One involves the phase separation of oxygen interstitials in $La_2CuO_{4+\delta}$. The interstitials are mobile near room temperature, giving a homogeneous phase, while cooling to lower temperatures can lead to segregation into phases with distinct oxygen concentrations. Early work focussed on samples with $\delta < 0.05$, in which case the two phases present at low temperature are the nearly-stoichiometric antiferromagnet (described in Chapter 4) and an oxygen-rich superconducting phase. For larger δ, multiple superconducting phases have been discovered, with the T_c varying with δ. It has also been shown that, within a particular oxygen-rich phase, the interstitials order in a layered fashion. Neutron diffraction has played a crucial role in establishing the nature of phase separation and order in $La_2CuO_{4+\delta}$, and that work is the subject of the first half of this chapter.

Another type of inhomogeneity that has been observed involves a periodic modulation of the charge density within the CuO_2 planes. A static form of this modulation has been clearly identified in the system $La_{1.6-x}Nd_{0.4}Sr_xCuO_4$. The experimental evidence appears to be consistent with a picture in which the holes doped into the planes segregate into stripes that act as domain walls between antiferromagnetic domains, with a π phase shift across the domain walls. A dynamical form of this modulation provides one model for interpreting the spin correlations in $La_{2-x}Sr_xCuO_4$

and $YBa_2Cu_3O_{6+x}$ (see Chapters 3 and 4). The experimental evidence for charge stripe correlations will be discussed in the second half of the chapter.

Two model systems that have been valuable for understanding the cuprates are $La_2NiO_{4+\delta}$ and $La_{2-x}Sr_xNiO_4$. The former exhibits phase separation and oxygen ordering similar to that found in $La_2CuO_{4+\delta}$, while the two systems together provided the first examples of charge-stripe order. Some of the relevant work on the nickelates will also be described.

1.1. BASIC STRUCTURE VARIATIONS AND NOTATION

Before continuing, it may be useful to briefly review some of the basic structural variations and the relevant notation. The variously doped versions of La_2CuO_4 and La_2NiO_4 are commonly referred to as "214" compounds. The structures of the undoped parent compounds are essentially identical, consisting of MO_2 (M = Cu, Ni) planes separated by La_2O_2 layers. Within an MO_2 plane, the metal ions form a square lattice with oxygen atoms bridging the nearest-neigbor sites. The unit-cell vectors a_1 and a_2 are parallel to nearest-neighbor M–O bonds within the planes, and a_3 is perpendicular to the planes. There are two MO_2 planes per unit cell, and they are related by the basis vector $\frac{1}{2}a_1 + \frac{1}{2}a_2 + \frac{1}{2}a_3$. For each M ion there is one out-of-plane oxygen directly above (along a_3) and one below, effectively completing a tetragonally-distorted octahedron of oxygens. La ions sit above and below the centers of the squares formed by the M ions.

The simple structure described above is known as the High Temperature Tetragonal (HTT) phase or the K_2NiF_4 structure. The restoring forces for rotations of the MO_6 octahedra are small, and variations on the basic structure can be characterized by distinct tilt patterns of the octahedra [2]. To describe the different structures it is convenient to define order parameters Q_1 and Q_2 such that Q_1 is the magnitude of the octahedral tilt about a [110] axis and Q_2 is the tilt magnitude about [1$\bar{1}$0]. If only one of the two order parameters is nonzero, the resulting phase is called the Low Temperature Orthorhombic (LTO). In the LTO structure, the in-plane unit-cell vectors rotate by 45° relative to those of the HTT, and the corresponding lattice parameters increase by $\sqrt{2}$. The orthorhombic strain $2(b-a)/(b+a)$ is typically $\lesssim 1\%$. In the Low Temperature Less Orthorhombic (LTLO) phase, Q_1 and Q_2 are both finite but unequal. Finally, if $Q_1 = Q_2 \neq 0$ the structure is called Low Temperature Tetragonal (LTT). All of the phases with finite tilts can be detected by superstructure reflections in a diffraction measurement. The different structural variations and their space groups are summarized in Table 1. Occasionally diffraction studies have determined a structure to be orthorhombic, but have not identified superstructure reflections. The space group $Fmmm$ has been used to characterize the effective structure in such cases. (Note that without some sort of distortion such as octahedral tilts there is no reason for the crystal symmetry to deviate from tetragonal.)

TABLE 1. Summary of commonly observed structural variants in 214 compounds. Notation is defined in the text.

Phase	Space group	Unit cell	Tilts
HTT	$I4/mmm$	$a \times a \times c$	$Q_1 = Q_2 = 0$
LTO	$Bmab$	$\sqrt{2}a \times \sqrt{2}b \times c$	$Q_1 \neq 0, Q_2 = 0$
			or $Q_1 = 0, Q_2 \neq 0$
LTLO	$Pccn$	$\sqrt{2}a \times \sqrt{2}b \times c$	$Q_1 \neq 0, Q_2 \neq 0, Q_1 \neq Q_2$
LTT	$P4_2/ncm$	$\sqrt{2}a \times \sqrt{2}a \times c$	$Q_1 = Q_2 \neq 0$
tentative	$Fmmm$	$\sqrt{2}a \times \sqrt{2}b \times c$	undetermined

Throughout this chapter, a, b, and c will refer to lattice parameters measured with respect to the HTT unit cell. In section 2, it will be convenient to specify reciprocal lattice vectors with respect to the orthorhombic cell in units of $(2\pi/\sqrt{2}a, 2\pi/\sqrt{2}b, 2\pi/c)$, whereas in section 3, the reference will be the HTT cell with units of $(2\pi/a, 2\pi/b, 2\pi/c)$.

2. Phase Separation in $La_2CuO_{4+\delta}$

2.1. OXYGEN INTERCALATION AND PHASE SEPARATION

In the system first studied by Bednorz and Müller [3], $La_{2-x}Ba_xCuO_4$ and the related $La_{2-x}Sr_xCuO_4$, the superconducting transition temperature, T_c, drops toward zero as x is reduced to 0.05, and antiferromagnetic order is observed for $x < 0.02$ (see Chapter 4). Thus, it came as a surprise when it was first reported that a small volume fraction of superconductivity had been detected in samples of nominally pure La_2CuO_4 [4, 5, 6]. It was soon demonstrated that the superconducting volume fraction, with a $T_c \sim 32$ K could be increased to roughly 50% by annealing in high pressure (a few kbar) oxygen [7, 8].

The question of how antiferromagnetism and superconductivity might coexist in $La_2CuO_{4+\delta}$ was largely resolved in a neutron powder diffraction study reported by Jorgensen et al. [9]. At low temperatures their samples contained two very similar orthorhombic phases, a primary one with $\delta \approx 0$, and a second oxygen-rich phase that was superconducting. The nearly-stoichiometric phase had the usual LTO structure, whereas the superlattice peaks characteristic of the LTO phase were missing for the oxygen-rich phase, and hence it was analyzed as $Fmmm$.

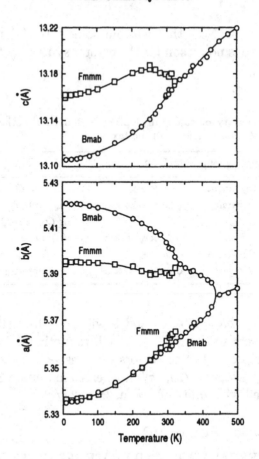

Figure 1.   Lattice constants vs. temperature for a $La_2CuO_{4+\delta}$ sample annealed in 0.1 kbar O_2 at 580°C. Two phases (denoted *Bmab* and *Fmmm*) were detected below 320 K. Above 320 K there is a single *Bmab* phase, which transforms to $I4/mmm$ above 430 K. Here *a* and *b* are lattice constants of the orthorhombic cell. From Jorgensen *et al.* [9].

The measurements by Jorgensen *et al.* [9] also showed that only a single phase was present at high temperatures, with a reversible phase separation occurring (in one sample) at 320 K. The temperature dependence of the lattice parameters for a sample annealed in 100 bar O_2 are shown in Fig. 1. Similar results were obtained in a combined neutron and synchrotron x-ray powder diffraction study by Zolliker *et al.* [10].

Evidence that the oxygen-deficient phase in phase-separated samples is antiferromagnetic was initially provided by muon spin rotation (μSR) [11] and nuclear magnetic resonance (NMR) [12] studies. The first neutron diffraction study to directly verify that it is indeed the *Bmab* phase that is antiferromagnetically ordered was performed by Vaknin *et al.* [13] on a single crystal. Similar results were obtained in another single-crystal study by Koga et al. [14], in which X-ray and neutron diffraction measurements of the cell parameters and magnetic order were compared with resistivity and magnetization results. In both of the latter diffraction studies [13, 14], although the oxygen concentrations of the samples were different, the phase-separation temperature was 260–265 K, with antiferromagnetic order appearing at 245–250 K.

The location of the interstitials within the lattice was determined in single-crystal neutron diffraction studies by Chaillout et al. [15, 16]. The Q resolution was limited by the choice of a very short neutron wavelength (0.484 Å), and hence the sample was initially assumed to be single phase [15]. The same data were later analyzed in terms of two phases [16]. In both cases the interstitials were found to occupy a position close to $(\frac{1}{4}, \frac{1}{4}, \frac{1}{4})$ with respect to the $Fmmm$ unit cell. This position is located halfway between neighboring CuO_2 planes, in the approximate center of a tetrahedron formed by 4 apical oxygens. Essentially the same interstitial position was determined for $La_2NiO_{4+\delta}$ by Jorgensen et al. [17] by neutron powder diffraction.

The maximum value of δ that can be achieved with high-pressure techniques is ~ 0.03. A substantially larger range of δ can be accessed through chemical [18, 19] or electrochemical [20, 21] oxidation. Besides the superconducting phase with $T_c \sim 32$ K on the oxygen-rich side of the miscibility gap, the electrochemically-prepared samples provided evidence for a higher-δ phase with $T_c \sim 45$ K [20, 21, 22, 23]. One complication is that it has proven difficult to obtain single-phase samples with large δ. Feng et al. [23] have argued that this is because of nonequilibrium conditions in samples electrochemically oxidized at room temperature, a situation which is exacerbated by high ionic currents. They have observed changes in the superconducting properties after annealing samples at temperatures up to 110°C (with no oxygen loss). There is evidence for an intermediate phase with $T_c \sim 15$ K [23, 24, 19]. It has been argued [23, 24] that this low-T_c phase is analogous to the depressed-T_c phase found in $La_{2-x}Ba_xCuO_4$ at $x \approx \frac{1}{8}$ [25].

Another complication involves the determination of δ. The amount of excess oxygen is typically measured by either thermogravimetric analysis (TGA) or iodometric titration. Both techniques are capable of providing accurate results, but each has possible pitfalls. Titration can miss oxygen if it escapes in bubbles due to an excessively rapid dissolution of the sample [26]. In TGA, the sample's weight loss during heating can be precisely measured, but an accurate estimate of δ depends on knowing how much of the weight loss actually corresponds to oxygen. Discrepancies between results obtained with the two methods on identical samples have frequently been reported. For example, Chou and coworkers [21, 27] found

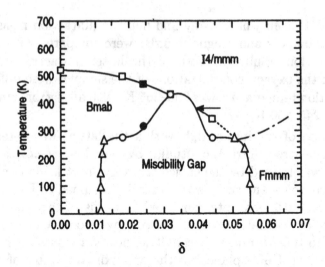

Figure 2.   Phase diagram of $La_2CuO_{4+\delta}$ in the region of the miscibility gap, as determined by neutron powder diffraction studies. From Radaelli *et al.* [29].

significant differences between TGA and titration measurements on their electrochemically-oxidized polycrystalline samples, with the titrated values of δ being much smaller than the TGA results. In comparison, Li *et al.* [28] have reported consistent results for $\delta \leq 0.03$, but a systematic deviation at larger δ. Thus, one must be cautious in comparing results on different samples with nominally identical values of δ.

Neutron diffraction studies have provided important information on the nature of interstitial order in the large δ phases, and we will discuss those results shortly. First, we return to the problem of the phase separation at low δ. Radaelli *et al.* [29] mapped out the miscibility gap by studying a series of electrochemically-oxidized polycrystalline samples with neutron diffraction. Their resulting phase diagram is shown in Fig. 2. They found that the low-δ *Bmab* phase is stable for $\delta \lesssim 0.01$, and that the high-δ edge of the miscibility gap is at $\delta \approx 0.055$.

Fairly recently Balagurov *et al.* [30] have reported a neutron diffraction study of two single crystals oxygenated by a high-pressure technique. In a crystal with $\delta = 0.03$, which should be in the middle of the miscibility gap, they observed no phase separation. This sample was superconducting, but T_c was only 12 K. These results appear to conflict with the those obtained by Vaknin *et al.* [13] on a nominally similar crystal. There is no objective way to compare sample quality without further measurements; however, it is interesting to note that phase separation appears to be suppressed in $La_{2-x}Bi_xCuO_{4+\delta}$ with $x = 0.05$ [31, 32].

2.2. LESSONS ON INTERSTITIAL ORDER FROM La$_2$NiO$_{4+\delta}$

Structurally, La$_2$NiO$_{4+\delta}$ is quite similar to the cuprate, but the greater ease with which oxygen can be intercalated into the nickelate makes it a useful model system for studying interstitial order [33, 17]. Annealing in air yields δ as high as ~ 0.15 [34, 35, 36], and it is possible to raise δ up to ~ 0.18 by an appropriate annealing in 1 atm. of O$_2$ [17]. Oxygen can also be intercalated electrochemically [35, 37, 38, 39, 40], and a value of δ as large as 0.25 has been reported [35, 37]. With the exception of electron, X-ray, and neutron diffraction studies of the $\delta \approx 0.25$ phase by Demourgues et al. [37, 41], most structural studies have focussed on the $\delta \lesssim 0.18$ regime.

As in the cuprate, the $\delta = 0$ phase has the LTO structure at room temperature; however, several neutron [42, 43, 44] and X-ray [36] diffraction studies have shown that below ~ 70 K the structure transforms to a nearly-tetragonal LTLO phase. Jorgensen et al. [17] performed initial neutron diffraction measurements on powder samples with $\delta \approx 0.07$ and 0.18. They found a single phase for the latter sample, but the former contained two phases. A very broad miscibility gap, analogous to the one identified in La$_2$NiO$_{4+\delta}$ [9], was suggested [17]. From Rietveld refinement of the diffraction data, the average position of the oxygen interstitials was determined to be the $(\frac{1}{4}, \frac{1}{4}, \frac{1}{4})$ site in the orthorhombic $Fmmm$ structure. Essentially identical conclusions concerning the structure were reached in a similar study by Rodríguez-Carvajal et al. [34].

The possibility that the phase diagram involved more than simply two phases separated by a wide miscibility gap was first suggested in an electron-diffraction investigation by Hiroi et al. [45]. A variety of superstructures was observed in samples with δ ranging from 0.04 to 0.20. It was suggested that the superstructures corresponded to ordered arrangements of interstitials with large unit cells [45]; however, the same superstructures have never been detected in bulk samples by neutron diffraction. It is possible that the reported structures are stable only in the extremely thin specimens required for electron microscopy.

Although the specific superstructures were not confirmed, the existence of several distinct phases was determined in X-ray diffraction studies on a large number of samples with δ covering the range $0 \leq \delta \leq 0.18$ [46, 47]. Rice and Buttrey [46] showed that, at room temperature, there exists an LTT phase in the narrow range $0.02 < \delta < 0.03$, and that an HTT phase occurs for $0.055 \lesssim \delta \lesssim 0.14$. A distinct orthorhombic phase ($Fmmm$) was identified at $\delta \approx 0.17$. Very similar results were reported by Tamura et al. [47], with the main difference being that within the range $0.05 < \delta < 0.09$ the structure was identified as orthorhombic rather than tetragonal. Hosoya et al. [48] found indirect evidence for several phases from a neutron diffraction study of antiferromagnetic order in single-crystal samples. Mehta and Heany [49] reported an X-ray diffraction study on a powder sample with a nominal δ of 0.18, in which they advocated a structure refinement with space group *Bbcm*. That choice allows a rotation of the NiO_6 octahedra about the *c*-axis; however, their analysis is not supported by observations of any unique superstructure reflections, and it appears to be inconsistent with the results described below.

The nature of the interstitial order in the range $0.05 \lesssim \delta \lesssim 0.14$ has been worked out subsequently in a series of neutron diffraction studies on single crystals [50, 51, 52, 53]. This work focussed on superlattice peaks that occur in the $(0kl)$ zone of reciprocal space (based on an orthorhombic unit cell). To understand the results, consider first the tilt pattern of the NiO_6 octahedra in the LTO phase. It results in superlattice peaks within the $(0kl)$ zone at positions such that k is an odd integer and l is even. (The structure factors for these peaks depend predominantly on the displacements of the oxygen ions, so that in detecting the superstructure, neutrons have a relative sensitivity advantage over X-rays.) The modulation wave vector for the LTO tilt pattern is $\mathbf{K} = (0, 1, 0)$. In oxygen-intercalated crystals with δ in the range $0.05 \lesssim \delta \lesssim 0.11$, the characteristic LTO peaks are absent, and instead one observes incommensurate peaks at positions $(0, k, l \pm \Delta)$ with k odd, l even, and $\frac{1}{4} < \Delta \leq \frac{1}{2}$. The interstitial-induced modulation wave vectors are then $\mathbf{K}_{\pm} = (0, 1, \pm\Delta)$.

The first observation of the incommensurate superstructure involved a crystal with $\delta = 0.105$, for which $\Delta \approx \frac{1}{2}$ [50]. On the assumption that the interstitial order had been properly characterized by electron diffraction [45], the significance of the neutron observations was initially misinterpreted [50]. Eventually, it became apparent that the systematic variation of Δ with δ could be explained in terms of a simple model involving a one-dimensional ordering of the interstitials [51]. Suppose that an interstitial is placed at a $(\frac{1}{4}, \frac{1}{4}, \frac{1}{4})$ site within the LTO structure. The extra oxygen ion will tend to repel neighboring apical oxygen ions, and this repulsion can be accomodated conveniently if the octahedra in the layers above the interstitial site reverse their tilt direction. Such a change, which can be described as an antiphase domain boundary, creates an entire layer of sites that are favorable for occupation by interstitials (see Fig. 3). The density of interstitials within a single layer will be limited by Coulomb interactions. When multiple layers are occupied, long-range interactions will favor periodic spacing of the antiphase domain boundaries.

If there is only a 1D ordering of the interstitials, with no regular order within the occupied layers, then the model described is quite similar to the "staging" order observed when alkali ions are intercalated into graphite [54]. It is convenient to adopt the staging language to describe the present

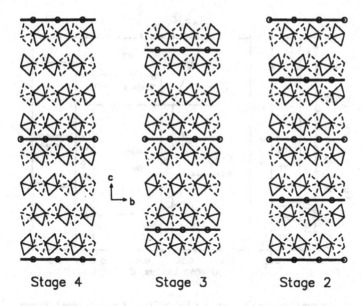

Stage 4   Stage 3   Stage 2

Figure 3. Schematic tilt patterns of the NiO$_6$ octahedra for several possible stagings of the interstitial oxygen layers. The solid octahedra intersect the plane of the figure, while the dashed octahedra are displaced perpendicular to the plane by $\frac{1}{2}a$. The possible interstitial positions (only a fraction of which are occupied), indicated by open circles, are displaced by $\frac{1}{4}a$. From Ref. [51].

situation. A structure in which the interstitial layers are separated by n NiO$_2$ layers is described as stage n. Examples of stage-4, -3, and -2 orderings are illustrated in Fig. 3. The modulation wave vectors for a stage n structure are $(0, 1, \pm\frac{1}{n})$. Thus, the parameter Δ in the observed modulation wave vector should be equal to $1/n$. The fact that measured values of Δ deviate from precisely commensurate fractions can be understood in terms of defects in the staging structure. Measured peak positions and asymmetric lineshapes can be quantitatively modelled using the formulas of Hendricks and Teller [55] for a system containing a random mixture of 2 distinct layer spacings [51].

In the samples that exhibit staging order, the high temperature structure, in which the interstitials have no long-range order, is HTT. At the transition to the ordered state, the symmetry changes from tetragonal to orthorhombic, consistent with the LTO-like tilting of the octahedra. In the original study of the $\delta = 0.105$, stage-2 sample, the low-temperature measurements suggested that the sample had phase separated into tetragonal and orthorhombic phases. It was later shown that the tetragonal component corresponded to disordered HTT phase that had failed to transform because of the slow kinetics of the transition [52]. The ordering transition occurs by a nucleation and growth mechanism, and the transformation rate

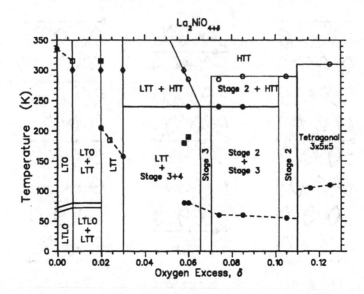

Figure 4. Approximate phase diagram for $La_2NiO_{4+\delta}$. Open circles (diamonds) indicate phase boundaries determined by neutron single-crystal (X-ray powder) diffraction. Solid circles (squares) denote Néel temperatures of primary (secondary) phase; open squares indicate magnetic transitions translated to the appropriate values of δ. From Ref. [51].

drops to zero at the transition temperature. It is possible to quench in a considerable amount of disorder by rapidly cooling the sample. In fact, Nakajima *et al.* [56] have shown that interstitial order can be completely inhibited by a sufficiently rapid quench.

An approximate phase diagram is shown in Fig. 4 [51]. Evidence has been observed for stage-2 and stage-3 phases, although a pure stage-3 phase has not been observed yet. Crystals at intermediate δ's exhibit distinct stage-2 and stage-3 superlattice peaks when slowly cooled. Crystals with $0.03 < \delta \lesssim 0.06$ phase separate into LTT plus a phase corresponding to a random stacking of stage-3 and stage-4 domains. The ratio of the δ values for the stage-2 and stage-3 phases is consistent with a fixed density of interstitials within the occupied layers.

For $\delta \gtrsim 0.11$, 3D order of the interstitials appears. Studies of crystals with $\delta = 0.125$ have revealed two types of superlattice reflections, characterized by $\mathbf{K}_1 = (\frac{1}{3}, 0, 1)$ and $\mathbf{K}_2 = (0, \pm\frac{4}{5}, \frac{4}{5})$ [53, 56]. The results can be understood in terms of a large orthorhombic unit cell of size $3\sqrt{2}a \times 5\sqrt{2}b \times 5c$ (relative to the HTT cell), although an orthorhombic splitting of the lattice parameters was not observed [53]. Analysis of the superstructure intensities yielded a specific model for the interstitial order, corresponding to an ideal oxygen excess of $\delta = \frac{2}{15}$. The accuracy of this ideal density estimate appears to be confirmed by a recent study on a crystal with δ very close to $\frac{2}{15}$ [57]. No cooling-rate dependence was observed for the oxygen-order peaks in the latter sample, in contrast to the $\delta = 0.125$ samples [53, 56].

2.3. STAGING ORDER IN $La_2CuO_{4+\delta}$

We have seen that a number of distinct phases occur in $La_2NiO_{4+\delta}$. Evidence for at least 3 different phases in the oxygen-rich cuprate was provided by a synchrotron X-ray diffraction study on a series of electrochemically-intercalated powder samples [58]. The high Q resolution obtainable with synchrotron X-rays made it possible to resolve the small differences in lattice parameters for the 3 coexisting phases. With increasing δ, both the orthorhombic strain and the c lattice parameter were found to systematically increase. One puzzle in this study was that the values of δ determined by iodometric titration and TGA measurements appear to be at least a factor of 2 smaller than those for corresponding phases determined by other groups.

Determination of the interstitial order requires the study of single crystals by neutron diffraction. Because of the substantial effort required to electrochemically intercalate crystals, measurement of δ by a destructive technique such as titration or TGA is not practical. Instead, one can measure the total charge per La_2CuO_4 formula unit that flows in the electro-chemical circuit by integrating the current. According to Grenier et al. [22] δ should be equal to the charge divided by 2, provided no reactions other than oxygen intercalation occur in the cell. In early studies, this latter constraint was not always satisfied.

Radaelli et al. [27] performed the first neutron diffraction study on a single crystal, utilizing a sample with $\delta \sim 0.1$ intercalated by Chou et al. [59]. Superlattice peaks characterized by modulation wave vectors $\mathbf{K}_1 = (\frac{1}{10}, \frac{1}{10}, \frac{1}{6})$ and $\mathbf{K}_2 = (\frac{1}{5}, 0, \frac{1}{3})$ were observed, but no model for the interstitial order was developed. Such modulation wave vectors suggest 3D correlations of the interstitials. The c lattice parameter reported for this crystal would be consistent with a δ substantially greater than the nominal value of 0.1.

Somewhat different results were obtained in a study of several electrochemically-oxidized crystals by Wells et al. [60]. Neutron diffraction measurements revealed superlattice peaks very similar to those found in $La_2NiO_{4+\delta}$, indicative of staging order. Examples of scans near the point $(0,1,4)$ (the position of a superlattice peak of the LTO phase) are shown in Fig. 5. The staging peaks appear at $(0, 1, 4 \pm \Delta)$, with Δ ranging from $\approx \frac{1}{6}$ to $\frac{1}{2}$. Finite intensity at $(0,1,4)$ indicates coexistence of the oxygen-poor LTO phase. In sample C, 3 different phases coexist, corresponding to stages 4, 3, and 2. The staging model would also appear to be consistent with the

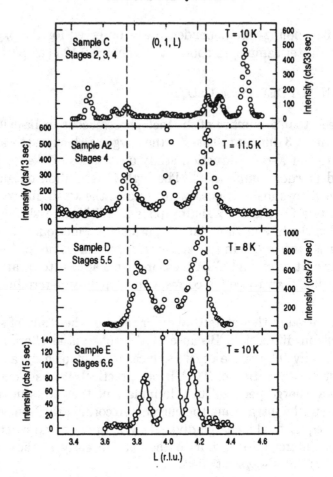

Figure 5. Neutron scattering scans along $\mathbf{Q} = (0, 1, L)$, showing the superlattice peaks due to interstitial order for several different crystals. The peak at $(0,1,4)$ is a superlattice reflection from the oxygen-poor LTO phase. The vertical, dashed reference lines are at $L = 4 \pm \frac{1}{4}$. From Wells *et al.* [60].

distribution of octahedral tilts inferred from La and Cu NMR measurements [61, 62]. The indication from the NMR studies [62] that the largest tilts occur adjacent to the interstitial sites appears to be compatible with the nature of the oxygen order; however, the absence of higher-harmonic satellites in the neutron diffraction measurements requires some degree of disorder with respect to the average relative positions of the occupied layers [60].

While the nature of the ordered state appears to be similar to the staging order found in the nickelate, the temperature dependence of the ordering is different. Above ~ 200 K, the staging order parameter in the cuprate crystals decreases continuously as the temperature is increased [60]. After the staging order disappears, the orthorhombic distortion survives up to a significantly higher temperature (~ 400 K [29]). This is in contrast to the nickelate, where the transitions to interstitial order and an orthorhombic cell are coincident and first order. The continuous development of staging order in the cuprate may explain why the order parameter does not appear to depend on the cooling rate [60], at least for temperatures above ~ 200 K.

There does exist a different kind of cooling-rate dependence that has been observed in $La_2CuO_{4+\delta}$. In a number of studies on the superconducting transition in samples with δ in the miscibility gap it was found that the value of T_c depends on how the sample is cooled [63, 64, 65]. In particular, it was shown that annealing in the neighborhood of 200 K could raise T_c by several degrees. A structural response associated with this annealing effect was discovered by Xiong et al. [66]. Using neutron diffraction, they studied the temperature dependence of the $(0, 1, 4 + \Delta)$ staging-order peak ($\Delta \approx 0.15$) in the oxygen-rich phase of a $\delta \sim 0.015$ crystal previously characterized by Koga et al. [14]. Monitoring the peak intensity while cooling the crystal slowly, they observed an abrupt drop in intensity as the temperature decreased through 200 K. Further investigation revealed that the apparent intensity loss occurred because the peak split in two along h, with positions $(\Delta', 1, 4 + \Delta)$. Representative scans along $(h, 1, 4 + \Delta)$ from Wochner et al. [67] are shown in Fig. 6. Similar results have been reported by Birgeneau et al. [68].

To explain the peak splitting along h, Wochner et al. [67] have suggested that a domain boundary in the b–c plane at which the positions of the occupied interstitial layers shift by one layer position along the c direction will involve a π phase shift in the octahedral tilt pattern. A roughly periodic distribution of such antiphase domain boundaries along the a direction would yield the observed peak splitting. The resulting staggering of the staging domains, as indicated in the inset of Fig. 6, is similar to the domain model originally suggested by Daumas and Herold for intercalated graphite [54].

The results obtained from a series of single-crystal samples [60, 68, 69] have led to the proposed phase diagram shown in Fig. 7 [70]. Recent measurements by Blakeslee et al. [69] on carefully oxidized crystals confirm the result of Radaelli et al. [29] that the value of δ for the first oxygen-rich phase is 0.055 ± 0.005; however, the phase separation temperatures in the first miscibility gap are systematically lower for single-crystals compared to polycrystalline samples. The ratio of δ values for the stage-6 and stage-4 phases appears to be consistent with a fixed interstitial density per occupied layer [69], as in the nickelate [51], but that density must be significantly greater than in the nickelate. The phase diagram suggests that in-plane ordering of the staging domains [66, 67] occurs in both stage-6 and -4 phases;

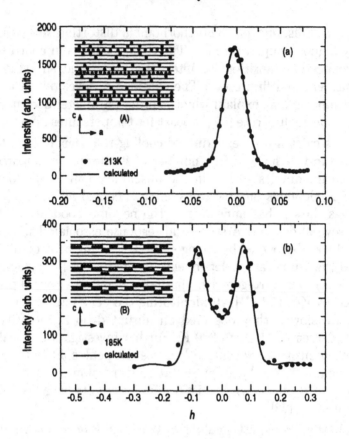

Figure 6. Neutron scattering scans along $\mathbf{Q} = (h, 1, 4.16)$ at temperatures above (a) and below (b) the transition to in-plane ordering of the staged-domains, measured in a crystal of $La_2CuO_{4+\delta}$ with $\delta \approx 0.015$. The solid lines are fits using the respective models indicated schematically in the insets. From Wochner *et al.* [67].

however, that transition has only been observed in the stage-6 phase so far. One unresolved issue concerns the relationship between the modulation wave vectors observed by Radaelli *et al.* [27] and the staging order. Ongoing studies are likely to provide some insight in the near future [71].

2.4. IMPLICATIONS

The similarity between the phase-separation temperature (in the first miscibility gap) and the Néel temperature of the oxygen-poor phase led experimentalists to speculate that the phase separation might be driven by magnetic fluctuations [72, 12]. A more specific mechanism was proposed by Emery and Kivelson [73] who argued, based on an analysis of the $t-J$ model [74], that holes doped into an antiferromagnet will tend to phase separate. Such behavior would normally be frustrated by the long-range part of the

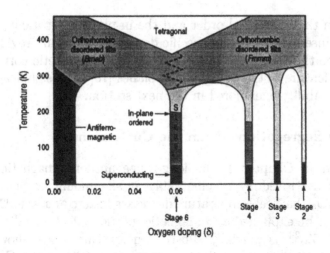

Figure 7.   Proposed phase diagram of La₂CuO₄₊δ based on single-crystal diffraction studies. From Wells *et al.* [70].

Coulomb interaction, but mobile dopants, as in $La_2CuO_{4+\delta}$, could move with the holes and screen the repulsion.

The original arguments focussed on the concept of a single miscibility gap. Of course, we now know that the phase diagram is considerably more complicated. Does the existence of multiple miscibility gaps invalidate the idea that electronic interactions play a role in the phase separations? Not necessarily. If staging order of interstitials involves constant-density layers, then electronic energies may be important for determining that density. The interstitial order appears to adjust to maintain the same local interstitial density, even as the average density changes. It would be quite interesting if one could test to see whether the hole density per CuO_2 plane varies with distance from the interstitial layers.

One might also ask whether the doping mechanism (interstitial oxygen in $La_2CuO_{4+\delta}$ versus elemental substition in $La_{2-x}Sr_xCuO_4$) modifies the nature of the electronic correlations within the CuO_2 planes. One approach to this issue is to study the magnetic correlations by inelastic neutron scattering. Such measurements have been performed recently by Wells *et al.* [70], following the successful electrochemical oxidation of large (2–3 g) single crystals. (Note that it was necessary to use D_2O rather than H_2O as solvent, in order to minimize the incoherent scattering from residual hydrogen in the samples.) They find that the magnetic response at low excitation energies (2–4 meV) exhibits an incommensurate Q dependence essentially identical to that found in $La_{2-x}Sr_xCuO_4$ (see Chapter 4). The wave vector describing the incommensurability is rotated by 45° with respect to the octahedral tilt direction, so that there appears to be no special connection between the interstitial order and the in-plane magnetic correlations. In fact, the insensitivity to the specific doping mechanism provides important evidence that the incommensurability of the magnetic correlations is an intrinsic feature of the doped CuO_2 planes [70]. The significance of the incommensurability is explored in the next section.

3. Charge Segregation within the CuO_2 planes

As discussed in Chapter 4, the long-range antiferrromagnetic order of La_2CuO_4 is destroyed by a remarkably small amount of Sr doping. In $La_{2-x}Sr_xCuO_4$ the Néel temperature decreases to zero at $x \approx 0.02$. This behavior cannot be explained as a magnetic dilution effect—studies in which nonmagnetic Zn^{2+} is partially substituted for Cu^{2+} have shown that it takes at least 10 times more Zn (i.e., more than 20% of the Cu replaced by Zn) to destroy the long-range order [75, 76]. Furthermore, studies using techniques sensitive to the local hyperfine field, such as muon spin rotation (μSR) and nuclear quadrupole resonance (NQR), indicate that the average locally-ordered Cu moment changes relatively little over the range $0 \le x \lesssim 0.02$ [77, 78, 79, 80, 81]. Thus, doping with a low density of holes has a drastic effect on long-range correlations, but has relatively little effect on local magnetic interactions.

With increased doping, neutron scattering studies show that the magnetic correlation length becomes finite and decreases with x (see Chapters 3 and 4). At the same time, the energy scale of the magnetic excitations changes relatively little, even as the character of the material changes from an antiferromagnetic insulator to an unusual metal. Again, these results suggest that the holes effectively break up a CuO_2 plane into regions that behave as if they are locally antiferromagnetic. But what is the nature of the inhomogeneity, and how are the spatial distributions of charge and spin related?

An important clue was the discovery that the magnetic scattering in $La_{2-x}Sr_xCuO_4$ with $x > 0.05$ is incommensurate. Some indication of the incommensurability was provided by early neutron scattering measurements by Birgeneau and coworkers [82, 83], but the true nature of the Q-dependent cross section was first established by Cheong et al. [84] in an inelastic scattering study. Shortly thereafter, Hayden et al. [85] discovered incommensurate magnetic scattering in $La_{1.8}Sr_{0.2}NiO_{4-\delta}$. Although the orientation of the peaks about the antiferromagnetic wave vector was rotated with respect to that of the cuprate, intriguing similarities were noted.

The nickelate system has proven to be a useful model for studying charge segregation. The undoped compound is an antiferromagnetic insulator just like the cuprate, but on doping it remains insulating throughout the regime of hole densities where the cuprate is superconducting [86]. Neutron scattering studies have gradually established a rather clear picture of the charge and spin correlations in this system, and this work will be described in the next subsection. Evidence for similar charge segregation in the cuprates has been obtained in studies of the system $La_{1.6-x}Nd_{0.4}Sr_xCuO_4$, in which an anomalous suppression of T_c is found for $x \sim \frac{1}{8}$. Following a description of that work, results on other cuprates that indicate the generality of charge inhomogeneity will be discussed.

3.1. STRIPE ORDER IN $La_{2-x}Sr_xNiO_{4+\delta}$

Following the initial discovery of incommensurate magnetic scattering in a nickelate crystal by Hayden et al. [85], a second set of superlattice peaks, indicative of charge order, was detected in a series of $La_{2-x}Sr_xNiO_{4+\delta}$ samples by electron diffraction [87]. The proper relationship between the two sets of peaks was determined in a neutron diffraction study on a crystal of $La_2NiO_{4+\delta}$ with $\delta = 0.125$, in which the magnetic and charge-order superstructure peaks were observed simultaneously [88, 53]. As will be explained below, the observed superstructure provides clear evidence for a highly correlated state in which the dopant-induced holes segregate into periodically-spaced stripes that separate antiferromagnetic domains.

To describe the superlattice peaks, it is convenient to use an indexing based on the unit cell of the HTT phase. With this choice, the antiferromagnetic wave vector is $\mathbf{K}_{AF} = (\frac{1}{2}, \frac{1}{2}, 0)$. The positions of the superlattice reflections within the $(hk0)$ zone of reciprocal space are indicated schematically in Fig. 8(a). Four magnetic peaks are found, split about $\mathbf{K}_{AF}$ along the [110] and [1$\bar{1}$0] directions. The peaks can be characterized by two modulation wave vectors, $\mathbf{K}_\epsilon^{Ni} = (\frac{1}{2} - \frac{\epsilon}{2}, \frac{1}{2} - \frac{\epsilon}{2}, 0)$ and $\tilde{\mathbf{K}}_\epsilon^{Ni} = (-\frac{1}{2} + \frac{\epsilon}{2}, \frac{1}{2} - \frac{\epsilon}{2}, 0)$. The two pairs of peaks about a given antiferromagnetic point are not equivalent, and they have quite different intensities in a sample with long-range order. The intensity difference can be understood in terms of the sensitivity of the neutron cross section to the angle between $\mathbf{Q}$ and spin direction $\mathbf{S}$ [see Eq. (53) in Chapter 1]. Using measurements in many different Brillouin zones, it has been shown that the intensities can be quantitatively explained if one assumes that each modulation wave vector corresponds to a distinct twin domain, with the Ni spins oriented perpendicular to the modulation direction [53]. The two types of domains are illustrated in Fig. 8(b) and (c). The splitting of the peaks about $\mathbf{K}_{AF}$ indicates an antiphase domain structure [85, 89]; that is, the Ni spins order in antiferromagnetic domains of constant width, with the phase of the spin direction shifting by π on crossing a domain wall.

In measuring the charge order, the neutrons do not interact with the

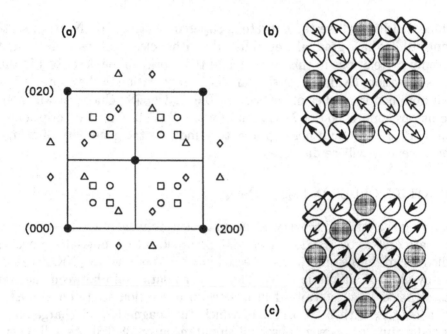

Figure 8.   (a) Schematic diagram of the $(hk0)$ zone in reciprocal space for an NiO_2 plane. Circles (squares) and diamonds (triangles) indicate positions of magnetic-order and charge-order peaks, respectively, corresponding to the real space domain shown in (b) [(c)]. (b) and (c) show models of spin and charge modulations for $n_h = \frac{1}{4}$. Only Ni sites are shown; filled circles indicate locations of holes (1 per site, with no ordered moment). Arrows indicate orientation of ordered magnetic moments, and shading distinguishes antiphase domains. Double line outlines the magnetic unit cell.

electrons themselves, but instead scatter from the small nuclear displacements induced by the modulated charge density. The positions of the charge-order peaks are approximately characterized by the modulation wave vectors $\mathbf{K}_{2\epsilon}^{Ni} = (\epsilon, \epsilon, 1)$ and $\tilde{\mathbf{K}}_{2\epsilon}^{Ni} = (-\epsilon, \epsilon, 1)$. The in-plane component of the modulations corresponds to a peak splitting twice that of the magnetic peaks. It follows that the period of the charge modulation is half of the magnetic period, which corresponds to the domain wall spacing. Although the diffraction measurements provide no information about the relative phases of the magnetic and charge modulations, it is logical to associate the maximum of the charge density with the domain wall positions (i.e., the nodes of the magnetic modulation). The out of plane component indicates that the positions of the charge stripes in neighboring layers are staggered in a body-centered fashion, as one might expect due to the Coulomb repulsion between the stripes. As a result, the charge-order peaks appearing in the $(hk0)$ zone are actually split about Bragg points above or below the plane. (Actually, it has been shown that, at least in certain cases, the body-centering is not perfectly achieved due to coupling of the charge stripes to the lattice [57].)

The superlattice peaks discussed so far provide evidence only for sinusoidal modulations of the spin and charge densities, whereas the caricatures of stripe order in Fig. 8(b) and (c) suggest narrow charge stripes and sharply defined magnetic domains. The deviations from a simple sine wave require higher Fourier components, and such components have been observed (at least for the spin density) as higher-order harmonics in the $\delta = \frac{2}{15}$ phase of $La_2NiO_{4+\delta}$ [57]. The most intense observed magnetic harmonic, the third, is only 1.5% of the first harmonic; nevertheless, quantitative modelling shows that the observed harmonic intensities are consistent with sharp magnetic domains. Analysis of the relative intensities of first harmonic peaks indicates that the maximum ordered Ni moment is > 80% of that observed in undoped La_2NiO_4 [53]. Some of the moment reduction may be associated with the larger zero-point fluctuations expected due to the anisotropy in effective spin-spin interactions induced by the domain walls [90, 91]. Thus, the large moments are consistent with a strong segregation of the charge into relatively narrow stripes.

The caricatures in Fig. 8 also show the charge stripes centered on rows of Ni atoms. The actual phase of the stripes with respect to the lattice cannot be determined in an absolute way directly from the diffraction measurement. Nevertheless, in the case of the $\delta = \frac{2}{15}$ phase, it was possible to show from the l-dependence of the magnetic peak intensities that the stripes must have a preferred registry with the lattice [57]. Furthermore, ϵ is temperature dependent in that sample, with the stripe density increasing with temperature. The value of ϵ exhibits lock-in plateaus at certain rational fractions. From an analysis of that behavior, together with an unusual ferrimagnetic response above the magnetic ordering temperature, it has been possible to show that the stripes are centered on rows of O atoms at high temperature, shifting to a dominant Ni-centering at low temperature [57, 92].

The stripe order found in $La_2NiO_{4+\delta}$ occurs only for $\delta \gtrsim 0.11$ [88, 53, 93, 56, 57], where the interstitials exhibit 3D order. It now appears that the first such phase has an ideal interstitial density of $\delta = \frac{2}{15}$ [53, 57], and that samples with $\delta = 0.125$ [88, 53, 93, 56] correspond to the same interstitial order, but with vacancies. For $\delta \lesssim 0.11$, the magnetic order remains commensurate [34, 48, 93, 51].

The long-range 3D stripe order found in the oxygen-doped crystals has made it possible to study fine details of the spin and charge correlations. On the other hand, one can obtain a more continuous variation of the hole density via Sr doping. So far, neutron scattering studies have also been performed on crystals of $La_{2-x}Sr_xNiO_4$ with x in the range of 0.135 to 0.333 [85, 94, 95, 96]. The results for ordering temperatures and incommensu-

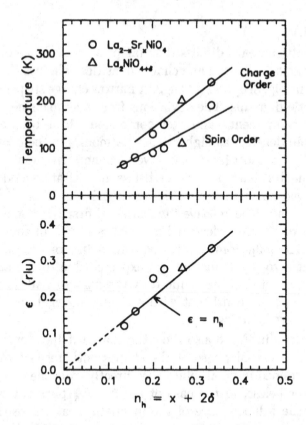

Figure 9. Summary of transition temperatures (top) and incommensurability ϵ (bottom) vs. net hole concentration for stripe-ordered phases in nickelate samples studied by neutron diffraction. References are given in the text.

rability (at low temperature) as a function of the net hole concentration, $n_h = x + 2\delta$, are summarized in Fig. 9. In all cases, the charge orders at a higher temperature than the spins. Such behavior is incompatible with the transitions being driven by the spin-density order parameter [97], such as occurs in Cr and Cr alloys [98]. The systematic increase in both the charge and spin ordering temperatures with n_h indicates the important role of the charge correlations. The bottom panel in Fig. 9 shows that $\epsilon \approx n_h$, which indicates that the hole density within the stripes (at low temperature) stays roughly constant at 1 hole/Ni site, consistent with Hartree-Fock calculations by Zaanen and Littlewood [99]. This means that the increasing average density of holes is accomodated by decreasing the spacing between stripes.

The highest transition temperatures occur for the $x = 0.33$ sample studied by Lee and Cheong [96]. The peak splitting parameter ϵ locks in at the commensurate value of $\frac{1}{3}$ in this case, independent of temperature. Optical

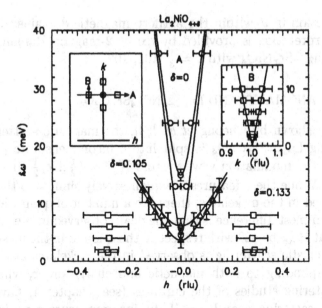

Figure 10. Low-energy spin-wave dispersions measured on crystals of $La_2NiO_{4+\delta}$ with $\delta = 0$ (circles), 0.105 (triangles), and 0.133 (squares). Left inset indicates directions A and B in the $(h, k, 0)$ plane along which the dispersion has been characterized. [Note that the coordinate system is based on an orthorhombic unit cell, so that $K_{AF} = (0, 1, 0)$.] Main panel shows dispersion along A; results of scans along B are shown in the right inset. Bars indicate measured peak widths (except at $h = 0$ for $\delta = 0$), with no correction for resolution. From Ref. [108].

spectra measured on a similar sample indicate that a large charge gap opens up below the charge-ordering temperature, reaching 0.26 eV at 10 K [100]. Raman-scattering studies indicate that a number of new Raman-active phonons appear at the same transition [101, 102]. Temperature-dependent and incommensurate peak splitting, similar to that found in the $\delta = \frac{2}{15}$ phase, has been observed in an $x = 0.225$ crystal by both neutron [95] and X-ray [103] scattering studies.

The doping dependence of the effective magnetic exchange has been probed through measurements of the spin-wave velocities. A number of inelastic neutron scattering studies have characterized the spin-wave dispersion in Néel-ordered $La_2NiO_{4+\delta}$ [104, 105, 106, 107, 108]. At $\delta = 0$, the spin-wave velocity is approximately half the value found in La_2CuO_4, and corresponds to a superexchange energy J of approximately 30 meV [106, 107]. The low-energy spin-waves soften dramatically with doping, as illustrated in Fig. 10. Interestingly, the dispersion hardens substantially once the stripe-order regime is reached. As indicated in Fig. 10, the effective spin-wave velocity in a stripe-ordered $\delta = \frac{2}{15}$ sample is $\sim 60\%$ of that in the $\delta = 0$ phase [108]. Much of the velocity reduction is expected to come from a weakened effective coupling across domain walls, rather than from a reduction in J within the antiferromagnetic domains. Support for such an interpretation is provided by recent 2-magnon Raman scattering studies on $La_{2-x}Sr_xNiO_4$ with $x = \frac{1}{3}$ [101, 102].

3.2. THE "$\frac{1}{8}$ ANOMALY" AND $La_{1.6-x}Nd_{0.4}Sr_xCuO_4$

As originally shown by Cheong et al. [84], the magnetic scattering in superconducting $La_{2-x}Sr_xCuO_4$ is split into 4 incommensurate peaks characterized by the modulation wave vectors $\mathbf{K}_\epsilon^{Cu} = (\frac{1}{2} \pm \epsilon, \frac{1}{2}, 0)$ and $\tilde{\mathbf{K}}_\epsilon^{Cu} = (\frac{1}{2}, \frac{1}{2} \pm \epsilon, 0)$. While these features are qualitatively similar to the magnetic peaks observed in the nickelates, there are a number of quantitative differences. The simplest difference is that the peaks observed in the cuprates are rotated about $\mathbf{K}_{AF}$ by 45° with respect to those found in the nickelates. Going further, elastic scattering is observed in the nickelates at superlattice points corresponding to both magnetic and charge order, whereas most neutron-scattering studies of the cuprates (see Chapter 4) have focussed on inelastic scattering near $\mathbf{K}_{AF}$. If the incommensurate peaks found in $La_{2-x}Sr_xCuO_4$ are associated with charge-stripe correlations, then these stripes would be part of a very unusual dynamically correlated phase.

An important test of the stripe interpretation for the cuprates would be to detect a significant charge-density modulation. Detection of such a modulation is simplified if it can be made static. Such static order might occur as a result of commensurability between the stripe period and the lattice spacing. The first likely candidate for such a pinned stripe phase is associated with the "$\frac{1}{8}$ anomaly" first discoved in $La_{2-x}Ba_xCuO_4$ [25]. To appreciate the significance of this anomaly, we need to briefly review a number of results.

Moodenbaugh et al. [25] were the first to observe that there is a strong depression of T_c in $La_{2-x}Ba_xCuO_4$ within a narrow region of x centered at $x \approx \frac{1}{8}$. This depression is anomalous when contrasted with the x dependence of T_c in $La_{2-x}Sr_xCuO_4$. For Ba-doped samples with x close to $\frac{1}{8}$, anomalies are also observed in various transport properties at temperatures below ~ 60 K [25, 109]. Experiments involving counter-doping with 4+ ions such as Ce and Th established that the local minimum in T_c is associated with a net hole concentration of $\frac{1}{8}$ [110, 111]. The minimum can also be shifted by varying the oxygen stoichiometry [112].

Using X-ray scattering on powder samples of $La_{2-x}Ba_xCuO_4$, Axe et al. [113] and Suzuki and Fujita [114] discovered that the crystal structure begins to transform below ~ 80 K from the usual LTO structure to an LTT phase for $0.05 < x < 0.20$. Rietveld analysis of neutron powder diffaction measurements confirmed the nature of the low-temperature structure and the fact that that transformation was not 100% [115]. In retrospect, it ap-

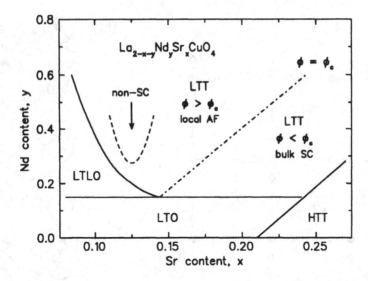

Figure 11. Low-temperature phase diagram for $La_{2-x-y}Nd_ySr_xCuO_4$, after Büchner *et al.* [123] and Moodenbaugh *et al.* [124]. ϕ is the average tilt angle of the CuO_6 octahedra, and ϕ_c is the critical value [123].

pears that the structural transition was first detected, but not recognized, in an earlier neutron diffraction study by Paul *et al.* [116]. Although the same structural transition does not occur in $La_{2-x}Sr_xCuO_4$, a recent electron diffraction study has provided evidence for narrow LTT-like regions at the boundaries between LTO twin domains [117].

μSR studies [118, 119] provided evidence for local magnetic order, with Cu moments of ~ 0.3 μ_B, at the $\frac{1}{8}$ concentration in Ba-doped samples, which would make them a tempting target for neutron scattering studies except for the fact that all attempts to grow single crystals have failed. Fortunately, there is an alternative. Crawford *et al.* [120] showed that one can induce the LTT structure in $La_{2-x}Sr_xCuO_4$, with a concomitant T_c anomaly at $x \approx \frac{1}{8}$, by partial substitution of Nd^{3+} for La^{3+}. The occurrence of the LTT phase is purely an ionic size effect [121, 122], and the rare-earth substitution has no effect on the charge density.

Büchner *et al.* [125, 123] have established the low-temperature phase diagram of $La_{2-x-y}Nd_xSr_yCuO_4$ (see Fig. 11) through neutron and X-ray powder diffraction studies. A minimum Nd concentration of $y \approx 0.18$ is required to obtain an LTT (or LTLO) phase. Within the LTT phase regime, the average tilt angle of the CuO_6 octahedra varies with x and y. Büchner *et al.* [123] have identified a critical tilt angle, such that in samples with an average tilt angle exceeding the critical value, anomalies are observed in the resistivity, and local magnetic order occurs [126, 127]. They also argued, on the basis of rather conservative criteria, that no bulk superconductivity

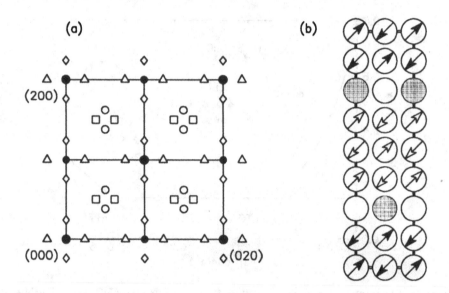

Figure 12.   (a) Schematic diagram of the $(hk0)$ zone in reciprocal space for a CuO_2 plane. (b) Real-space model of spin and charge modulations for $n_h = \frac{1}{8}$. Meaning of symbols is same as in Fig. 8. In (b), filled and empty circles indicate hole density of 1 per 2 Cu sites in domain walls.

occurs above the critical tilt line; however, recent studies have provided reasonable evidence that the boundary for suppression of superconductivity is shifted to somewhat higher Nd concentrations [124, 128]. In any case, the existence of a critical tilt boundary for local magnetism seems intuitively consistent with the idea of pinning stripe correlations.

Crystals of $La_{1.6-x}Nd_{0.4}Sr_xCuO_4$ were successfully grown by Nakamura and Uchida [129], who also characterized the anisotropic transport properties of samples with $x = 0$, 0.12, and 0.20. Somewhat later, the $x = 0.12$ crystal, which had its superconducting transition depressed below 5 K, was selected for a neutron diffraction study. If charge and spin modulations were pinned due to commensurability with the lattice, then one would expect to see elastic magnetic peaks at positions $\mathbf{K}_\epsilon^{Cu}$ and $\tilde{\mathbf{K}}_\epsilon^{Cu}$ with $\epsilon = \frac{1}{8}$. The experiment did indeed reveal elastic peaks with $\epsilon = 0.12$, shifted only slightly from the expected commensurate positions [130, 131]. The peaks were rather sharp within the $(hk0)$ zone, but diffuse along l. Elastic scans near the (200) Bragg point yielded weak, temperature-dependent peaks at positions displaced by $\mathbf{K}_{2\epsilon}^{Cu} = (\pm 2\epsilon, 0, 0)$ [130, 131]. These peaks correspond to charge order, as confirmed by high-energy X-ray diffraction [132]. The positions of the observed peaks, and equivalent superlattice points (assuming rods of scattering perpendicular to the plane), are indicated in Fig. 12(a). No intensity was detected at the positions $(2, \pm 2\epsilon, 0)$. Keeping in mind that the intensity of the 2ϵ peaks should be due to atomic displacements induced by the charge modulation, the anisotropy of the intensities is consistent with displacments that are parallel to the modulation wave vector [131].

The observed superlattice peaks provide direct evidence for spatial modulations of the spin and charge densities in real space, with the period of the magnetization-density modulation twice that of the charge density. It is natural to imagine, by analogy with the nickelates, that the modulations are one-dimensional, and that the appearance of peaks displaced in orthogonal directions is due to the presence of two types of domains. A simple model for a commensurate phase with $\epsilon = \frac{1}{8}$ is shown in Fig. 12(b). The hole concentration corresponds to $\frac{1}{2}$ per unit cell of the charge-ordered system. It is important to note that the phase of the modulations with respect to the lattice has not been determined experimentally. Furthermore, the observed peaks provide evidence only for sinusoidal modulations. The densities shown in the model deviate from sinusoidal form, and the deviations should result in extra superlattice peaks at higher-harmonic positions; however, it has been shown in the case of $La_2NiO_{4.133}$ that such higher-harmonic peaks are quite weak [57], and would be difficult to detect in the present case.

Another important issue concerns the amplitude of the modulations. The charge density is not directly measured, and hence cannot be determined directly. The amplitude of the spin density can be determined from the intensity of the magnetic scattering, but the uncertainties are considerable due to the problem of integrating over diffuse scattering along l. Instead, a useful measure is provided by μSR studies, which indicate that the local magnetic field seen by the muons has an amplitude half that found in antiferromagnetically-ordered La_2CuO_4 [133, 127, 81]. Such a static (or quasi-static) value is quite large, considering the lack of true long-range order and the considerable quantum spin fluctuations one would expect in such a modulated structure [90, 91]. The μSR results appear to justify the strong modulation indicated in the model.

The 1D stripe modulation provides a natural explanation for the fact that static spin and charge order (and a strong depression of T_c) occur in the LTT, but not the LTO, phase. Figure 13 compares the modulations of the CuO_2 planes in the two structures. The displacement pattern of the oxygens changes from a diagonal orientation in the LTO phase to horizontal (alternating with vertical in every other plane) in the LTT structure. It is intuitively obvious that the stripe model shown in Fig. 12(b) can be pinned by the horizontal pattern, but not by the diagonal pattern. Furthermore, one would expect the pinned stripe correlations to rotate by 90° from one LTT layer to the next, following the structural modulation. The resulting

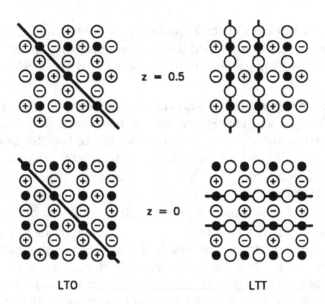

Figure 13. Displacement patterns within the CuO_2 planes for the LTO and LTT structures. Open (solid) circles represent oxygen (copper) atoms. Oxygen atoms are displaced out of the plane (+ or −) by local rotations of square planar CuO_4 units about tilt axes indicated by doubled lines. In the LTT structure, the tilt axes rotate by 90° from $z = 0$ to $z = 0.5$, where z is the height along the c axis in lattice units. From Ref. [130].

stacking pattern of the stripe layers is shown in Fig. 14(b). [This contrasts with the body-centered stacking of parallel stripes found in the nickelates and illustrated in Fig. 14(a).] The orthogonal charge stripes in neighboring layers would have no preferred relative alignment, and this is consistent with the extremely limited magnetic correlations along the c axis detected in the neutron scattering work [131]. The X-ray scattering study indicated a staggered alignment of parallel stripes in next-nearest-neighbor layers, which is also consistent with the model [132].

The possibility that the spin and charge modulations are locally 2D rather than 1D has not yet been ruled out, but it seems rather unlikely. If the modulations were 2D, then one could imagine commensurability with the LTO, as well as the LTT, structure, which is not observed. Also, the magnetic unit cell of a 2D modulation would be rotated by 45° with respect to the charge unit cell. Such a rotation is not consistent with the superlattice peaks that have been observed so far.

The transition from the LTO to the LTT structure occurs near 70 K, and the charge-order peaks appear at approximately the same temperature [130, 131, 132]. The magnetic signal detected by neutron scattering appears at a somewhat lower temperature, ~ 50 K, and μSR measurements indicate that the magnetic correlations are not truly static until the temperature drops to ~ 30 K [133]. As in the case of the nickelates, the higher temperature for

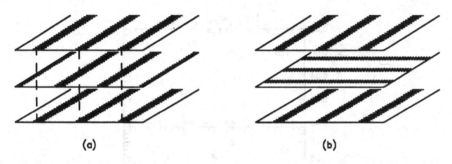

Figure 14. Sketch of the stacking of stripe-ordered layers in (a) $La_{2-x}Sr_xNiO_{4+\delta}$, where Coulomb repulsion determines the stacking, and (b) $La_{1.6-x}Nd_{0.4}Sr_xCuO_4$, where the rotation of the stripe direction between layers is controlled by pinning to the LTT structure. From Ref. [134].

charge ordering is incompatible with a spin-density-wave mechanism [97]. Below ~ 3 K, the Nd moments begin to order in phase with the neighboring Cu moments, resulting in an order of magnitude increase in the intensity of the magnetic peaks. Inelastic neutron scattering investigations of the Nd fluctuations at higher temperatures have been performed by Roepke *et al.* [135] on samples with varying Sr and Nd concentrations.

The stripe order observed in $La_{1.6-x}Nd_{0.4}Sr_xCuO_4$ at $x = 0.12$ occurs in a sample with strongly suppressed superconductivity. What happens to the stripe order as x is increased? Neutron scattering measurements on crystals with $x = 0.15$ and $x = 0.20$ showed that the elastic signal at the split magnetic peak positions survives, although the relative intensities and the ordering temperatures decrease substantially as x deviates from 0.12 [136]. A recent high-field magnetization study on a piece of the $x = 0.15$ crystal provides clear evidence of bulk superconductivity [128], so that it appears that static (or quasi-static) stripe order can coexist with superconductivity. Some of the results obtained for $La_{1.6-x}Nd_{0.4}Sr_xCuO_4$ are summarized in Fig. 15. As one can see, the values of ϵ and T_c both increase with x for the crystals studied, so that the coexistence phenomenon is not consistent with a mixture of a unique stripe-ordered phase and a distinct superconducting phase.

Also presented in Fig. 15 are the results for T_c and ϵ obtained from a series of $La_{2-x}Sr_xCuO_4$ crystals by Yamada *et al.* [137]. Although their neutron scattering measurements were inelastic rather than elastic, the incommensurability is essentially identical to that found in the Nd-doped crystals where the Sr concentrations overlap. Given this result, there can be little doubt that the same mechanism must be responsible for the incommensurability in the two systems. Furthermore, the fact that the stripe order in the Nd-doped, $x = 0.12$ sample is charge driven then implies that there must also be a substantial dynamically-correlated charge modulation

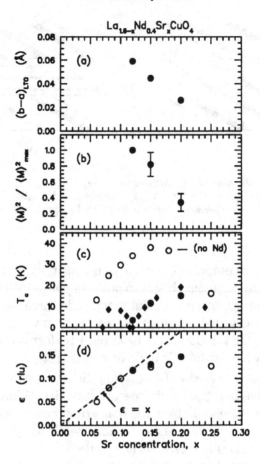

Figure 15. Comparison of results as a function of x: (a) difference between a and b lattice parameters in the LTO phase measured just above the transition to the LTT phase, (b) square of the low-temperature staggered magnetization, normalized to the $x = 0.12$ result, (c) superconducting transition temperature, and (d) incommensurate splitting ϵ. Filled symbols: $La_{1.6-x}Nd_{0.4}Sr_xCuO_4$; open symbols: $La_{2-x}Sr_xCuO_4$ (Ref. [137].) Circles: single-crystal samples; diamonds: ceramic samples. From Ref. [136].

in the $La_{2-x}Sr_xCuO_4$ samples. Note that the linear variation of ϵ with x for $x \lesssim \frac{1}{8}$ is similar to the behavior found in the nickelates (see Fig. 9), although ϵ clearly saturates for $x \gtrsim \frac{1}{8}$. Thus, it appears that dynamic stripe correlations are compatible with superconductivity, but that static stripe order competes the superconducting order.

While it is clear that the hole concentration at $x = \frac{1}{8}$ is special, another feature that changes with x is the magnitude of the octahedral tilts. A useful measure of the tilts, as demonstrated by Büchner *et al.* [123], is the orthorhombic splitting $(b - a)_{LTO}$ just above the transition to LTT. As one can see in the Fig. 15(a), $(b - a)_{LTO}$ is decreasing as x increases, and for $x = 0.20$ it drops just below the critical value (≈ 0.035 Å) for local magnetic order [123]. The change in buckling as well as the variation in hole concentration may contribute to the increase in T_c as x deviates from $\frac{1}{8}$.

The importance of the LTT structure for suppressing T_c was made especially clear in a study by Katano *et al.* [138]. They used neutron diffraction to determine the low-temperature structure of T_c-suppressed $La_{2-x}Ba_xCuO_4$ with $x = 0.125$ as a function of pressure. A substantial increase in T_c was observed when the LTT phase was converted to LTO at pressures above 0.5 GPa.

Dabrowski *et al.* [139] have presented evidence from neutron powder diffraction studies that, even within the LTO phase, reducing the average octahedral tilts is correlated with an increase in T_c. As discussed above, the static modulations of the plane in the LTO structure are not expected to pin stripes, so what is the significance of the observed correlation? As discussed in Chapter 5, there exists in the LTO phase a low-energy phonon mode that corresponds to LTT-like displacements. This mode has an energy of 1.5–2 meV in superconducting crystals of $La_{2-x}Sr_xCuO_4$, as determined by inelastic neutron scattering studies [140, 141]. A coupling between slowly fluctuating charge stripes and the soft LTT-like mode might have a negative impact on T_c. That impact should be decreased by a reduction of the average static octahedral tilt angle.

Very recently, Suzuki *et al.* [142] have reported an observation by neutron diffraction of weak elastic scattering at the split magnetic peak positions in an orthorhombic single crystal of $La_{1.88}Sr_{0.12}CuO_4$. In interpreting this result it is important to note that the presence of an LTT-like distortion at the LTO twin boundaries has been detected in $La_{2-x}Sr_xCuO_4$ with $x = 0.115$ by electron diffraction [117]. Furthermore, a high resolution X-ray diffraction study has found evidence for an LTT component at the 10% level in an $x = 0.12$ powder sample [143]. It seems likely that the elastic magnetic peaks found with neutrons are associated with a minority LTT phase.

3.3. IS THERE EVIDENCE FOR STRIPES IN OTHER SYSTEMS?

We have seen that modifying the octahedral tilt pattern in $La_{2-x}Sr_xCuO_4$ from the usual LTO to the LTT structure can induce static modulations of the charge and spin densities. The modulations have been interpreted as evidence for charge segregation in the form of stripes. Removing the LTT distortion tends to eliminate the static character of the modulations, but the observed incommensurability of the spin correlations suggests the presence of dynamical charge stripes. The physics of doped CuO_2 planes should be common to all of the copper-oxide superconductors, so if the stripe interpretation is correct, it should be possible to find evidence of stripes in cuprates without an LTT-like structure.

A likely alternative method for pinning stripes is the substitution of a small amount of nonmagnetic Zn for Cu. An earlier indication of this possibility was provided in a study by Koike et al.[144], in which it was shown that substitution of 1% Zn in $La_{2-x}Sr_xCuO_4$ causes a drastic dip in T_c near $x = 0.115$, very similar to the behavior in $La_{2-x}Ba_xCuO_4$ [25]. Recently, Hirota et al. [145] reported on a neutron diffraction study of a single crystal with $x = 0.14$, 1.2% Zn, and $T_c = 19$ K. They found that the Zn does not modify the incommensurability of the magnetic scattering, but that it does shift spectral weight to lower energies and induces a sharp (in Q) elastic component at temperatures below ~ 20 K.

As discussed in Chapter 3, there have also been several neutron scattering studies of Zn-doped $YBa_2Cu_3O_{6+x}$ crystals [146, 147, 148]. Although no elastic magnetic scattering has been reported, the results are otherwise qualitatively similar to those found in the 214 system. The Zn-doping does not modify the Q dependence of the inelastic magnetic scattering, but it does shift spectral weight from higher energies to the low-energy region, making the frequency dependence in the low-energy region appear like that of a disordered antiferromagnet.

The Q dependence of the inelastic magnetic scattering in pure $YBa_2Cu_3O_{6+x}$ has long been viewed as distinct from that in $La_{2-x}Sr_xCuO_4$. As discussed in Chapter 3, the low-energy scattering is peaked commensurately at K_{AF}. Although flat-topped peak shapes have been reported in samples with $x \sim 0.5$ [149, 150], no resolved incommensurability had been seen, until now. Dai, Mook, and Doğan [151] have recently found evidence for an incommensurate splitting at energies near 25 meV in a crystal with $x = 0.6$ and $T_c = 62.7$ K. While this result suggests that incommensurate spin fluctuations may be a common feature of layered cuprates, further investigations are required to test the extent to which the stripe model may be applicable.

3.4. THEORETICAL MODELS

One class of theoretical models attributes the magnetic correlations in superconducting cuprates to the scattering of electrons across the Fermi surface. Some of this work is discussed in Chapter 4. If the correlations become strong enough, a transition to a spin-density-wave state can occur. We have noted above several features that appear to be incompatible with an SDW instability. For example, the static stripe order is driven by the charge, rather than the spin, density. Also, static magnetic order can be induced by the presence of Zn impurities, whereas one might expect impurities to weaken any instability dependent on a sharp Fermi surface and translational symmetry. Nevertheless, calculations of stripe phases induced by Fermi-surface effects have had a strong influence on experimental work, and hence require some discussion.

The 2D single-band Hubbard model contains the electronic kinetic energy and a locally repulsive interaction that are commonly believed to be of dominant importance for understanding the cuprates. Spin and charge modulations associated with an electronic instability can be evaluated using the Hartree-Fock approximation, provided that the repulsive interaction is not too strong. The first identification of a spin and charge stripe-phase solution was made in a numerical analysis by Zaanen and Gunnarsson [152], and similar results have been obtained by many other groups [153, 154, 155, 156, 157]. Stripes have also been found in variational [158] and fixed-node [159] Monte Carlo analyses of the single-band Hubbard model, Hartree-Fock treatments of 3-band Hubbard [160, 161] and Peierls-Hubbard [162] models, and in a model of dilute ferromagnetic bonds in an antiferromagnet [163]. Both vertical and diagonal domain walls have been found, the relative stability of each depending on the parameter values. The analysis by Schulz [154] makes clear that the mean-field stripe phase is stabilized by the opening of a gap on parts of the Fermi surface. A consequence of this is that the hole concentration in a stripe will be exactly one per Cu site, and the ordered state will be an insulator. Both of these results are inconsistent with experimental results for the cuprates.

An alternative view of stripe origins developed out of studies of the so-called t-J model, which is a simplified, strong-interaction version of the Hubbard model. In a controversial paper, Emery, Kivelson, and Lin [74] argued that for a given hole concentration there is a ratio of J/t beyond which the holes will segregate into hole-rich and no-hole phases. Recent numerical calculations on large lattices provide support for the claim that charge segregation occurs within this model for hole concentrations and parameter regimes relevant to the cuprates [164]. Coulomb interactions, which are not included in the t-J model, should frustrate the phase separation in the real materials, leading to large amplitude fluctuations of the hole density on intermediate length scales [73]. Analysis of a simple model with competing interactions and long-range repulsion demonstrated the likelihood of obtaining stripe phases [165]. Haas et $al.$ [166] found a variety of charge-ordered solutions when a long-range repulsive interaction was added to the t-J model. Castellani, Di Castro, and Grilli [167] have explored the consequences of an incommensurate charge-density-wave instability driven by Coulomb-frustrated phase separation.

Other groups have explored inhomogeneous solutions of the original t-J

model. Prelovšek and Zotos [168] were the first to identify a tendency of holes to form diagonal domain walls within the disordered antiferromagnet. Viertiö and Rice [169] found that such domain walls should become disordered due to quantum fluctuations. Tsunetsugu, Troyer, and Rice [170] discussed vertical, insulating charge stripes, and argued that they should melt at relevant values of J/t. More recently, using the density matrix renormalization group, White and Scalapino [171] have stabilized vertical stripes with the experimentally observed charge density by applying a staggered magnetic field at the open boundaries of the lattice.

Some groups have returned to the Hubbard model to consider stripes. Nayak and Wilczek [172, 173] have proposed a "minimal domain wall" hypothesis. Alternatively, Seibold et al. [174] have used the Gutzwiller variational approach to evaluate the energies of half- and quarter-filled domain walls in a 2D Hubbard model with a long-range interaction included.

Some studies have started from the assumption that charge stripes exist, and have considered their implications. For example, Zaanen and coworkers have considered the effect of classical charge-stripe fluctuations on low-frequency spin fluctuations [175], and the problem of treating stripe fluctuations quantum mechanically [176]. Castro Neto and Hone [177] have used a stripe model to explain the destruction of Néel order at low hole concentrations. Several recent papers have investigated the possibility of hole pairing due to interactions between a metallic stripe and an antiferromagnetic domain [178, 179], the problem of superconducting phase coherence in a stripe phase [180], and the importance of transverse stripe fluctuations for preventing an insulating charge-ordered state at low temperatures [181].

4. Concluding Remarks

Neutron scattering has played a crucial role in elucidating the nature of interstitial order in $La_2CuO_{4+\delta}$, and of spin and charge modulations within the CuO_2 planes. Studies of closely related model compounds have provided valuable insights along the way. There is no question that neutron scattering techniques will continue to be essential tools as we strive to improve our understanding of these fascinating materials.

Acknowledgments

In preparing this review, I have benefitted from discussions with numerous colleagues and collaborators. I would especially like to thank D. J. Buttrey, V. J. Emery, G. Shirane, and P. Wochner for critical readings of the manuscript. This work is supported by the U.S. Department of Energy, Division of Materials Sciences, under Contract No. DE-AC02-98CH10886.

References

1. K. A. Müller, M. Takashige, and J. G. Bednorz, Phys. Rev. Lett. **58** (1987) 1143.
2. J. D. Axe and M. K. Crawford, J. Low Temp. Phys. **95** (1994) 271.
3. J. G. Bednorz and K. A. Müller, Z. Phys. B **64** (1986) 189.
4. J. Beille et al., C. R. Acad. Sc. Paris, Série II **304** (1987) 1097.
5. P. M. Grant et al., Phys. Rev. Lett. **58** (1987) 2482.
6. K. Sekizawa et al., Jap. J. Appl. Phys. **26** (1987) L840.
7. J. E. Schirber et al., Physica C **152** (1988) 121.
8. G. Demazeau et al., Physica C **153–155** (1988) 824.
9. J. D. Jorgensen et al., Phys. Rev. B **38** (1988) 11337.
10. P. Zolliker et al., Phys. Rev. B **42** (1990) 6332.
11. E. J. Ansaldo et al., Phys. Rev. B **40** (1989) 2555.
12. P. C. Hammel et al., Phys. Rev. B **42** (1990) 6781.
13. D. Vaknin et al., Phys. Rev. B **49** (1994) 9057.
14. K. Koga, M. Fujita, K. Ohshima, and Y. Nishihara, J. Phys. Soc. Jpn. **64** (1995) 3365.
15. C. Chaillout et al., Physica C **158** (1989) 183.
16. C. Chaillout et al., Physica C **170** (1990) 87.
17. J. D. Jorgensen, B. Dabrowski, S. Pei, and D. G. Hinks, Phys. Rev. B **40** (1989) 2187.
18. P. Rudolf and R. Schöllhorn, J. Chem. Soc., Chem. Commun. (1992) 1158.
19. E. Takayama-Muromachi, T. Sasaki, and Y. Matsui, Physica C **207** (1993) 97.
20. J.-C. Grenier et al., Physica C **173** (1991) 139.
21. F. C. Chou, J. H. Cho, and D. C. Johnston, Physica C **197** (1992) 303.
22. J.-C. Grenier et al., Physica C **202** (1992) 209.
23. H. H. Feng et al., Phys. Rev. B **51** (1995) 16499.
24. J.-S. Zhou, H. Chen, and J. B. Goodenough, Phys. Rev. B **50** (1994) 4168.
25. A. R. Moodenbaugh et al., Phys. Rev. B **38** (1988) 4596.
26. D. J. Buttrey, private communication.
27. P. G. Radaelli et al., Phys. Rev. B **48** (1993) 499.
28. Z. G. Li et al., Phys. Rev. Lett. **77** (1996) 5413.
29. P. G. Radaelli et al., Phys. Rev. B **49** (1994) 6239.
30. A. M. Balagurov, V. Y. Pomjakushin, V. G. Simkin, and A. A. Zakharov, Physica C **272** (1996) 277.
31. S. Wakimoto et al., J. Phys. Soc. Jpn. **65** (1996) 581.
32. M. Kato, H. Chizawa, Y. Ono, and Y. Koike, Physica C **256** (1996) 253.
33. D. J. Buttrey et al., J. Solid State Chem. **74** (1988) 233.
34. J. Rodríguez-Carvajal, M. T. Fernández-Díaz, and J. L. Martínez, J. Phys.: Condens. Matter **3** (1991) 3215.
35. A. Demourgues et al., Physica C **109** (1992) 425.
36. A. Hayashi, H. Tamura, and Y. Ueda, Physica C **216** (1993) 77.
37. A. Demourgues et al., J. Solid State Chem. **106** (1993) 317.
38. J. F. DiCarlo, I. Yazdi, S. Bhavaraju, and A. J. Jacobson, Chem. Mater. **5** (1993) 1692.
39. I. Yazdi et al., Chem. Mater. **6** (1994) 2078.
40. S. Bhavaraju et al., Solid State Ionics **86–8** (1996) 825.
41. A. Demourgues et al., J. Solid State Chem. **106** (1993) 330.
42. J. Rodríguez-Carvajal, J. L. Martínez, J. Pannetier, and R. Saez-Puche, Phys. Rev. B **38** (1988) 7148.
43. G. H. Lander, P. J. Brown, J. Spałek, and J. M. Honig, Phys. Rev. B **40** (1989) 4463.
44. T. Kajitani et al., Physica C **185** (1991) 579.
45. Z. Hiroi et al., Phys. Rev. B **41** (1990) 11665.
46. D. E. Rice and D. J. Buttrey, J. Solid State Chem. **105** (1993) 197.

47. H. Tamura, A. Hayashi, and Y. Ueda, Physica C **216** (1993) 83.
48. S. Hosoya *et al.*, Physica C **202** (1992) 188.
49. A. Mehta and P. J. Heaney, Phys. Rev. B **49** (1994) 563.
50. J. M. Tranquada, D. J. Buttrey, and D. E. Rice, Phys. Rev. Lett. **70** (1993) 445.
51. J. M. Tranquada *et al.*, Phys. Rev. B **50** (1994) 6340.
52. J. E. Lorenzo, J. M. Tranquada, D. J. Buttrey, and V. Sachan, Phys. Rev. B **51** (1995) 3176.
53. J. M. Tranquada, J. E. Lorenzo, D. J. Buttrey, and V. Sachan, Phys. Rev. B **52** (1995) 3581.
54. S. Safran, Solid State Phys. **40** (1987) 183.
55. S. Hendricks and E. Teller, J. Chem. Phys. **10** (1942) 147.
56. K. Nakajima *et al.*, J. Phys. Soc. Jpn. **66** (1997) 809.
57. P. Wochner, J. M. Tranquada, D. J. Buttrey, and V. Sachan, Phys. Rev. B **57** (1998) 1066.
58. M. K. Crawford *et al.*, J. Phys. Chem. Solids **56** (1995) 1459.
59. F. C. Chou, D. C. Johnston, S.-W. Cheong, and P. C. Canfield, Physica C **216** (1993) 66.
60. B. O. Wells *et al.*, Z. Phys. B **100** (1996) 535.
61. P. C. Hammel *et al.*, Phys. Rev. Lett. **71** (1993) 440.
62. B. W. Statt *et al.*, Phys. Rev. B **52** (1995) 15575.
63. J. Ryder *et al.*, Physica C **173** (1991) 9.
64. E. T. Ahrens *et al.*, Physica C **212** (1993) 317.
65. R. K. Kremer *et al.*, Z. Phys. B **91** (1993) 169.
66. X. Xiong *et al.*, Phys. Rev. Lett. **76** (1996) 2997.
67. P. Wochner, X. Xiong, and S. C. Moss, J. Superconductivity **10** (1997) 367.
68. R. J. Birgeneau *et al.*, in *Proc. of the 10th Anniversary HTS Workshop on Physics, Materials, and Applications*, edited by B. Batlogg *et al.* (World Scientific, Singapore, 1996), pp. 421–4.
69. P. Blakeslee *et al.*, Phys. Rev. B (submitted).
70. B. O. Wells *et al.*, Science **277** (1997) 1067.
71. Y. S. Lee, private communication.
72. M. F. Hundley *et al.*, Phys. Rev. B **41** (1990) 4062.
73. V. J. Emery and S. A. Kivelson, Physica C **209** (1993) 597.
74. V. J. Emery, S. A. Kivelson, and H. Q. Lin, Phys. Rev. Lett. **64** (1990) 475.
75. B. Keimer *et al.*, Phys. Rev. B **45** (1992) 7430.
76. P. Carretta, A. Rigamonti, and R. Sala, Phys. Rev. B **55** (1997) 3734.
77. J. I. Budnick *et al.*, Europhys. Lett. **5** (1988) 651.
78. Y. J. Uemura *et al.*, Physica C **153–155** (1988) 769.
79. F. C. Chou *et al.*, Phys. Rev. Lett. **71** (1993) 2323.
80. F. Borsa *et al.*, Phys. Rev. B **52** (1995) 7334.
81. C. Niedermayer *et al.*, Phys. Rev. Lett. (in press).
82. R. J. Birgeneau *et al.*, Phys. Rev. B **39** (1989) 2868.
83. T. R. Thurston *et al.*, Phys. Rev. B **40** (1989) 4585.
84. S.-W. Cheong *et al.*, Phys. Rev. Lett. **67** (1991) 1791.
85. S. M. Hayden *et al.*, Phys. Rev. Lett. **68** (1992) 1061.
86. R. J. Cava *et al.*, Phys. Rev. B **43** (1991) 1229.
87. C. H. Chen, S.-W. Cheong, and A. S. Cooper, Phys. Rev. Lett. **71** (1993) 2461.
88. J. M. Tranquada, D. J. Buttrey, V. Sachan, and J. E. Lorenzo, Phys. Rev. Lett. **73** (1994) 1003.
89. P. J. Brown *et al.*, Physica B **180 & 181** (1992) 380.
90. D. Hone and A. H. Castro Neto, J. Superconductivity **10** (1997) 349.
91. C. N. A. van Duin and J. Zaanen, Phys. Rev. Lett. **80** (1998) 1513.
92. J. M. Tranquada, P. Wochner, A. R. Moodenbaugh, and D. J. Buttrey, Phys. Rev. B **55** (1997) R6113.

93. K. Yamada et al., Physica C 221 (1994) 355.
94. V. Sachan et al., Phys. Rev. B 51 (1995) 12742.
95. J. M. Tranquada, D. J. Buttrey, and V. Sachan, Phys. Rev. B 54 (1996) 12318.
96. S.-H. Lee and S.-W. Cheong, Phys. Rev. Lett. 79 (1997) 2514.
97. O. Zachar, S. A. Kivelson, and V. J. Emery, Phys. Rev. B 57 (1998) 1422.
98. E. Fawcett et al., Rev. Mod. Phys. 66 (1994) 25.
99. J. Zaanen and P. B. Littlewood, Phys. Rev. B 50 (1994) 7222.
100. T. Katsufuji et al., Phys. Rev. B 54 (1996) R14230.
101. G. Blumberg, M. V. Klein, and S.-W. Cheong, Phys. Rev. Lett. 80 (1998) 564.
102. K. Yamamoto, T. Katsufuji, T. Tanabe, and Y. Tokura, Phys. Rev. Lett. 80 (1998) 1493.
103. A. Vigliante et al., Phys. Rev. B 56 (1997) 8248.
104. G. Aeppli and D. J. Buttrey, Phys. Rev. Lett. 61 (1988) 203.
105. T. Freltoft et al., Phys. Rev. B 44 (1991) 5046.
106. K. Yamada et al., J. Phys. Soc. Jpn. 60 (1991) 1197.
107. K. Nakajima et al., J. Phys. Soc. Jpn. 62 (1993) 4438.
108. J. M. Tranquada, P. Wochner, and D. J. Buttrey, Phys. Rev. Lett. 79 (1997) 2133.
109. M. Sera et al., Solid State Commun. 69 (1989) 851.
110. Y. Koike et al., Solid State Commun. 79 (1991) 155.
111. Y. Maeno, N. Kakehi, M. Kato, and T. Fujita, Phys. Rev. B 44 (1991) 7753.
112. A. R. Moodenbaugh, U. Wildgruber, Y. L. Wang, and Y. Xu, Physica C 245 (1995) 347.
113. J. D. Axe et al., Phys. Rev. Lett. 62 (1989) 2751.
114. T. Suzuki and T. Fujita, Physica C 159 (1989) 111.
115. D. E. Cox et al., Mat. Res. Symp. Proc. 156 (1989) 141.
116. D. M. Paul et al., Phys. Rev. Lett. 58 (1987) 1976.
117. Y. Horibe, Y. Inoue, and Y. Koyama, Physica C 282–287 (1997) 1071.
118. G. M. Luke et al., Physica C 185-189 (1991) 1175.
119. K. Kumagai et al., Hyperfine Int. 86 (1994) 473.
120. M. K. Crawford et al., Phys. Rev. B 44 (1991) 7749.
121. T. Suzuki, M. Sera, T. Hanaguri, and T. Fukase, Phys. Rev. B 49 (1994) 12392.
122. K. Yoshida et al., Physica C 230 (1994) 371.
123. B. Büchner, M. Breuer, A. Freimuth, and A. P. Kampf, Phys. Rev. Lett. 73 (1994) 1841.
124. A. R. Moodenbaugh, L. H. Lewis, and S. Soman, Physica C 290 (1997) 98.
125. B. Büchner et al., Europhys. Lett. 21 (1993) 953.
126. M. Breuer et al., Z. Phys. B 92 (1993) 331.
127. W. Wagener et al., Phys. Rev. B 55 (1997) R14 761.
128. J. E. Ostenson et al., Phys. Rev. B 56 (1997) 2820.
129. Y. Nakamura and S. Uchida, Phys. Rev. B 46 (1992) 5841.
130. J. M. Tranquada et al., Nature 375 (1995) 561.
131. J. M. Tranquada et al., Phys. Rev. B 54 (1996) 7489.
132. M. v. Zimmermann et al., Europhys. Lett. (in press).
133. G. M. Luke et al., Hyp. Int. 105 (1997) 113.
134. J. M. Tranquada, Ferroelectrics 177 (1996) 43.
135. M. Roepke et al., J. Phys. Chem. Solids (in press).
136. J. M. Tranquada et al., Phys. Rev. Lett. 78 (1997) 338.
137. K. Yamada et al., Phys. Rev. B 57 (1998) 6165.
138. S. Katano et al., Phys. Rev. B 48 (1993) 6569.
139. B. Dabrowski et al., Phys. Rev. Lett. 76 (1996) 1348.
140. M. Braden et al., Z. Phys. B 94 (1994) 29.
141. C.-H. Lee et al., Physica C 257 (1996) 264.
142. T. Suzuki et al., Phys. Rev. B 57 (1998) R3229.
143. A. R. Moodenbaugh et al., (preprint).
144. Y. Koike et al., Solid State Commun. 82 (1992) 889.

145. K. Hirota, K. Yamada, I. Tanaka, and H. Kojima, Physica B (in press).
146. K. Kakurai *et al.*, Phys. Rev. B **48** (1993) 3485.
147. H. Harashina *et al.*, J. Phys. Soc. Jpn. **62** (1993) 4009.
148. Y. Sidis *et al.*, Phys. Rev. B **53** (1996) 6811.
149. J. M. Tranquada *et al.*, Phys. Rev. B **46** (1992) 5561.
150. B. J. Sternlieb *et al.*, Phys. Rev. B **50** (1994) 12915.
151. P. Dai, H. A. Mook, and F. Doğan, Phys. Rev. Lett. **80** (1998) 1738.
152. J. Zaanen and O. Gunnarsson, Phys. Rev. B **40** (1989) 7391.
153. D. Poilblanc and T. M. Rice, Phys. Rev. B **39** (1989) 9749.
154. H. J. Schulz, Phys. Rev. Lett. **64** (1990) 1445.
155. M. Kato, K. Machida, H. Nakanishi, and M. Fujita, J. Phys. Soc. Jpn. **59** (1990) 1047.
156. J. A. Vergés *et al.*, Phys. Rev. B **43** (1991) 6099.
157. M. Inui and P. B. Littlewood, Phys. Rev. B **44** (1991) 4415.
158. T. Giamarchi and C. Lhuillier, Phys. Rev. B **42** (1990) 10641.
159. G. An and J. M. J. van Leeuwen, Phys. Rev. B **44** (1991) 9410.
160. G. Seibold, J. Seidel, and E. Sigmund, Phys. Rev. B **53** (1996) 5166.
161. T. Mizokawa and A. Fujimori, Phys. Rev. B **56** (1997) 11920.
162. Z. G. Yu, J. Zang, J. T. Gammel, and A. R. Bishop, Phys. Rev. B **57** (1998) R3241.
163. N. M. Salem and R. J. Gooding, Europhys. Lett. **35** (1996) 603.
164. C. S. Hellberg and E. Manousakis, Phys. Rev. Lett. **78** (1997) 4609.
165. U. Löw, V. J. Emery, K. Fabricius, and S. A. Kivelson, Phys. Rev. Lett. **72** (1994) 1918.
166. S. Haas, E. Dagotto, A. Nazarenko, and J. Riera, Phys. Rev. B **51** (1995) 5989.
167. C. Castellani, C. Di Castro, and M. Grilli, Phys. Rev. Lett. **75** (1995) 4650.
168. P. Prelovšek and X. Zotos, Phys. Rev. B **47** (1993) 5984.
169. H. E. Viertiö and T. M. Rice, J. Phys.: Condens. Matter **6** (1994) 7091.
170. H. Tsunetsugu, M. Troyer, and T. M. Rice, Phys. Rev. B **51** (1995) 16456.
171. S. R. White and D. J. Scalapino, Phys. Rev. Lett. **80** (1998) 1272.
172. C. Nayak and F. Wilczek, Int. J. Mod. Phys. B **10** (1996) 2125.
173. C. Nayak and F. Wilczek, Phys. Rev. Lett. **78** (1997) 2465.
174. G. Seibold, C. Castellani, C. Di Castro, and M. Grilli, Report No. cond-mat/9803184.
175. J. Zaanen, M. L. Horbach, and W. van Saarloos, Phys. Rev. B **53** (1996) 8671.
176. H. Eskes, R. Grimberg, W. van Saarloos, and J. Zaanen, Phys. Rev. B **54** (1996) R724.
177. A. H. Castro Neto and D. Hone, Phys. Rev. Lett. **76** (1996) 2165.
178. V. J. Emery, S. A. Kivelson, and O. Zachar, Phys. Rev. B **56** (1997) 6120.
179. Y. A. Krotov, D.-H. Lee, and A. V. Balatsky, Phys. Rev. B **56** (1997) 8367.
180. A. H. Castro Neto, Phys. Rev. Lett. **78** (1997) 3931.
181. S. A. Kivelson, E. Fradkin, and V. J. Emery, Nature (in press).

MAGNETIC 2-D AND 3-D ORDERING PHENOMENA IN RARE-EARTH BASED COPPER-OXIDE SUPER-CONDUCTORS AND RELATED SYSTEMS

P. FISCHER AND M. MEDARDE
Laboratory for Neutron Scattering, ETHZ Zurich & Paul Scherrer Institute
CH-5232 Villigen PSI, Switzerland

1. Introduction

Magnetic ordering phenomena in high-T_c compounds of rare earths R and Y of type $R_1Ba_2Cu_3O_{7-\delta}$ (R1-2-3), $R_1Ba_2Cu_4O_8$ (R1-2-4), $R_2Ba_4Cu_7O_{15-\delta}$ (R2-4-7) and of related systems, mainly based on elastic neutron scattering experiments, are the subject of the present review. Compared to previous summaries such as published in references [1-11] a rather complete, updated survey of mostly generally available publications is attempted for these classes of important layered perovskite compounds with fascinating physical properties including high-temperature superconductivity. Primary information comes from nuclear and magnetic neutron scattering. This is due to the sensitivity of thermal neutrons to light atoms such as oxygen, which form together with copper ions superconducting CuO_2 planes as well as Cu-O chains providing a charge reservoir. Because of the magnetic moment of the neutron interacting with atomic magnetic moments in neutron scattering experiments, magnetic ordering phenomena and magnetic interactions may be studied by this method in a rather direct way on an atomic scale also in such relatively new materials containing magnetic Cu and rare earth ions. Investigations of phase diagrams as well as of crystal-field effects and magnetic excitations of such high-T_c superconducting compounds are covered in chapters 2, 9 and 3 of this volume, respectively.

After a discussion of the layer type crystal structures of these isostructural rare-earth and Y systems, magnetic ordering due to copper ions in such compounds will be described. More detailed structural information on high-temperature cuprate superconductors or more specifically on Y1-2-3 compounds is available from the recent review published by Park and Snyder [12] and from chapter 2 of this volume, respectively. The main part of the present contribution will then consist of a review of two-dimensional (2-D) to 3-D magnetic ordering due to rare-earth ions in these systems. Then the influences of crystal-field effects, of dipole and exchange interactions will be discussed, and finally a summary will be presented.

In copper oxide superconductors magnetic interactions between magnetic moments of copper ions (Cu^{2+} with $3d^9$ electron configuration) and/or rare earth ions (generally due to 4f electrons of R^{3+}, replacing Y^{3+}) are important in relation to superconductivity, in particular concerning the question of the pairing mechanism and the possibility of coexistence of magnetic ordering and superconductivity. Simultaneous long-range rare-earth antiferromagnetism and superconductivity was previously found in ternary compounds of rare earths such as Chevrel phases like $DyMo_6S_8$ or for borides and stannides, e.g. $TmRh_4B_4$ and $ErRh_{1.1}Sn_{3.6}$ [13] and recently for borocarbides [11, 14]. A similar coexistence of antiferromagnetism due to magnetic moments of rare earth ions R^{3+} and superconductivity is found to exist in R1-2-3 high-T_c materials and related compounds.

2. Crystal Structure Aspects

Average layer type structures of the 'prototype' systems Y1-2-3, Y1-2-4 and Y2-4-7 are illustrated in figure 1 similar to ref. [15]. Possible influences of phase separation and charge-stripe order effects are discussed in chapters 6 and 9 of this volume. The T_c = 90 K Y1-2-3$O_{x \approx 7}$ superconductor is characterised by CuO_2 conduction planes perpendicular to the [0, 0, 1] direction and single CuO chains parallel to the b-axis which serve as charge reservoirs. This ortho-I layer structure is orthorhombic, corresponding to space group Pmmm with lattice parameters a $\approx$ 3.81 Å, b $\approx$ 3.88 Å and c $\approx$ 11.66 Å at 5 K [16]. In Y1-2-3O_x it is of course now well known that for approximately x < 6.4 the compounds are nonsuperconducting and tetragonal with space group symmetry P4/mmm and at first glance surprisingly increased lattice parameter c $\approx$ 11.82 Å and a $\approx$ 3.85 Å for x = 6 at 5 K [16]. In the sense of average structures we do not consider here superstructures due to oxygen ordering such as the ortho-II or Herringbone phases which may exist e. g. for x = 6.5 [17, 18], as usually they are not long-range ordered. As illustrated in figure 2, showing the unit cell of Tm1-2-3O_7, one may replace Y^{3+} by trivalent rare earths without major changes in superconductivity [19] and structure type. However - as shown in detail by Buchgeister [19] - the T_c curves versus oxygen concentration x of light rare earths show considerably smaller plateau ranges compared to those of heavy rare earths which resemble the Y case. Systematic trends of the R1-2-3O_7 crystal structures associated with lanthanide contraction etc. were published by Guillaume et al. [20]. In particular it was found that the chain copper Cu(1) - apex oxygen O(1) distance remains within error limits constant throughout the R1-2-3O_7 series, whereas the distance Cu(2)-O(1) increases. Moreover the existence of a critical puckering angle in the CuO_2 planes was established. Another aspect to be considered is the reported phase separation in Y1-2-3O_x for x > 6.9 [21], yielding a splitting of the c-lattice parameter.

On the other hand the 80 K Y1-2-4 superconductor has Ammm symmetry and a c-lattice parameter of approximately 27 Å, associated with double CuO chains [22, 23]. Structure parameters of similar superconducting R1-2-4 compounds of heavy rare earths R, refined by neutron diffraction, were published in refs. [24-26]. T_c decreases systematically with the increase of the ionic radius of the rare earth ion.

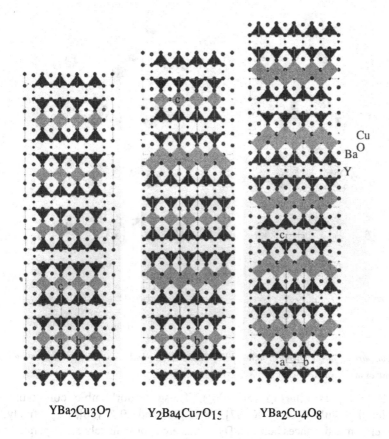

YBa₂Cu₃O₇   Y₂Ba₄Cu₇O₁₅   YBa₂Cu₄O₈

Figure 1. Comparison of the $YBa_2Cu_3O_7$, $Y_2Ba_4Cu_7O_{15}$ and $YBa_2Cu_4O_8$ crystal structures, projected along the a-axis. Parallel to the horizontal b-axis the single and double Cu-O chains are illustrated, light-coloured. Y is located between superconducting CuO_2 planes forming pairs. Unit cells are indicated by lines and letters.

The Y2-4-7 structure with the largest c-lattice parameter of about 50 Å consists (see figure 1) of a regular stacking of 1-2-3 blocks sandwiched by mirror symmetric 1-2-4 units, corresponding to space group Ammm [27, 28]. Depending on oxygen concentration and sample perfection (also Cu nonstoichiometry) T_c of Y2-4-7O_y may vary between approximately 14 K and 70 K [29, 30] to 93 K [31] and 95 K [28, 32]. Oxygen may be mainly removed from the single Cu-O chains. For the highest T_c values apparently a small number of stacking faults is more important than oxygen chain order. In particular for samples prepared by sol-gel techniques the chain disorder may be remarkably large [33]. High-resolution neutron diffraction structural studies of similar R2-4-7 compounds with $T_c \leq 80$ K were published by Currie et al. [34], whereas recently by sol-gel techniques $T_c = 89$ K was realised for a Er2-4-7O_{15} sample by Böttger et al. [35].

Concerning magnetic rare-earth ordering it is instructive to compare the rare-earth sublattices of the R1-2-3, R1-2-4 and R2-4-7 systems. Figure 3 illustrates this for the

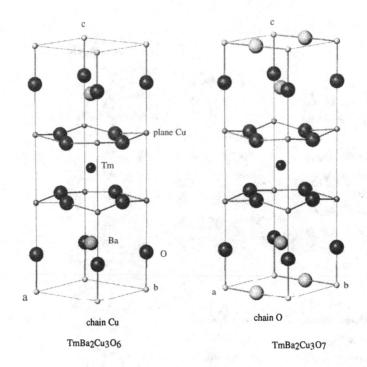

Figure 2. *Comparison of the chemical unit cells of TmBa$_2$Cu$_3$O$_6$ and TmBa$_2$Cu$_3$O$_7$, illustrating CuO$_2$ planes and Cu or CuO chains.*

case of R = Dy (similar to ref. [36]). These orthorhombic compounds are superconducting with T_c = 91 K [37], 78 K [24] and 60 K [36], respectively. The shortest interatomic distances between Dy^{3+} ions are approximately 3.8 Å parallel to the a- and b-axes in each case, but parallel to the c-axis the Dy layer distances are at least three times larger: 11.7, 13.6, 11.4 and 13.6 Å for Dy1-2-3, Dy1-2-4 and Dy2-4-7, respectively. Thus one expects dominant exchange interactions within (a,b) planes: J_{ab} » J_c and rather 2-D than 3-D magnetic ordering.

Another aspect of importance is the Ammm symmetry of the R1-2-4 and R2-4-7 systems [24], in contrast to the Pmmm or P4/mmm space groups of the R1-2-3 compounds. The A-lattice implies translation b/2 + c/2. Therefore in case of antiferromagnetic coupling along a-, b- and c-directions considerable frustration results in nearest neighbour co-ordination etc. below and above a particular magnetic ion (in each next layer one spin parallel and one antiparallel, resulting in net interaction zero, see figure 3). As suggested by Zhang et al. [36, 38] this is the main reason for expecting 2-D antiferromagnetism in most R1-2-4 systems such as Dy1-2-4 [24, 38, 39]. Frustration tendencies exist also to a certain extent in case of R2-4-7, where in addition the crystal structure favours the formation of magnetically coupled bilayers due to the shorter R-R distance in the R1-2-3 block along the c-axis, see figure 3 [36].

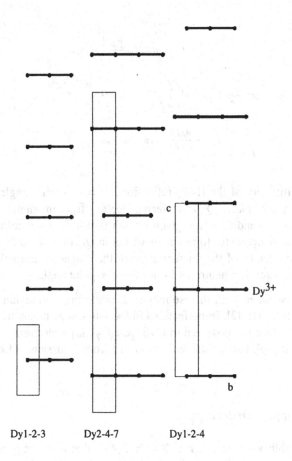

Dy1-2-3   Dy2-4-7   Dy1-2-4

Figure 3. Comparison of rare-earth sublattices for $DyBa_2Cu_3O_7$, $Dy_2Ba_4Cu_7O_{15}$ and $DyBa_2Cu_4O_8$, view along the a-axis. The rectangles indicate unit cells.

3. Magnetic Neutron Diffraction

For magnetic 2-D or 3-D ordering the theory of elastic neutron scattering has been summarised e. g. in refs. [36, 38-40] (see also chapter 1 of this volume) and is indicated for the simplest 2-D case in formula (1).

2-D antiferromagnetic ordering corresponds to magnetic rods in reciprocal space instead of magnetic Bragg points for 3-D ordering, which yield a superposition of Gaussian Bragg peaks in case of usual constant wavelength neutron powder diffraction. In such a measurement one scans reciprocal space radially, i. e. suddenly the first magnetic rod gets touched, resulting in a steep increase of magnetic intensity. As the rods extend in principle to infinity, the 2-D peaks of magnetic neutron intensity decrease at higher scattering angles only gradually. A typical example is illustrated in figure 14 for Dy1-2-4 [39]. In case of magnetic correlations between neighbouring planes modulations of the magnetic rod neutron intensity result which imply modulations of

$$I_{hk}^{nM} (2-D) = const. M_{hk} \frac{|\vec{F}_M|^2}{sin\theta} F(\theta),$$

$$\vec{F}_M = \vec{e}_Q x (\vec{e}_\mu x \vec{e}_Q) \sum_{j=1}^{N} < \mu_z >_j f_j(Q) e^{-W_j} e^{i\vec{Q}\vec{r}_j}, \qquad (1)$$

$$F(\theta) = \int_{0}^{\pi/2} e^{-\frac{4\pi L^2}{\lambda^2}(sin\theta cos\varphi - sin\theta_{hk})^2} d\varphi,$$

where M_{hk} = multiplicity of the (hk0) reflection, 2θ = scattering angle, θ_{hk} = Bragg angle of the 2-D peak (hk0), $\vec{Q}$ = scattering vector, $\vec{\mu}$ = magnetic moment with expectation value $<\mu_z>$ and direction given by unit vector $\vec{e}_\mu$ for a collinear magnetic structure, f = neutron magnetic form factor of the magnetic ions, e^{-W} = temperature factor, $\vec{r}$ = position vector of the magnetic ion in the magnetic unit cell, L = domain size within the 2-D layer, λ = neutron wavelength, φ = polar angle.

the 2-D peaks. Associated with diffuse magnetic scattering Lorentzian peak shape is discussed in e. g. ref. [41, 42]. Imperfections in the long-range magnetic ordering such as stacking faults as recently observed in $HoBa_2Cu_3O_7$ [43] yield further complications such as anisotropic peak halfwidths and variations from Gaussian to Lorentzian peak shape.

4. Magnetic Copper Ordering

It is now well established (see e.g. Fig. 8 in ref. [44]) that long-range magnetic order due to Cu ions cannot coexist with superconductivity in R1-2-3 systems. However dynamical spin correlations survive (cf. e. g. [44, 45]). On the other hand in nonsuperconducting $Y1$-2-$3O_{6+\delta}$ 3-D long-range antiferromagnetic Cu ordering was found with very high Néel temperatures T_{N1} up to approximately 500 K (in recent phase diagrams [44] up to approximately 420 K). Neutron diffraction investigations performed on tetragonal ceramic $Y1$-2-$3O_{6+\delta}$ powder samples at Brookhaven National Laboratory [46] and on both powders and single crystals by French groups [44, 47, 48] as well as later by Casalta et al. on pure single crystals [49] yielded clear evidence for magnetic ordering of approximately divalent Cu ions in CuO_2 planes with a maximum ordered magnetic moment of Cu of 0.66 μ_B, oriented perpendicular to the c-axis (see table Ia). The basic magnetic structure is shown in figure 4.

The low values of ordered magnetic copper moments are due to quantum spin fluctuations and covalency effects. Within the CuO_2 planes the magnetic coupling is antiferromagnetic along the (tetragonally equivalent) a- and b-directions, corresponding to distances Cu - Cu of about 3.8 Å. Parallel to the c-axis neighbouring CuO_2 planes are also antiferromagnetically coupled. The Cu - Cu distance perpendicular to the planes is approximately 3.4 Å. These antiferromagnetic bilayers are separated by usually nonmagnetic ($\approx Cu^+$) chain copper layers, resulting in an approximately 8.3 Å distance

TABLE Ia. Neutron diffraction results on magnetic ordering in $YBa_2Cu_3O_x$ compounds. P = powder, SC= single crystal specimen, T_N = Néel temperature, $\vec{k}$ = propagation vector, μ_{sat} = ordered magnetic moment at saturation, MO = magnetic order. a) oxygen superstructure, b) absence of a second antiferromagnetic transition at T ≥ 2 K for pure crystals.

compound/ sample	$T_N[K]$	$\vec{k}$	$\mu_{sat}[\mu_B]$	$\mu_{direction}$	ref.-year	remarks
$YBa_2Cu_3O_{6.40}$/P	$T_{N1}(Cu2) \approx$ 220	[1/2,1/2,0]	$\mu_{Cu2} \approx$ 0.38	$\perp$ [0,0,1]	[46]-1988	
$YBa_2Cu_3O_{6.38}$/SC	$T_{N1}(Cu2) =$ 245(10)	[1/2,1/2,0]	$\mu_{Cu2} =$ 0.28(5)	$\perp$ [0,0,1]	[48]-1988	
$YBa_2Cu_3O_{6.35}$/SC	$T_{N1}(Cu2) \approx$ 405(5); $T_{N2}(Cu1) \approx$ 40	[1/2,1/2,0] (Cu2)& [1/2,1/2,1/2]	$\mu_{Cu2} =$ 0.50(7); $\mu_{Cu1} =$ 0.012(7)	$\perp$ [0,0,1]*	[50]-1988	* $\vec{\mu}_{Cu2}$ $\perp$ $\vec{\mu}_{Cu1}$
$YBa_2Cu_3O_{6.35}$/SC	$T_{N1}(Cu2) =$ 390(5)	[1/2,1/2,0]	$\mu_{Cu2} =$ 0.40(5)	$\perp$ [0,0,1]	[18]-1991	a)
$YBa_2Cu_3O_{6.25}$/SC	$T_{N1}(Cu2) =$ 290(10)	[1/2,1/2 0]	$\mu_{Cu2} =$ 0.45(5)	$\perp$ [0,0,1]	[48]-1988	$\mu_{Cu1} < .05 \mu_B$
$YBa_2Cu_3O_{6.18}$/SC	$T_{N1}(Cu2) =$ 368(1)	[1/2,1/2 0]	$\mu_{Cu2} =$ 0.44(3)	$\perp$ [0,0,1]	[49, 51]-1994	b)
$YBa_2Cu_3O_{6.15}$/P	$T_{N1}(Cu2) \approx$ 400	[1/2,1/2,0]	$\mu_{Cu2} =$ 0.50(5)	$\perp$ [0,0,1]	[46]-1988	
$YBa_2Cu_3O_{6.13}$/P	$T_{N1}(Cu2) \approx$ 500	[1/2,1/2,0]	$\mu_{Cu2} =$ 0.64(6)	$\perp$ [0,0,1]	[52]-1988	
$YBa_2Cu_3O_{6.1}$/SC	$T_{N1}(Cu2) =$ 411(1)	[1/2,1/2 0]	$\mu_{Cu2} =$ 0.55(3)	$\perp$ [0,0,1]	[49, 51]-1994	b)
$YBa_2Cu_3O_{6.0}$/P	$T_{N1}(Cu2) \approx$ 400	[1/2,1/2,0]	$\mu_{Cu2} =$ 0.48(8)	$\perp$ [0,0,1]	[46]-1988	
$YBa_2Cu_3O_{6.0}$/P	$T_{N1}(Cu2) \approx$ 500	[1/2,1/2,0]	$\mu_{Cu2} =$ 0.48(5)	‖ [0,0,1]	[53]-1988	‖ [0,0,1] ?
$YBa_2Cu_3O_{6.0}$/SC	$T_{N1}(Cu2) =$ 420(10)	[1/2 1/2 0]	$\mu_{Cu2} =$ 0.60(5)	$\perp$ [0,0,1]	[47]-1988	no Cu1 MO
$YBa_2Cu3O_{6.0}$/SC&P	$T_{N1}(Cu2) =$ 420(10)	[1/2,1/2,0]	$\mu_{Cu2} =$ 0.61(5)	$\perp$ [0,0,1]	[48]-1988	
$YBa_2Cu_3O_{5.94}$/P	$T_{N1}(Cu2) \approx$ 500	[1/2,1/2,0]	$\mu_{Cu2} =$ 0.66(6)	$\perp$ [0,0,1]	[46]-1988	

between Cu ions of neighbouring bilayers. According to inelastic neutron scattering studies of magnetic Cu spin excitations of Y1-2-3O_{6+x} [44, 45] the dominant exchange interaction is the nearest-neighbour superexchange in the CuO_2 plane: $J_{Cu-Cu} = J \approx 120$ meV. There exists also a large antiferromagnetic coupling between two neighbouring CuO_2 layers: $J_\perp \approx 10$ meV which persists in the metallic and superconducting state. However the interaction between bilayers (J' $\approx 3.5 \times 10^{-5}$J) and planar anisotropy ($\Delta J/J \approx 2 \times 10^{-4}$) were found to be weak. The coupling of different CuO_2 layers yields neutron intensity modulations along the magnetic ridge (1/2,1/2,l) [44, 54]. Also the critical

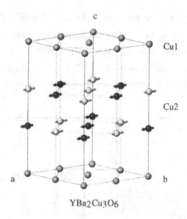

Figure 4. Magnetic Cu ordering in pure $YBa_2Cu_3O_6$ with easy directions of magnetisation [1,0,0] or [0,1,0] [44].

exponent $\beta = 0.25(3)$ [44] indicates a quasi 2-D behaviour with planar (XY) anisotropy.

More details concerning the dependence of the ordered magnetic copper moment on oxygen concentration and temperature are summarised in the review by Rossat-Mignod et al. [44] and in table Ia. The remarkable reentrant behaviour of the Cu sublattice magnetization in Y1-2-3O$_x$ at low temperatures was explained by Korenblit and Aharony [55] by a model which assumes that localized holes reduce the sublattice magnetisation more strongly than mobile holes.

Several elastic neutron scattering investigations have provided evidence that also chain copper may order magnetically at low temperatures (generally $T_{N2} < T_{N1}$, e. g. $T_{N1} = 430$ K and $T_{N2} = 80$ K for Nd1-2-3O$_{6.1}$ [56]) both in case of the nonsuperconducting compounds Y1-2-3O$_{6.35}$ [50], Pr1-2-3O$_{6+\delta}$ [57] (cf. also table Ie) and Nd1-2-3O$_{6+\delta}$ (see tables Ig and Ih). Magnetic moments on chain copper Cu1 sites tend to a doubling of the magnetic unit cell also along [0, 0, 1] and to noncollinear magnetic ordering (cf. e. g. ref. [56]). Also neutron diffraction studies on nonsuperconducting Pr1-2-3O$_{6+\delta}$ crystals with δ up to about 0.65 [57] yielded at small δ values $T_{N1} \approx 370$ K due to simultaneous ordering of both plane and chain Cu spins. For $x \approx 6.6$ the Cu planes still order at $T_{N1} \approx 370$ K, but the chain copper ordering is reduced in temperature to $T_{N2} \approx 160$ K. As illustrated e. g. in references [56, 57] the chain copper ordering results in a temperature dependent reorientation of the magnetic Cu moments with approximately perpendicular orientation of magnetic chain and plane copper moments perpendicular to [0, 0, 1]. Moreover simultaneous magnetic ordering of Cu and rare-earth ions as in the system Nd1-2-3O$_{6+\delta}$ [41, 56] enhances the tendency to noncollinearity of the magnetic ordering. At saturation (cf. fig. 12 in ref. [56]) the chain copper moment attains in case of Nd1-2-3O$_{6.1}$ only about half the value of the plane copper moment. This is one of the reasons why the magnetic chain copper ordering was rarely observed in neutron powder measurements, but it could be also due to e. g. higher chemical oxygen ordering in crystals [56] or perhaps more likely caused by impurities introduced during crystal growth. A new noncollinear magnetic structure was suggested

TABLE Ib. Neutron diffraction results on magnetic ordering in La and Al substituted $YBa_2Cu_3O_x$. a) sample dependence of transition temperatures with equal nominal composition.

compound/ sample	$T_N[K]$	$\vec{k}$	$\mu_{sat}\,[\mu_B]$	$\mu_{direction}$	ref.- year	remarks
$YBa_{1.1}La_{0.9}Cu_3O_{7+y}$/P	$T_N(Cu) \approx$ 430	[1/2,1/2,0]			[58]- 1991	
$YBa_{1.2}La_{0.8}Cu_3O_{7+y}$/P	$T_N(Cu) \approx$ 419	[1/2,1/2,0]			[58]- 1991	
$YBa_{1.3}La_{0.7}Cu_3O_{7+y}$/P	$T_N(Cu) \approx$ 400	[1/2,1/2,0]			[58]- 1991	
$YBa_{1.4}La_{0.6}Cu_3O_{7+y}$/P	$T_N(Cu) \approx$ 180	[1/2,1/2,0]			[58]- 1991	
$YBa_2Cu_{2.81}Al_{0.19}O_{6.28}$/ SC	$T_{N1}(Cu2) =$ 407(1) $T_{N2(Cu)} =$ 17(1)	[1/2,1/2,0] [1/2,1/2,1/2]	$\mu_{Cu2} =$ 0.58(2) at 20 K $\mu_{Cu2} =$ 0.55(1) at 4.2 K	⊥ [0,0,1]	[59]- 1995	a) $\mu_{Cu1} \approx 0$
$YBa_2Cu_{2.86}Al_{0.14}O_{6.25}$/ SC	$T_{N1}(Cu2) =$ 410(1) $T_{N2\,(Cu)} =$ 8(1)	[1/2,1/2,0] [1/2,1/2,1/2]			[49, 51]- 1994	
$YBa_2Cu_{2.86}Al_{0.14}O_{6.25}$/ SC	$T_{N1}(Cu2) =$ 404(1) $T_{N2} = 4(1)$				[59]- 1995	
$YBa_2Cu_{2.90}Al_{0.10}O_{6.21}$/ SC	$T_{N1}(Cu2) =$ 411(1) $T_{N2} = 8(1)$				[59]- 1995	
$YBa_2Cu_{2.93}Al_{0.07}O_{6.18}$/ SC	$T_{N1}(Cu2) =$ 403(1) $T_{N2} = 12(1)$				[59]- 1995	
$YBa_2Cu_{2.94}Al_{0.06}O_{6.14}$/ SC	$T_{N1}(Cu2) =$ 411(1) $T_{N2} = 8(1)$				[59]- 1995	

recently by Boothroyd et al. [60] for $PrBa_2Cu_3O_{6+x}$, implying essential magnetic coupling of the Cu2 and Pr sublattices, whereas Uma et al. [61] observed only a modest 3d-4f coupling. Particularly complex magnetic Cu and Pr ordering with temperature dependent reorientations of magnetic moments were also reported for crystals of $PrBa_2Cu_{3-y}Al_yO_{6+x}$ by Longmore et al. [62].

As summarised in tables Ib, Ic and Id, substitution of Cu by e. g. Co or Fe stabilises magnetic chain ordering which is associated with mainly occupation of chain sites. A similar effect was established in single crystal studies of the compounds $Nd_{1.5}Ba_{1.5}Cu_3O_x$ (x = 6.4 and 6.9) containing excess of Nd [63]. The often complex, noncollinear magnetic ordering in substituted Y1-2-3O_{6+x} was found to depend essentially on sample preparation such as deoxygenation at high or low temperatures and on effects of impurity clustering, cf. e. g. the recent neutron diffraction investigations on $YBa_2(Cu_{1-x}Fe_x)_3O_y$, in particular by Mirebeau et al. [64, 65], where also evidence

TABLE Ic. Neutron diffraction results on magnetic ordering in Co and Fe ($x_O > 6.4$) substituted $YBa_2Cu_3O_x$. a) $\vec{\mu}(M2)$ antiparallel $\vec{\mu}(M1)$, b) angle $[\vec{\mu}(M2);\vec{\mu}(M1)] = 110(20)°$, c) spin glass transition below 70 K.

compound/ sample	T_N[K]	$\vec{k}$	μ_{sat} [μ_B]	$\mu_{direction}$	ref.-year	remarks
$YBa_2Cu_{2.2}Co_{0.8}O_{6.91}$/P	$T_N(M) = $ 435(5)	[1/2,1/2,1/2]	$\mu_M = $ 0.68(6)	$\perp$ [0,0,1]	[66]-1988	a)
$YBa_2Cu_{2.8}Co_{0.2}O_{6.75}$/P	$T_{N1}(M2)\approx$ 226; $T_{N2}(M1) \approx$ 40	[1/2,1/2,0] (M2) & [1/2,1/2,1/2]			[67]-1989	No MO for $O_{7.0}$.
$YBa_2Cu_{2.8}Co_{0.2}O_{6.65}$/P	$T_{N1}(M2) \approx$ 334; $T_{N2}(M1) \approx$ 40	[1/2,1/2,0] (M2) & [1/2,1/2,1/2]			[67]-1989	
$YBa_2Cu_{2.8}Co_{0.2}O_{6.48}$/P	$T_{N1}(M2) \approx$ 374; $T_{N2}(M1) \approx$ 75	[1/2,1/2,0] (M2) & [1/2,1/2,1/2]			[67]-1989	
$YBa_2Cu_{2.8}Co_{0.2}O_{6.45}$/P	$T_{N1}(M2) = $ 415(7); $T_{N2}(M1) = $ 211(6)	[1/2,1/2,0] (M2) & [1/2,1/2,1/2]	$\mu_{M1} = $ 0.62(6); $\mu_{M2} = $ 0.18(2)	$\perp$ [0,0,1]	[66]-1988	b)
$YBa_2Cu_{2.8}Co_{0.2}O_{6.41}$/P	$T_{N1}(M2) \approx$ 414; $T_{N2}(M1) \approx$ 210	[1/2,1/2,0] (M2) & [1/2,1/2,1/2]			[67]-1989	
$YBa_2Cu_{2.8}Co_{0.2}O_{6.32}$/P	$T_N(M) \approx 376$	[1/2,1/2,1/2]			[67]-1989	
$YBa_2Cu_{2.7}Fe_{0.3}O_{6.67}$P	$T_N(M) = 295$	[1/2,1/2,1/2]	$\mu(M2) = $ 0.36	$\perp$ [0,0,1]	[68]-1994	$\mu(M1) \approx 0.4\mu$ (M2)
$YBa_2Cu_{2.7}Fe_{0.3}O_{6.60}$P	$T_N(M) = 388$	[1/2,1/2,1/2]	$\mu(M2) = $ 0.40	$\perp$ [0,0,1]	[68]-1994	$\mu(M1) \approx 0.4\mu$ (M2)
$YBa_2Cu_{2.64}Fe_{0.36}O_{6.5}$/P	$T_N(M) = 430$	[1/2,1/2,1/2]	$\mu_M = 0.22$		[69]-1991	c)
$YBa_2Cu_{2.7}Fe_{0.3}O_{6.42}$/P	$T_N(M) = 440$	[1/2,1/2,1/2]	$\mu(M2) = $ 0.46	$\perp$ [0,0,1]	[68]-1994	$\mu(M1) \approx 0.4\mu$ (M2)

for a kind of spin glass transition at low temperatures was obtained. Most of these observations may be explained by the recently by Andersen and Uimin [70] published model for magnetic ordering in doped $YBa_2Cu_3O_{6+x}$, suggesting ferromagnetic coupling via free spins in CuO_x chain layers.

Both unpolarised [3] and polarised neutron diffraction studies of magnetisation densities of R1-2-3O_x (cf. e. g. [71-73]) did not reveal substantial covalence in the sense of a significant magnetic moment on oxygen ions, i. e. confirmed the essential 3d character of the neutron magnetic Cu form factor. However also on chain copper Cu1 a small magnetic field induced magnetic moment is observed. In addition newer polarised

TABLE Id. Neutron diffraction results on magnetic ordering in Fe substituted $YBa_2Cu_3O_{x \leq 6.4}$. a) samples deoxygenated by a low temperature process; partially spin glass like transition around T_{N2}, b) deoxygenated at high temperature, c) partially short-range ordered antiferromagnetic phase at $T < T^*$, noncollinear, certainly at temperatures below T_{N2}, d) T_f = freezing temperature of frustrated paramagnetic Fe spins (spin glass) on Cu1 sites.

compound/ sample	$T_N[K]$	$\vec{k}$	μ_{sat} $[\mu_B]$	$\mu_{direction}$	ref.- year	remarks
$YBa_2Cu_{2.64}Fe_{0.36}O_{6.4}$/P	$T_{N1} = 410$	[1/2,1/2,0]	$\mu_1(M2) \approx$ 0.58	$\perp$ [0,0,1]	[65]- 1997	a)
	$T_{N2} \approx 80$	[1/2,1/2,1/2]	$\mu_1(M1) \approx$ 0.26			
$YBa_2Cu_{2.52}Fe_{0.48}O_{6.4}$/P	$T_{N1} = 410$	[1/2,1/2,0]	$\mu_1(M2) \approx$ 0.56	$\perp$ [0,0,1]	[65]- 1997	a)
	$T_{N2} \approx 150$	[1/2,1/2,1/2]	$\mu_1(M1) \approx$ 0.16			
$YBa_2Cu_{2.97}Fe_{0.03}O_{6.3}$/P	$T_{N1} = 420$	[1/2,1/2,0]		$\perp$ [0,0,1]	[65]- 1997	a)
	$T_{N2} \approx 40$	[1/2,1/2,1/2]				
$YBa_2Cu_{2.94}Fe_{0.06}O_{6.3}$/P	$T_{N1} \approx 440$	[1/2,1/2,0]	$\mu_1(M2) \approx$ 0.54	$\perp$ [0,0,1]	[65]- 1997	a)
	$T_{N2} \approx 40$	[1/2,1/2,1/2]	$\mu_1(M1) \approx$ 0.06			
$YBa_2Cu_{2.99}Fe_{0.01}O_{6.2}$/P	$T_{N1} =$ 420(10)	[1/2,1/2,0]		$\perp$ [0,0,1]	[64]- 1994	b), c)
	$T^* \approx 200$					
	$T_{N2} = 40(10)$	[1/2,1/2,1/2]				
$YBa_2Cu_{2.98}Fe_{0.02}O_{6.2}$/P	$T_{N1} =$ 410(10)	[1/2,1/2,0]		$\perp$ [0,0,1]	[64]- 1994	b), c)
	$T^* \approx 175(25)$					
	$T_{N2} = 110$	[1/2,1/2,1/2]				
$YBa_2Cu_{2.97}Fe_{0.03}O_{6.2}$/P	$T_{N2} = 410$	[1/2,1/2,1/2]	$\mu(M2) =$ 0.45, $\mu(M1) =$ 0.04	$\perp$ [0,0,1]	[64]- 1994	b)
$YBa_2Cu_{2.94}Fe_{0.06}O_{6.2}$/P	$T_{N2} = 435$ $T_f \approx 60$	[1/2,1/2,1/2]		$\perp$ [0,0,1]	[64]- 1994	b), d)
$YBa_2Cu_{2.92}Fe_{0.08}O_{6.2}$/P	$T_{N2} = 443$ $T_f = 85$	[1/2,1/2,1/2]	$\mu(M2) =$ 0.55, $\mu(M1) =$ 0.18	$\perp$ [0,0,1]	[64]- 1994	b), d)
$YBa_2Cu_{2.98}Fe_{0.02}O_{6.1}$/P	$T_{N1} \approx T^* =$ 415(10)	[1/2,1/2,0]	$\mu(M2) =$ 0.54	$\perp$ [0,0,1]	[64]- 1994	b), c)
	$T_{N2} = 370$	[1/2,1/2,1/2]	$\mu(M1) =$ 0.04			

neutron studies on Y1-2-3O$_x$ by Boucherle et al. [74] confirm that the magnetic density is located mainly on copper sites.

A remarkable effect is the surprisingly large pressure change of the Néel temperature T_{N1}: $dT_{N1}(Cu2)/dp \approx 23$ K/kbar up to 4 kbars in $NdBa_2Cu_3O_{6.35}$, which was reported by Lynn et al. [75].

5. Magnetic Rare-Earth (R) Ordering

5.1. MAGNETIC ORDERING IN R1-2-3 COMPOUNDS

5.1.1. *Magnetic Ordering in R1-2-3O₇ Systems*
With respect to experiments bulk magnetic measurements performed on Gd1-2-3O₇ [76] clearly indicate an antiferromagnetic transition at $T_N \approx 2.2$ K. As the sample remains diamagnetic below the transition, superconductivity and antiferromagnetic ordering associated with magnetic Gd moments appear to coexist. Similar heat capacity measurements performed on the compounds R1-2-3O₇ of heavy rare earths R = Dy, Ho and Er (cf. e. g. [77]) also show well defined magnetic phase transitions with Néel temperatures $T_N < 1$ K. For a detailed determination of magnetic ordering phenomena neutron scattering experiments are necessary.

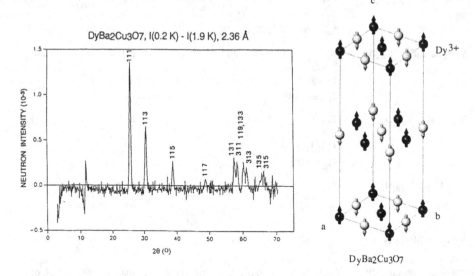

Figure 5. *Magnetic difference neutron diffraction pattern I(0.2 K) - I(1.9 K) of $DyBa_2Cu_3O_7$ [37].*

Figure 6. *Magnetic unit cell of $DyBa_2Cu_3O_7$, corresponding to antiferromagnetic Dy ordering with $\vec{k} = [1/2,1/2,1/2]$.*

Figure 5 shows as an illustrative example a corresponding magnetic difference neutron diffraction pattern for the case of Dy1-2-3O₇ [37]. Due to magnetic moments of Dy^{3+} ions one observes additional antiferromagnetic, sharp Bragg peaks, which prove the existence of long-range antiferromagnetic ordering despite the large separation between the magnetic ions along the c-axis. It is easy to index the new lines with a magnetic unit cell which is doubled along all three directions x, y, z with respect to the chemical cell. This magnetic structure with propagation vector $\vec{k}$ = [1/2,1/2,1/2] is illustrated in figure 6. It implies antiferromagnetic coupling along all basic translations. Magnetic moments of magnitude of about 7 μ_B at saturation are oriented parallel to the c-axis. In view of the approximately three times larger separation between rare earth ions along the c-axis compared to the (a,b)-plane it is surprising that such a three-dimensional antiferromagnetic configuration is typical for most R1-2-3O$_7$ compounds. It occurs not only for the heavy rare earth R = Dy [37, 78], but also in the S state Gd system [79] and in case of the light rare earths R = Nd [80, 81] and Pr [82] (see also tables Ie - Ik).

As already pointed out by Maple et al. [83] and shown in figure 7, the transition temperatures scale approximately with the de Gennes factor $(g-1)^2 J(J+1)$, except for the anomalous nonsuperconducting Pr1-2-3O$_7$ compound. This indicates the importance

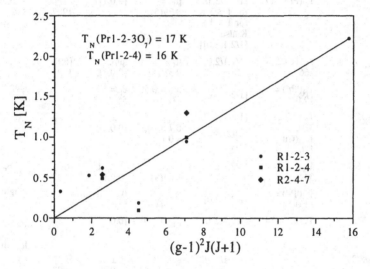

Figure 7. *De Gennes plot of measured Néel temperatures T_N associated with rare earths ions R^{3+} for R1-2-3O$_7$, R1-2-4 and R2-4-7 compounds.*

of exchange interactions such as superexchange in addition to dipole interactions. A modified RKKY mechanism was proposed by Liu for R1-2-3O$_x$ compounds [84]. On the other hand the Néel temperature T_N = 2.27(3) K of nonsuperconducting GdBa$_2$Cu$_3$O$_{6.1}$ [85] was found to be within error limits equal to T_N = 2.24 K of superconducting GdBa$_2$Cu$_3$O$_{\geq 6.8}$ [76]. Differences in ordering temperatures between other R1-2-3O$_7$ and R1-2-3O$_6$ compounds were recently attributed by Drössler et al. [86] to essential 4f-3d coupling in case of oxygen-deficient specimens.

In view of the small ordered magnetic moment at saturation of Pr of (0.74 ± 0.08) μ_B in the case of Pr1-2-3O$_7$ the exceptionally high Néel temperature T_N(Pr) ≈ 17 K [82] must be related to hybridisation effects (cf. e. g. [87, 88]). On the other hand from µSR [89] and NMR [90] investigations it is known that in semiconducting Pr1-2-3O$_7$ [82] the plane Cu spins order already at approximately T_{N1}(Cu2) ≈ 280 K, with approximate orientation perpendicular to the c-axis. It was therefore proposed [90] that the Pr-Pr exchange interaction might be also enhanced by a coupling via ordered Cu spins. Partial

TABLE Ie. Neutron diffraction results on magnetic ordering in $PrBa_2Cu_3O_x$ and in related compounds. a) noncollinear Cu spin arangement in the bilayer, θ_{Pr} = angle of the magnetic Pr moments to the (a,b)-plane, b) $\bar{\mu}(Cu2) \perp \bar{\mu}(Cu1)$, $\mu^z_{Cu1} = 0.09\ \mu_B$, c) $T_{N2}(Cu1) = T_{N1}(Cu2)$, $\bar{\mu}(Cu2) \perp \bar{\mu}(Cu1)$.

compound/ sample	T_N[K]	$\vec{k}$	μ_{sat} [μ_B]	$\mu_{direction}$	ref.-year	remarks
$PrBa_2Cu_3O_7$/P	$T_N(Pr) \approx 17$	[1/2,1/2,1/2]	$\mu_{Pr} = 0.74(8)$	[0,0,1]	[82]-1989	
$PrBa_2Cu_3O_{6.97}$/SC	$T_N(Pr) = 16.6$	[1/2,1/2,0], for T < 13.4 K [1/2,1/2,1/2]	$\mu_{Pr} \approx 0.8$, at 25 K $\mu_{Cu} = 0.62(2)$	$\theta_{Pr} = 30°$	[61]-1998	$T_N(Cu) = 281$ K, first-order trans. at 11 to 13.4 K
$PrBa_2Cu_3O_{6.93}$/P	$T_N(Pr) \approx 17$	[1/2,1/2,1/2] for T > 5 K, for T = 1.8 K also [1/2,1/2,0]]	$\mu_{Pr} = 0.79(5)$	[0,0,1]	[91]-1997	$T_N(Cu) \approx$ 330 K
$PrBa_2Cu_3O_{6.92}$/SC	$T_{NJ}(Cu2) = 266$; $T_N('Pr') = 19.0(5)$	[1/2,1/2,0]; [1/2,1/2,0]	$\mu_{Cu2} = 0.58(5)$; $\mu_{Pr} = 0.56(7)$	$\bar{\mu}_{Cu} \perp [0,0,1]$, $\theta_{Pr} = 55(20)°$	[60]-1997	a), Pr c-correl. length 160 Å
$PrBa_2Cu_3O_{6.65}$/SC	$T_{N1}(Cu2) = 370$; $T_{N2}(Cu1) = 170$	[1/2,1/2,0] (Cu2) & [1/2,1/2,1/2]	$\mu_{Cu2} \approx 0.7$; $\mu_{Cu1} \approx 0.1$	$\approx \perp$ [0,0,1]	[57]-1992	b)
$PrBa_2Cu_3O_{6.35}$/SC	$T_{NJ}(Cu2) = 347$; $T_N('Pr') = 11.0(5)$	[1/2,1/2,0]; [1/2,1/2,0]	$\mu_{Cu2} = 0.64(9)$; $\mu_{Pr} = 1.2(1)$	$\bar{\mu}_{Cu} \perp [0,0,1]$, $\theta_{Pr} = 38(6)°$	[60]-1997	a), Pr c-correl. length 10Å
$PrBa_2Cu_3O_{6.3}$/P	$T_N(Pr) \approx 14$	[1/2,1/2,1/2]	$\mu_{Pr} \approx 1.7$	[0,0,1]	[92]-1993	Pr c-correl. length 11Å
$PrBa_2Cu_3O_{6.2}$/SC		[1/2,1/2,1/2]		$\approx \perp$ [0,0,1]	[57]-1992	c)
$Pr_{1.05}Ba_{1.95}Cu_{2.57}Al_{0.12}O_{6.73}$/SC	$T_{N1}(Cu1) = 360$; $T_{N2}(CuII) = 11$; $T_N(Pr) \approx 11$	[1/2,1/2,0]; [1/2,1/2,1/2]; [1/2,1/2,0]	$\mu_{Cu2} = 0.49(1)$ at 40 K; $\mu_{Pr} = 0.50(4)$	$\approx \perp$ [0,0,1]; $\theta_c(Pr) = 59(3)°$	[62]-1996	Cu1 2-D MO ? Pr 3-D MO, c-correl. length = 34 (3) Å
$PrBa_2Cu_{2.8}Al_xO_{6.40}$/SC	$T_{N1}(Cu1) = 370$; $T_{N2}(CuII) = 100$; $T_N(Pr) \approx 8$	I: [1/2,1/2,0] II: [1/2,1/2,1/2]	$\mu_{Cu2} = 0.59(5)$, $\mu_{Cu1} = 0.04(1)$at 4 K		[61]-1996	Sample compos. ? Pr 2-D MO
$PrBa_2Cu_{2.96}Ga_{0.04}O_7$/P	$T_N(Pr) \approx 14$	[1/2,1/2,1/2]	$\mu_{Pr} \approx 0.7$	[0,0,1]	[93]-1994	
$PrBa_2Cu_{2.92}Ga_{0.08}O_7$/P	$T_N(Pr) \approx 10$	[1/2,1/2,0]	$\mu_{Pr} \approx 0.76(7)$	[0,0,1]	[93]-1994	
$PrBa_2Cu_{2.7}Zn_{0.3}O_7$/P	$T_N(Pr) \approx 17$	[1/2,1/2,0]	$\mu_{Pr} \approx 0.7(7)$	[0,0,1]	[94]-1993	$\beta = 0.49(9)$

TABLE If. Neutron diffraction results on magnetic ordering in Pr compounds related to $PrBa_2Cu_3O_x$.

compound/ sample	$T_N[K]$	$\bar{k}$	$\mu_{sat}[\mu_B]$	$\mu_{direction}$	ref.- year	remarks
$TlBa_2PrCu_2O_{7-y}$	$T_N(Cu) \approx$ 370 $T_N(Pr) \approx 8$	[1/2,1/2,0] [1/2,1/2,1/2] (Pr)	$\mu_{Cu} \approx 0.59$ $\mu_{Pr} \approx 1.05$	$\perp [0,0,1]$ [0,0,1]	[95]- 1994	3-D MO (Pr)
$PrBa_2Cu_2NbO_8/P$	$T_N(Pr) \approx$ 12.65(7) (Pr) $T_N(Cu) \approx$ 340(15) (Cu)	[1/2,1/2,1/2] (Pr) [1/2,1/2,0] (Cu)	$\mu_{Pr} \approx$ 1.2(1) $\mu_{PCu} \approx$ 0.5(1)	$\mu_{Pr} \parallel$ [0,0,1]	[96]- 1993	c-correl. length (Pr) $\approx 7(1)$ Å
$Pb_2Sr_2PrCu_3O_8$	$T_N(Pr) \approx 7$	[1/2,1/2,1/2] (Pr)		$\mu_{Pr} \parallel$ [0,0,1]	[97]- 1994	c-correl. length (Pr) ≈ 20 Å
$(Pr_{1.5}Ce_{0.5})Sr_2Cu_2NbO_{10-\delta}$	$T_N(Cu) \approx$ 200	[1/2,1/2,0]		$\bar{\mu}_{Cu} \perp$ [0,0,1]	[98]- 1997	several magnetic phases

substitution of Cu by Ga and Zn in $PrBa_2Cu_{3-x}Ga_xO_7$ causes an essential change of magnetic ordering from $\bar{k} = [1/2,1/2,1/2]$ to $[1/2,1/2,0]$ for $x = 0.08$ and 0.3, respectively, combined in the first case with a reduction of $T_N(Pr)$ to 10 K, as may be seen from table Ie.

Despite the three-dimensional antiferromagnetic ordering a substantial two-dimensional character of the magnetic interactions is visible for example in the temperature dependence of the order parameters. This has been shown in the case of Dy1-2-3O_7 already in 1989 by Allenspach et al. [99]. A 2-D Ising fit [100, 101] with S = 1/2 (ground state doublet) is in good agreement with the measured temperature dependence of the sublattice magnetisation, in contrast to a mean-field crystal field (CEF) calculation. On the other hand the calculated magnetic saturation moment of Dy of 6.8 to 7.1 μ_B, based on the crystal field parameters determined by inelastic neutron scattering, agrees well with the experimentally determined moment values of Dy: 6.8 [37] and 7.2 μ_B [78]. Also the easy direction of magnetisation [0,0,1] is caused by the crystalline electric field. Moreover 2-D antiferromagnetism was directly seen in single crystal neutron diffraction studies of Dy1-2-3O_7 at temperatures just above $T_N = 0.91(2)$ K [42, 102, 103]. There are indeed 2-D rods of magnetic neutron intensity, as was shown by scans across and parallel to the rods. In the case of the light rare-earth superconductor Nd1-2-3O_7 the temperature dependence of the order parameter indicates rather 3-D than 2-D character [80, 81]. Again the observed magnitude and direction of the magnetic saturation moment of Nd agree approximately with calculated values based on CEF measurements.

Different types of antiferromagnetic structures exist in the superconductors Ho1-2-3O_7 [43], Er1-2-3O_7 [104-111] and Yb1-2-3O_7 [112], corresponding to $\bar{k} = [0,1/2,1/2]$, $[1/2,0,1/2]$ and $[0,1/2,1/2]$, respectively (see also tables Ij and Ik). In the first neutron diffraction investigation at temperatures $T \geq 0.3$ K of Er1-2-3O_7 [104] this compound appeared as 2-D antiferromagnet due to magnetic moments of Er ions coupled antiferromagnetically along the a-direction and ferromagnetically along the b-axis, with

TABLE Ig. Neutron diffraction results on magnetic ordering in $NdBa_2Cu_3O_{x \geq 6.13}$ and in related systems. MO = magnetic ordering, LRMO = long-range magnetic ordering. a) 2-D magnetic correlations at T = 0.54 K, b) 2-D MO of Nd at T = 0.3 K, c) incomplete 3-D LRMO of Nd at low T, d) spin reorientation as f(T), e) $dT_{N1}(Cu2)/dp$ = 23(3) K/kbar up to 4kbars, f) LT MO not in powders, g) pressure dependence of $T_{N1}(Cu2)$.

compound/ sample	$T_N[K]$	$\bar{k}$	μ_{sat} [μ_B]	$\mu_{direction}$	ref.- year	remarks
$NdBa_2Cu_3O_{6.94}$/P	$T_N(Nd)$ = 0.53	[1/2,1/2,1/2]		[0,0,1]	[41, 42]- 1993,5	a)
$NdBa_2Cu_3O_{6.9}$/P	$T_N(Nd)$ = 0.55(1)	[1/2,1/2,1/2]	μ_{Nd} = 1.14(6)	[0,0,1]	[80]- 1989	
$NdBa_2Cu_3O_{6.9}$/P	$T_N(Nd) \approx 0.5$	[1/2,1/2,1/2]	μ_{Nd} = 1.07(7)	[0,0,1]	[81]- 1989	
$Nd_{1.5}Ba_{1.5}Cu_3O_{6.88}$/SC	$T_N(Cu)$ = 390	[1/2,1/2,0]			[63]- 1991	
$NdBa_2Cu_3O_{6.78}$/P	$T_N(Nd)$ = 1.5	[1/2,1/2]			[41, 42]- 1993,5	b)
$NdBa_2Cu_3O_{6.62}$/P		[1/2,1/2,0] & [1/2,1/2,1/2]			[112]- 1992	T ≤ 3 K
$NdBa_2Cu_3O_{6.45}$/P		[1/2,1/2,1/2]			[41, 42]- 1993,5	c)
$Nd_{1.5}Ba_{1.5}Cu_3O_{6.39}$/SC	$T_{N1}(Cu2)$ = 390; $T_{N2}(Cu1)$ = 150	[1/2,1/2,0] (Cu2) & [1/2 1/2 1/2]			[63]- 1991	
$NdBa_2Cu_3O_{6.35}$/SC	$T_{N1}(Cu2)$ = 230; $T_{N2}(Cu1)$ = 10	[1/2 1/2 0] (Cu2) & [1/2,1/2,1/2]		⊥ [0,0,1]	[113, 114]- 1988	d)
$NdBa_2Cu_3O_{6.35}$/SC	$T_{N1}(Cu2)$ = 230; $T_{N2}(Cu1)$ = 10	[1/2,1/2,0] (Cu2) & [1/2,1/2,1/2]			[75]- 1989	e)
$NdBa_2Cu_3O_{6.35}$/SC	$T_{N1}(Cu2)$ = 230; $T_{N2}(Cu1)$ = 10	[1/2,1/2,0] (Cu2) & [1/2,1/2,1/2]	μ_{Cu2} = 0.27; μ_{Cu1} = 0.23	⊥ [0,0,1]	[56]- 1990	d), f)
$NdBa_2Cu_3O_{6.3}$/P	$T_N(Nd) \approx 1.5$	[1/2,1/2,1/2]		$\vartheta_c = 45°$	[41, 42]- 1993,5	3-D MO of Nd at low T
$NdBa_2Cu_3O_{6.13}$/P	$T_N(Nd) \approx$ 1.75	[1/2,1/2,1/2]	μ_{Nd} = 0.85(4)	$\vartheta_c = 45°$	[41, 42]- 1993,5	3-D LRMO of Nd at low T

the magnetic moments oriented parallel to the latter direction. Later investigations on both powder and single crystal samples of $Er1$-2-$3O_7$ (see table Ik) at lower temperatures established the existence of 3-D antiferromagnetism in this compound, but the results are still to some extent inconsistent, presumably caused by differences in sample quality. In case of $Yb1$-2-$3O_7$ Mössbauer measurements first yielded evidence for

TABLE Ih. Neutron diffraction results on magnetic ordering in $RBa_2Cu_3O_x$, R = light rare earths Nd ($x_O \leq$ 6.1) and Sm and in related systems.[*] at room temperature. a) spin reorientation as f(T), b) pressure dependence of $T_{N1}(Cu2)$, c) LT MO not in powders.

compound/ sample	$T_N[K]$	$\vec{k}$	μ_{sat} [μ_B]	$\mu_{direction}$	ref.- year	remarks
$NdBa_2Cu_3O_{6.1}$/SC	$T_{N1}(Cu2) =$ 430; $T_{N2}(Cu1) =$ 80	[1/2,1/2,0] (Cu2) & [1/2,1/2,1/2]	$\mu_{Cu2} =$ 0.97(9); $\mu_{Cu1} =$ 0.46(6)	$\perp$ [0,0,1]	[113, 114]- 1988	a)
$NdBa_2Cu_3O_{6.1}$/SC	$T_N(Cu) =$ 385(2)	[1/2,1/2,1/2]	$\mu_{Cu2} =$ 0.40(2)[*]; $\mu_{Cu1} =$ 0.04(2)	$\perp$ [0,0,1]	[115]- 1988	
$NdBa_2Cu_3O_{6.1}$/SC	$T_N(Cu) =$ 385(2)	[1/2,1/2,1/2]	$\mu_{Cu2} =$ 0.40(2); $\mu_{Cu1} =$ 0.04(2)	$\perp$ [0,0,1]	[116]- 1989	
$NdBa_2Cu_3O_{6.1}$/SC	$T_{N1}(Cu2) =$ 430; $T_{N2}(Cu1) =$ 80	[1/2,1/2,0] (Cu2) & [1/2,1/2,1/2]			[75]- 1989	b)
$NdBa_2Cu_3O_{6.1}$/SC	$T_{N1}(Cu2) =$ 430; $T_{N2}(Cu1) =$ 80	[1/2,1/2,0] (Cu2) & [1/2,1/2,1/2]	$\mu_{Cu2} =$ 0.81; μ_{Cu1} = 0.34	$\perp$ [0,0,1]	[56]- 1990	a), c)
$NdBa_2Cu_2NbO_8$/P	$T_N(Na) \approx$ 1.69(5) $T_N(Cu) \approx$ 375(10)	[1/2,1/2,1/2] (Na) [1/2,1/2,0] (Cu)	$\mu_{Nd} \approx$ 0.74(5) $\mu_{Cu} \approx$ 0.5(1)	μ_{Nd} ‖ [0,0,1]	[96]- 1993	c-correl. length (Pr) $\approx$ 25 (10) Å
$SmBa_2Cu_3O_7$/P		[1/2,1/2,1/2] ?	$\mu_{Sm} \leq$ 0.2(2)	[0,0,1] ?	[117]- 1993	Sm isotopes

magnetic ordering due to magnetic moments of Yb^{3+} ions below 0.35 K, and a magnetic moment value of 1.7 μ_B was estimated [118]. Low-temperature neutron diffraction measurements performed at Saphir reactor, Würenlingen by Roessli et al. [112] yielded clear evidence for long-range antiferromagnetic ordering in Yb1-2-3O_7. Below approximately 330 mK the magnetic moments of Yb are aligned parallel to the b-axis, and attain at saturation the value 1.4 μ_B, in approximate agreement with the Mössbauer results and crystal field calculations ($\mu_{CEF} \approx 1.7$ μ_B, easy direction of magnetisation [0,1,0]) based on inelastic neutron spectroscopic results published by Guillaume et al. [119].

A particularly interesting case is Ho1-2-3O_7 [43] with singlet crystal-field ground state. The antiferromagnetic ordering below 190 mK is due to hyperfine interactions associated with nuclear polarisation of ^{165}Ho with large nuclear spin I = 7/2. Moreover the magnetic ordering was found to be imperfect according to magnetic stacking faults along the c-axis. This is illustrated in figures 8, 9, 11 and 12.

μSR experiments on HoBa$_2$Cu$_3$O$_7$ [120] were interpreted with $\vec{k}$ = [0,1/2,1/2] and magnetic Ho moment direction [0,0,1] below 100 mK. The ordered magnetic Ho moment of $\approx$ 2.6 μ_B agrees reasonably with the one determined by neutron diffraction. At about 100 mK the μSR data indicated a reorientation of the magnetic Ho moment

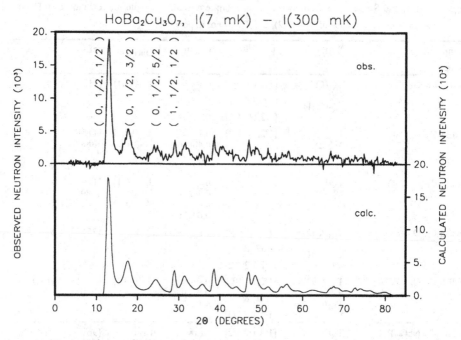

Figure 8. *Observed and calculated magnetic neutron diffraction patterns of $HoBa_2Cu_3O_7$ with magnetic ordering caused by Ho hyperfine interactions and magnetic stacking faults along the c-direction [43].*

direction to the [1,0,0]-axis at higher temperatures. Simultaneously a change of the magnetic 3-D ordering according to $\bar{k}$ = [1/2,1/2,1/2] was proposed to exist for temperatures up to approximately 5 K. Magnetic short-range correlations (probably 2-D) due to Ho were observed up to 50 K. Later neutron diffraction measurements performed by Staub and Ritter on $HoBa_2Cu_3O_7$ [121] established at 1.6 K short-range (of the order of 13 Å) 2-D antiferromagnetism according to $\bar{k}$ = [0,1/2], associated with magnetic Ho moments of 1.0(2) μ_B for this compound. Magnetic short-range interactions above 400 mK were shown to be due to the pecularities of induced magnetism in this particular singlet ground-state system.

To our knowledge the presently highest Néel temperature T_N = 5.4 K of a superconducting R1-2-3O_7 compound has been published recently by Li et al. for $TbSr_2Cu_{2.69}Mo_{0.31}O_7$ [122]. In this superconductor with T_c = 32 K antiferromagnetic Tb ordering according to $\bar{k}$ = [1/2,1/2,1/2] coexists with superconductivity. Mo replaces Cu in the Cu-O chain layers. At saturation the ordered magnetic moments attain a value of 7 μ_B which are oriented parallel to the direction [0,0,1]. Magnetic correlations between magnetic Tb moments were observed up to a temperature around 40 K.

As related systems with similar ordering temperatures we would like to mention $Pb_2Sr_2Tb_{1-x}Ca_xCu_3O_8$ [123, 124]. The Néel temperatures are also in the range of 5.3 K to 5.5 K for nonsuperconducting $Pb_2Sr_2TbCu_3O_8$, whereas for superconducting $Pb_2Sr_2Tb_{0.5}Ca_{0.5}Cu_3O_8$ with T_c = 71 K $T_N \approx$ 4 K was reported by Staub et al. [123].

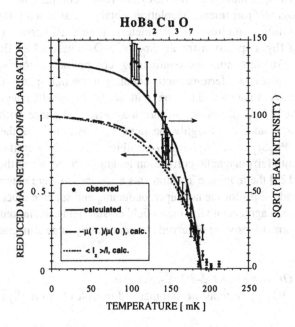

Figure 9. *The filled circles represent the temperature dependence of the observed and the line the one of the calculated square root of the magnetic neutron intensity of the $(0,1/2,1/2)$ peak of $HoBa_2Cu_3O_7$. The dot-dashed and dotted lines illustrate the corresponding calculated (reduced) electronic magnetisation $\mu(T)/\mu(0)$ and the nuclear polarisation $P = <I_x>/I$, respectively.*

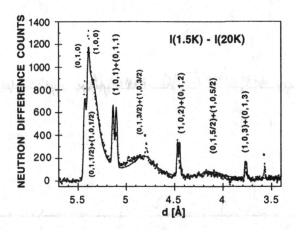

Figure 10. *Observed and calculated magnetic difference neutron diffraction pattern for $Pb_2Sr_2TbCu_3O_8$ obtained by time-of-flight neutron diffraction on IRIS/ISIS. The line corresponds to the fit of a quasi 2-D component (finite correlation length along c-direction) and a 3-D component with different ordering wave vectors. The two reflections marked by a star correspond to an impurity phase [123].*

Both the chemical and magnetic structures are rather complex in such compounds, requiring best possible instrumental resolution, obtainable at synchrotron X-ray sources and on high-resolution neutron diffractometers. Figure 10 shows a well resolved, complex time-of-flight spectrum for $Pb_2Sr_2TbCu_3O_8$ measured on IRIS at ISIS with cold neutrons. For this nonsuperconducting compound a dominant quasi 2-D antiferromagnetic phase (antiferromagnetic coupling in the (a,b)-plane, finite correlation length of about 32 Å along the c-direction) with antiferromagnetic coupling along the c-axis was observed. In addition an antiferromagnetic 3-D phase with ferromagnetic coupling along the c-axis and roughly the same T_N was found. On the other hand the superconductor $Pb_2Sr_2Tb_{0.5}Ca_{0.5}Cu_3O_8$ exhibits a quasi 2-D antiferromagnetic Tb ordering with finite ferromagnetic correlation length ($\approx$ 26 Å) along the c-direction. In contrast to ref. [124] the magnetic Tb moments at saturation were determined by Staub et al. as 9.1 μ_B and 8.8 μ_B for the nonsuperconducting and superconducting compounds, respectively, in good agreement with mean-field crystal field calculations. Magnetic 2-D short-range Tb correlations were observed in a remarkably large temperature range up to $8T_N$.

5.1.2. *Magnetic Ordering in R1-2-3$O_{x<7}$ Systems*
Results on R1-2-3$O_{x<7}$ systems are summarised in tables Ie to Ik. Systematic neutron

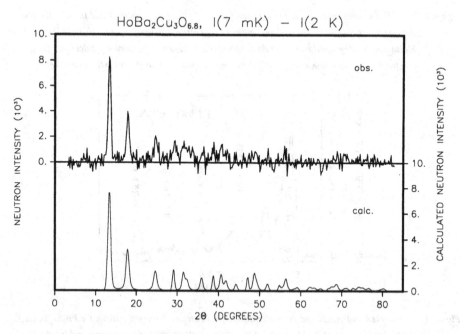

Figure 11.   *Observed and calculated magnetic neutron diffraction patterns of $HoBa_2Cu_3O_{6.8}$ with almost perfect 3-D long-range antiferromagnetic ordering induced by Ho hyperfine interactions [125].*

TABLE Ii. Neutron diffraction results on magnetic ordering in $RBa_2Cu_3O_x$ (R = heavy rare earths Gd, Tb) and in related systems. AF = antiferromagnetism, af. = antiferromagnetic. a) magnetic short-range Tb ordering up to ≈ 40 K, b) partially magnetic 3-D ordering at 1.4 K, 2-D correlation length = 550 Å at 6 K, c) antiferromagnetic coupling in (a,b)-plane,

compound/ sample	$T_N[K]$	$\bar{k}$	μ_{sat} $[\mu_B]$	$\mu_{direction}$	ref.- year	remarks
$^{160}GdBa_2Cu_3O_7$/P	$T_N(Gd)$ = 2.22(7)	[1/2,1/2,1/2]	μ_{Gd} = 7.4(6)	[0,0,1]	[79]- 1988	
$GdBa_2Cu_3O_{6.5}$/SC	$T_N(Gd)$ = 2.2	[1/2,1/2,0]	μ_{Gd} = 6.9(7)	[0,0,1]	[126]- 1988	
$^{160}GdBa_2Cu_3O_{6.4}$/P		[1/2,1/2, 1/2]			[85]- 1988	
$^{160}GdBa_2Cu_3O_{6.14}$/P	$T_N(Gd)$ = 2.27(3)	[1/2,1/2,1/2]			[85]- 1988	
$^{160}GdBa_2Cu_3O_{6.1}$/P	$T_N(Gd)$ = 2.3	[1/2,1/2,1/2] & [1/2,1/2,0]	μ_{Gd} = 5.2(5) & 3.2(8)	[0,0,1]	[127]- 1994	
$TbSr_2Cu_{2.69}Mo_{0.31}O_7$/P	$T_N(Tb)$ ≈ 5.4	[1/2,1/2,1/2]	μ_{Tb}= 6.98(7)	[0,0,1]	[122]- 1997	3-D AF, a)
$Pb_2Sr_2TbCu_3O_8$/P	$T_N(Tb)$ ≈ 5.5	[1/2,1/2] & [1/2,1/2,0]	μ_{Tb}= 7.43(2)	[0,0,1]	[124]- 1994	b)
$Pb_2Sr_2TbCu_3O_8$/P	$T_N(Tb)$ = 5.3	[1/2,1/2], [1/2,1/2,1/2], [1/2,1/2,1]			[128]- 1996	88 % 2-D AF + 3-D af. and ferromg. c-axis order
$Pb_2Sr_2TbCu_3O_8$/P	$T_N(Tb)$ = 5.3	[0,0,1/2] & 3-D: [0,0,0] (other chemical unit cell)	μ_{Tb}= 9.1(3)	[0,0,1]	[123]- 1997	c), nonsc., c-correl. length 32(6) Å, 2-D correl. up to 8T_N
$Pb_2Sr_2Tb_{0.5}Ca_{0.5}Cu_3O_8$/P	$T_N(Tb)$ ≈ 4	[0,0,0]	μ_{Tb}= 8.8(3)	[0,0,1]	[123]- 1997	c), T_c = 71 K, c-correl. length 26(6) Å

diffraction studies of magnetic ordering as a function of oxygen concentration have been first performed by Maletta et al. in the case of the heavy rare-earth system Er1-2-3O_x [109]. In the superconducting state 3-D antiferromagnetic ordering according to propagation vector $\bar{k}$ = [1/2,0,1/2] is found, whereas in the nonsuperconducting compounds only 2-D antiferromagnetism exists with $\bar{k}$ = [1/2,0]. With decreasing oxygen concentration T_N decreases from 620 to 460 mK, and the ordered magnetic moment changes from 4.8 to 3.7 μ_B, in reasonable agreement with CEF calculations based on inelastic neutron scattering results obtained by Mesot et al. [129], yielding μ_{Er} = 4.2 and 3.9 μ_B for x = 7 and 6.1, respectively.

In the case of Dy1-2-3O_x there is single crystal and powder neutron diffraction evidence for a change of the magnetic ordering from $\bar{k}$ = [1/2,1/2,1/2] ($T_N \approx 0.92$ K) for x = 7 to $\bar{k}$ = [1/2,1/2,0] (T_N = 0.87 K) for x = 6.54 [102, 103]. Neutron diffraction on a

TABLE Ij.   Neutron diffraction results on magnetic ordering in $RBa_2Cu_3O_x$, R = Dy, Ho. NIMO = nuclear induced magnetic order. a) first both $\bar{k}$ = [1/2,1/2,0] & [1/2,1/2,1/2] observed in the same crystal, b) 3-D 'short-range' magnetic order (c-correlation length ≈ 13 Å at 70 mK) below ≈ 2 K, c) NIMO & magn. stacking faults, d) 'broad' magnetic neutron peak.

compound/ sample	T_N[K]	$\bar{k}$	μ_{sat} [μ_B]	$\mu_{direction}$	ref.- year	remarks
$DyBa_2Cu_3O_7$/P	T_N(Dy) = 1.00(5)	[1/2,1/2,1/2]	μ_{Dy} = 7.2(6)	[0,0,1]	[78]- 1987	
$DyBa_2Cu_3O_7$/P	T_N(Dy) = 0.90(3)	[1/2,1/2,1/2]	μ_{Dy} = 6.8(1)	[0,0,1]	[37]- 1988	
$DyBa_2Cu_3O_7$/SC	T_N(Dy) ≈ 0.91	[1/2,1/2,1/2]			[102, 103]- 1991,2	a) 2-D MO (Er) at 0.93 K
$DyBa_2Cu_3O_7$/SC	T_N(Dy) = 0.93	[1/2,1/2,1/2]	μ_{Dy} ≈ 7	[0,0,1]	[42]- 1995	
$DyBa_2Cu_3O_{6.54}$/P	T_N(Dy) = 0.87	[1/2,1/2,0] [1/2,1/2]	μ_{Dy} ≈ 7	[0,0,1]	[42]- 1995	T_c = 50 K, 3-D LRMO, 2-D AF at 1 K
$DyBa_2Cu_3O_6$/P	T_N(Dy) ≈ 0.9	[1/2,1/2,0]	μ_{Dy} = 5.4(3)	[0,0,1]	[127]- 1994	c-axis correl. length 240 Å
$DyBa_2Cu_3O_6$/SC	T_N(Dy, 2-D) ≈ 0.55	[1/2,1/2,0]		[0 0 1]	[42]- 1995	b)
$HoBa_2Cu_3O_{7.00}$/P	T_N(Ho) = 0.19(1)	[0,1/2,1/2]	μ_{Ho} = 2.8(3)	[1,0,0]	[43]- 1993	c)
$HoBa_2Cu_3O_{7.00}$/P		[0,1/2] or [1/2,0]	μ_{Ho} = 1.0(2)	[1,0,0]	[121]- 1996	2-D AF at 1.6 K, correl. length 14(3) Å
$HoBa_2Cu_3O_{6.8}$/P	T_N(Ho) = 0.14(2)				[37]- 1988	d)
$HoBa_2Cu_3O_{6.77}$/P	T_N(Ho) = 0.10(1)	[0,1/2,1/2]	μ_{Ho} = 1.7(2)	[1,0,0]	[125]- 1994	NIMO & ≈ 3-D LRMO
$HoBa_2Cu_3O_{6.25}$/P					[125]- 1994	no LRMO for T ≥ 7 mK

$DyBa_2Cu_3O_6$ single crystal with T_c = 50 K [42] yielded below approximately 1.5 K 2-D antiferromagnetism with $\bar{k}$ = [1/2,1/2,0] and T_N = 0.55 K - together with magnetic 3-D short-range order along the c-axis. At 70 mK a finite correlation length of only about 13 Å was determined. A much larger c-correlation length of 240 Å was found in a powder neutron diffraction study of $DyBa_2Cu_3O_{6.1}$ by Guillaume et al. [127], confirming almost 3-D antiferromagnetism of type $\bar{k}$ = [1/2,1/2,0] for $DyBa_2Cu_3O_6$ with T_N ≈ 0.9 K, which is illustrated in figure 13. Thus oxygen removal in Dy1-2-3O_7 yields a transition from antiferromagnetic to ferromagnetic coupling of antiferromagnetic (a,b)-planes along the c-direction, as well as a reduced ordered magnetic Dy moment of 5.4 μ_B oriented parallel to the direction [0,0,1] at 10 mK.

TABLE Ik. Neutron diffraction results on magnetic ordering in $RBa_2Cu_3O_x$, R = heavy rare earths Er, Tm, Yb. a) composite sample of five crystals, b) β(LRMO) = 0.122(4) as for 2-D Ising system, 2-D MO at 0.67 K, c) Influence of magnetic Cu ordering ?

compound/ sample	T_N[K]	$\bar{k}$	μ_{sat} [μ_B]	$\mu_{direction}$	ref.- year	remarks
$ErBa_2Cu_3O_7$/P	T_N(Er, 2-D) ≈ 0.5	[1,2,0] or [0 1/2]	μ_{Er} = 2.9	[0,1,0]	[104]- 1987	2-D MO at T ≥ 0.3 K
$ErBa_2Cu_3O_7$/P		[1/2,0,0] & [1/2,0,1/2]	μ_{Er} = 4.9(2)	ϑ_c = 32°	[105]- 1988	3-D LRMO at 0.14 K
$ErBa_2Cu_3O_{7(-\delta)}$/ SC	T_N(Er) = 0.55(2)	[1/2,0,0] & [1/2,0,1/2]	μ_{Er} = 4.8(2)	[0,1,0]	[106]- 1989	a)
$ErBa_2Cu_3O_{7(-\delta)}$/ SC	T_N(Er) = 0.5	[1/2,0,0]	μ_{Er} = 4.8(2)	[0,1,0] (ϑ_c = 82(5)°)	[107]- 1989	T_c = 88 K
$ErBa_2Cu_3O_7$/P	T_N(Er) = 0.618	[1/2,0,1/2]; [1/2,0,0] & [1/2,0,1/2]			[108]- 1989 &[110]- 1991	b)
$ErBa_2Cu_3O_{7.00}$/P	T_N(Er) = 0.62		μ_{Er} = 4.8(2)	[0,1,0]	[109]- 1990	
$ErBa_2Cu_3O_{7.00}$/SC	T_N(Er) = 0.62	[1/2,0,1/2]	μ_{Er} = 4.8	[0,1,0]	[42]- 1995	
$ErBa_2Cu_3O_{6.87}$/P	T_N(Er) = 0.585(5)	[1/2,0,1/2]	μ_{Er} = 4.4(3)	[0,1,0]	[109]- 1990	
$ErBa_2Cu_3O_{6.64}$/P	T_N(Er) = 0.530(5)	[1/2,0,1/2]	μ_{Er} = 3.9(3)	[0,1,0]	[109]- 1990	
$ErBa_2Cu_3O_{6.6}$/SC	T_N(Er) = 0.48	[1/2,0,0]		[0,1,0]	[42]- 1995	
$ErBa_2Cu_3O_{6.53}$/P	T_N(Er) = 0.460(5)	[1/2,0,1/2]	μ_{Er} = 3.7(3)	[0,1,0]	[109]- 1990	
$ErBa_2Cu_3O_{6.40}$/P		[1/2,0]			[109]- 1990	
$ErBa_2Cu_3O_{6.20}$/P		[1/2,0]			[109]- 1990	
$ErBa_2Cu_3O_{6.11}$/P	T_N(Er, 2-D) < 0.8	[1/2,0]			[109]- 1990	
$ErBa_2Cu_3O_6$/SC	'T_N(Er)' ≈ 0.7	[1/2,0]		⊥ [0,0,1]	[42]- 1995	2-D AF with correl. length 15 Å at 60 mK, c)
$TmBa_2Cu_3O_7$/P					[130]- 1990	no LRMO for T ≥ 90 mK
$YbBa_2Cu_3O_7$/P	T_N(Yb) = 0.33(2)	[0,1/2,1/2]	μ_{Yb} = 1.44(6)	[0,1,0]	[112]- 1992	

On the other hand neutron diffraction investigations of a ^{160}GdBa$_2$Cu$_3$O$_x$ powder sample (see table Ii) did not show essential changes of both $\bar{k}$ = [1/2,1/2,1/2] and T_N. However a coexistence of $\bar{k}$ = [1/2,1/2,1/2] and [1/2,1/2,0] was detected by Guillaume et al. [127] in polycrystalline ^{160}GdBa$_2$Cu$_3$O$_{6.1}$, in contrast to a single crystal study for x = 6.5 [126] reporting $\bar{k}$ = [1/2,1/2,0]. For x = 6.14 a powder investigation [85] yielded the critical exponent β = 0.33(3), indicating 3-D long-range order.

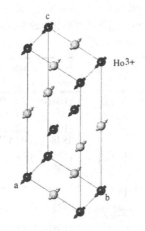

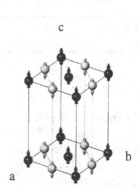

Figure 12.   Antiferromagnetic Ho ordering in   Figure 13.   Antiferromagnetic Dy ordering in
$HoBa_2Cu_3O_{6.8}$ [125]. $DyBa_2Cu_3O_6$ [127].

In the case of the light rare-earth system $NdBa_2Cu_3O_x$ powder neutron diffraction investigations of the oxygen dependence of magnetic Nd ordering [41, 42] showed a transition from 3-D antiferromagnetism with propagation vector $\bar{k} = [1/2,1/2,1/2]$ for x ≈ 6.94 to 2-D antiferromagnetic order with $\bar{k} = [1/2,1/2]$ for x = 6.78. At the same time a remarkable increase of the Néel temperature from 0.53 to 1.5 K was found, in agreement with specific heat measurements [42, 131]. For x = 6.3 and 6.13 the Nd magnetic ordering is again of 3-D character with $\bar{k} = [1/2,1/2,1/2]$, and $T_N(Nd)$ attains the remarkably high values of 1.5 K and 1.75 K, respectively. In these cases the Cu spins are ordered too (see table Ig and Ih).

Both in the nonsuperconducting compound Pr1-2-3O$_7$ and Pr1-2-3O$_6$ the Néel temperatures associated with antiferromagnetic Pr ordering are exceptionally high: 17 and 14 K, respectively (cf. also tables Ie and If for related Pr compounds).

Finally the special case of the oxygen concentration dependence of the magnetic Ho ordering induced by hyperfine interactions in $HoBa_2Cu_3O_x$ should be mentioned [37, 125] (cf. Figs. 8, 9, 11, 12). Whereas for x = 6.77 the antiferromagnetic Ho ordering below $T_N = 100$ mK is almost of 3-D type corresponding to $\bar{k} = [0,1/2,1/2]$ and ordered magnetic moment $\mu_{Ho} = 1.7(2)$ μ_B, imperfect antiferromagnetism with the same $\bar{k}$-vector and equal easy direction of magnetisation [1,0,0], but increased $T_N = 190$ mK and ordered magnetic moment $\mu_{Ho} = 2.8(3)$ μ_B at T = 7 mK, i. e. at saturation, are observed in case of x = 7.0. The magnetic stacking faults may be related to the low value of the Néel temperature T_N and to structural phase separation possible for x > 6.9 [21].

5.2. MAGNETIC ORDERING IN R1-2-4 COMPOUNDS

First neutron diffraction investigations of Dy1-2-4 at 50 mK by Zhang et al. [38] (see also table II) yielded an asymmetric magnetic peak indicating 2-D antiferromagnetism within (a,b)-planes with ordered magnetic Dy moments of magnitude 5.3 μ_B, oriented

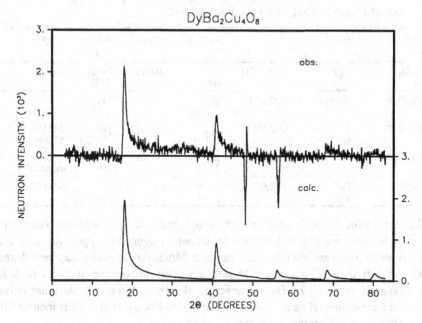

Figure 14. Observed and calculated magnetic neutron diffraction patterns of $DyBa_2Cu_4O_8$ with antiferromagnetic 2-D ordering due to Dy^{3+} ions for a temperature of 7 mK [39].

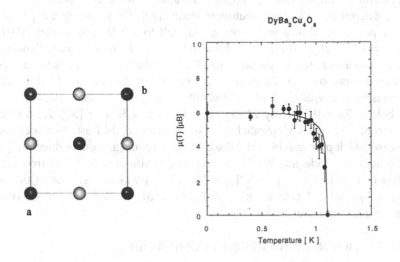

Figure 15. 2-D antiferromagnetic Dy ordering in $DyBa_2Cu_4O_8$. Dark and pale circles indicate antiparallel orientation of magnetic Dy^{3+} moments parallel to the c-axis.

Figure 16. Temperature dependence of the Dy sublattice magnetisation of $DyBa_2Cu_4O_8$.

TABLE II. Neutron diffraction results on magnetic ordering in $RBa_2Cu_4O_8$ compounds. * $\vartheta_c = 60^\circ$ in (b,c)-plane, a) 2-D MO due to frustration at T ≥ 50 mK.

compound/ sample	$T_N[K]$	$\vec{k}$	$\mu_{sat}[\mu_B]$	$\mu_{direction}$	ref.- year	remarks
$PrBa_2Cu_4O_8$/P	$T_N(Pr) \approx 16$	[1/2,1/2,1]	$\mu_{Pr} =$ 0.61(4)	[0,0,1]	[132]- 1997	3-D AF
$DyBa_2Cu_4O_8$/P	$T_N(Dy) \approx 0.9$	[1/2,1/2]	$\mu_{Dy} =$ 5.3(3)	[0,0,1]	[38]- 1990	a)
$DyBa_2Cu_4O_8$/P	$T_N(Dy) =$ 1.10(2)	[1/2,1/2]	$\mu_{Dy} =$ 5.9(5)	[0,0,1]	[24, 39] -1992,3	2-D MO at T ≥ 7 mK
$HoBa_2Cu_4O_8$/P	0.1	[0,1/2,0]	$\mu_{Ho}=$ 1.5(4)	[1,0,0]	[133]- 1994	c-correl. length 18 Å
$ErBa_2Cu_4O_8$/P	$T_N(Er) \approx$ 0.49	[1/2,0,1]	$\mu_{Er} =$ 3.9(2)	$\approx [0,1,0]$*	[38]- 1990	3-D LRMO

parallel to the c-axis. At our laboratory such powder measurements were extended with better resolution to 7 mK [24, 39]. Results are shown in figure 14. Apparently also at 7 mK the magnetic peaks are obviously asymmetric. Moreover the peaks may be indexed by (m/2, n/2, 0) with m and n = odd integers. This proves the stability of 2-D antiferromagnetic ordering of Dy moments as shown in figure 15. A quantitative analysis of the experimental neutron data performed by Roessli et al. [39] is included in figure 14, showing good agreement with the measurements.

Thus it was possible to determine the Dy sublattice magnetisation and the correlation length ('domain size') as function of temperature. The saturation moment agrees perfectly with the value of 6.0 μ_B calculated for the easy direction [0,0,1] from CEF data determined by inelastic neutron scattering [39]. The temperature dependence of the order parameter shown in figure 16 fits well to a 2-D Ising model [100] with exchange constant $2J_{ab} = 0.9(1)$ K, and the correlation length increases with decreasing temperature to about 500 Å. Although 'ideal' 2-D antiferromagnetism is to be expected because of frustration effects for most R1-2-4 systems tending to $\vec{k} = [1/2,1/2,1/2]$, 3-D antiferromagnetism with $\vec{k} = [1/2,0,1]$ and ordered magnetic moments $\mu_{Er} = 3.9$ μ_B was found below $T_N \approx 0.49$ K by Zhang et al. in case of R = Er [38]. As the nearest neighbours of a Er ion are all parallel along the b-direction, the frustration does not hold in this case which presumably is stabilised by the easy magnetisation direction [0,1,0]. Recently nonsuperconducting Pr1-2-4 was reported to also show 3-D antiferromagnetism according to $\vec{k} = [1/2,1/2,1]$ with $T_N = 16$ K [132] which is similar to the one of $PrBa_2Cu_3O_7$. Ho1-2-4 showed $\vec{k} = [0,1/2,0]$ antiferromagnetism with finite c-correlation length of about 18 Å [133].

5.3. MAGNETIC ORDERING IN R2-4-7 COMPOUNDS

Concerning magnetic ordering to our knowledge (table III) previously only elastic neutron scattering results for the 60 K superconductor $Dy_2Ba_4Cu_4O_{15}$ were published by Zhang et al. [4, 36, 134]. Below about 1.3 K 2-D, antiferromagnetic (a,b)-layers couple along the c-axis predominantly ferromagnetically to bilayers due to magnetic moments of Dy ions oriented parallel to the [0,0,1]-direction. This ordering appears to develop

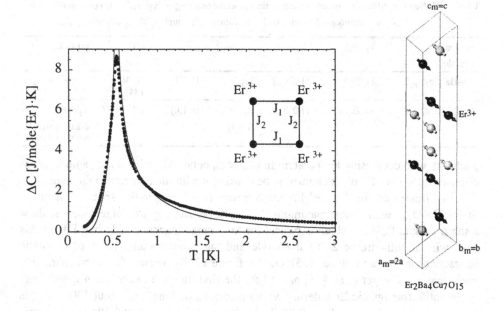

Figure 17. Measured and calculated magnetic specific heat of $Er_2Ba_4Cu_7O_{14.9}$. The inset shows the resulting coupling parameters in the (a,b)-plane [135], $J_1 = 0.039(1)$ meV, $J_2 = -0.009(1)$ meV (or reversed signs).

Figure 18. Magnetic unit cell of $Er_2Ba_4Cu_7O_{15}$ with $T_c = 89$ K, illustrating quasi-3-D antiferromagnetic Er ordering (c-axis correlation length ≈ 130 Å).

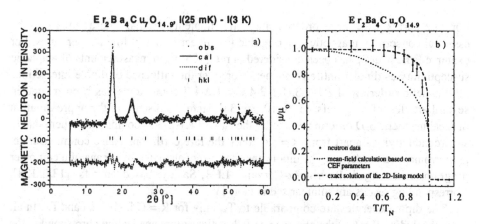

Figure 19. a) Magnetic observed, calculated and difference neutron diffraction patterns of $Er_2Ba_4Cu_7O_{15}$ and b) corresponding measured and calculated temperature dependencies of the ordered magnetic Er moment.

TABLE III. Neutron diffraction results on magnetic rare-earth ordering in $R_2Ba_4Cu_7O_y$ compounds. a) 2-D bilayer antiferromagnetic order with $\approx$ ferromagnetic coupling along direction [0,0,1].

compound/ sample	T_N [K]	$\vec{k}$	μ_{sat} [μ_B]	$\mu_{direction}$	ref.- year	remarks
$Dy_2Ba_4Cu_7O_{15}$/P	$T_N(Dy) \approx 1.3$	[1/2,1/2]	$\mu_{Dy} \approx 7$	[0,0,1]	[4, 36, 134]-1992	$T_c \approx 60$ K, a)
$Er_2Ba_4Cu_7O_{14.9}$/P	$T_N(Er) =$ 0.54	[1/2,0,0]	$\mu_{Er} =$ 3.7(1)	[0,1,0]	[35]-1997	$T_c = 89$ K, c-axis correl. length $\approx$ 130 Å

gradually with decreasing temperature in this compound. An ordered magnetic moment of Dy of the order of 7 μ_B is claimed to be consistent with the experimental results.

As illustrated in figure 17 measurements of magnetic specific heat of $Er_2Ba_4Cu_7O_{14.9}$ with almost maximum $T_c = 89$ K, prepared by sol-gel techniques, show a sharp peak at 0.54 K, which is typical for long-range magnetic ordering. Fits of this peak with an anisotropic 2-D Ising model indicate essential anisotropy of magnetic interactions in the (a,b)-plane [135], cf. also figure 17. By means of neutron diffraction investigations Böttger et al. [35] proved for the first time the existence of quasi long-range antiferromagnetic Er ordering with a c-correlation length of about 130 Å in this high-T_c compound below $T_N = 0.54$ K. As shown in figures 18 and 19a the magnetic ordering corresponds to $\vec{k} = [1/2,0,0]$. In agreement with mean-field calculations based on crystal field parameters determined by inelastic neutron scattering the ordered magnetic Er moments are oriented along the b-axis and attain a value of 3.7(1) μ_B at 25 mK. The corresponding temperature dependence of the ordered magnetic Er moment shown in figure 19b indicates more 2-D than 3-D character.

6. Dipolar Interactions and Crystal-Field Effects

For rare earths with large ordered magnetic moments dipolar interactions may be essential concerning magnetic ordering due to rare earth ions at low temperatures. For example $ErBa_2Cu_3O_6$ has been considered in ref. [136] from measurements of magnetic susceptibility as dipolar antiferromagnet. Moreover the influence of dipole interactions on magnetic ordering of R1-2-3, R1-2-4 and R2-4-7 compounds has been treated in several articles, cf. e. g. refs. [5, 6, 10, 36, 38]. Here we discuss in the approximation of ideal magnetic 3-D ordering explicitly the case of R1-2-3O_7 compounds where due to superconductivity - apart from R = Pr - only the rare earth ions cause commensurate, antiferromagnetic ordering. The dipole energies were calculated by means of a computer program written by J. Rodriguez-Carvajal, LLB, Saclay, based on refs. [137, 138]. Corresponding results are summarised in table IV.

The dipole energies are comparable to T_N only for R = Gd, Dy, Ho and Er. In all cases (excluding R = Gd) the observed easy directions of magnetisation correspond to the expectations from the crystalline electric field. Moreover table IV shows that one should expect for purely dipolar interactions in all cases $\vec{k}$ either [0,1/2,1/2] or [0,1/2,0] and the magnetic moment orientation along direction [1,0,0]. This configuration is only

TABLE IV. Calculated dipole energies in mK per magnetic rare-earth ion for R1-2-3O$_7$ compounds for different propagation vectors $\bar{k}$ (k_x,k_y,k_z) and different directions of the magnetic moments, using the observed magnetic moment values given in previous tables and lattice parameters from ref. [20]. The observed propagation vectors and easy directions of magnetisation (as deduced from CEF measurements) are indicated by arrows (←) and dots (•), respectively. Configurations with lowest dipole energies are marked with an asterisk (*).

R [CEF ref.]	Pr [139]	Nd [140]	Gd	Dy [99]	Ho [141]	Er [129]	Yb [119]
[1/2,1/2,1/2]	←	←	←	←			
x	4.2	10.2	440.2	422.2	64.3	159.8	17.3
y	3.4	8.1	343.6	325.4	49.1	121.1	13.0
z	-7.7•	-18.3•	-783.9•	-747.6•	-113.4	-281.0	-30.3
[1/2,1/2,0]							
x	4.3	10.2	440.3	422.3	64.3	159.9	17.3
y	3.4	8.1	343.7	325.5	49.1	121.2	13.0
z	-7.7	-18.3	-784.1	-747.8	-113.5	-281.0	-30.3
[1/2,0,1/2]							←
x	17.5	41.7	1785.4	1702.8	258.3	639.9	69.0
y	-14.5	-34.7	-1483.7	-1413.8	-214.4	-530.7	-57.2
z	-2.9	-7.0	-301.7	-289.1	-44.0	-109.3	-11.8
[0,1/2,1/2]*					←		←
x	-15.0*	-35.8*	-1537.6*	-1467.8*	-222.8*•	-552.3*	-59.6*
y	17.5	41.7	1786.2	1703.6	258.5	640.2	69.0•
z	-2.5	-5.9	-248.6	-235.8	-35.6	-88.0	-9.4
[1/2,0,0]							
x	17.5	41.7	1788.7	1705.9	258.8	641.1	69.1
y	-14.5	-34.7	-1483.7	-1413.8	-214.4	-530.7	-57.2
z	-3.0	-7.1	-305.0	-292.2	-44.5	-110.4	-12.0
[0,1/2,0]*							
x	-15.0*	-35.8*	-1537.6*	-1467.8*	-222.8*	-552.3*	-59.6*
y	17.5	41.8	1790.0	1707.2	259.0	641.6	69.2
z	-2.5	-6.0	-252.3	-239.4	-36.2	-89.3	-9.6
[0,0,1/2]							
x	-13.4	-32.1	-1380.8	-1318.9	-200.1	-496.1	-53.6
y	-12.7	-30.4	-1297.5	-1235.5	-187.0	-462.7	-49.8
z	26.2	62.4	2678.4	2554.4	387.2	958.7	103.4
[0,0,0]							
x	-13.4	-32.1	-1379.4	-1317.6	-200.1	-496.2	-53.6
y	-12.7	-30.3	-1296.1	-1234.1	-187.0	-462.8	-49.8
z	14.0	33.6	1443.7	1377.8	209.0	517.9	55.9

observed in case of R = Ho.

Concerning more realistic treatments of dipolar 2-D and 3-D magnetism with exchange interactions we refer to publications by MacIsaac et al. [142, 143] and to the previously mentioned references.

7. Discussion and Conclusions

In this review mainly magnetic ordering phenomena of rare earth ions R^{3+} in the high-T_c superconductors R1-2-3, R1-2-4, R2-4-7, R = rare earth and Y, and in related compounds, established by means of low-temperature neutron-diffraction experiments, are summarised and discussed. Concerning related specific heat data we refer to the new review to be published by Allenspach [144]. Neutron scattering experiments have provided basic results on magnetic ordering phenomena in mainly R1-2-3O_x systems. Because of the more difficult sample preparation more limited information is presently available on R1-2-4 and R2-4-7 compounds which also show very interesting physical properties such as large changes induced by pressure and the fact that the former system crystallises usually without twinning. Further neutron scattering efforts will depend essentially on the possibilities to produce such sufficiently large, well characterised polycrystalline and untwinned, pure and homogeneous single crystal samples. Apart from high-pressure synthesis sol-gel techniques provide in this respect new possibilities.

Due to the layer type crystal structures of the discussed systems with large separations R - R between the layers their magnetic properties are predominantly of 2-D character. 3-D long-range magnetic ordering due to copper spins does not exist in the superconducting compounds. On the other hand generally 3-D antiferromagnetic rare-earth ordering and superconductivity coexist in metallic R1-2-3$O_{x \approx 7}$ superconductors with $T_c \leq 96$ K at low temperatures. The Néel temperatures T_N are < 2.3 K, except for anomalous, nonsuperconducting Pr and Tb compounds such as nonsuperconducting Pr1-2-3O_7 with $T_N = 17$ K and antiferromagnetic, superconducting $TlSr_2Ca_{0.4}Tb_{0.6}Cu_2O_7$ [145], $TbSr_2Cu_{2.69}Mo_{0.31}O_7$ [122] and $Pb_2Sr_2Tb_{0.5}Ca_{0.5}Cu_3O_8$ [123] with $T_c = 62$ K, 32 K and 71 K, respectively and $T_N = 6.2$ K, 5.4 K and 4 K, respectively. To a large extent the ordered magnetic moments and easy directions of magnetisation of the rare earths are determined by crystal-field effects. Apart from usually small dipole energies, exchange interactions are important. In particular concerning temperature dependencies of the order parameters 2-D aspects are also present in case of 3-D ordering. Neutron scattering experiments on superconducting $PrBa_2Cu_3O_x$ crystals with $T_c \approx 85$ K as published recently by Zou et al. [146] should verify wether indeed magnetic Cu and Pr ordering do not exist in this case.

In similar nonsuperconducting compounds with oxygen concentration $x \approx 6$ antiferromagnetism due to Cu ions with high Néel temperatures up to approximately 500 K dominates over rare-earth ordering which tends to two-dimensional antiferromagnetism for $6 < x < 7$. Generally the magnetic rare-earth ordering depends essentially on oxygen concentration. In the case of Nd1-2-3O_x a remarkable increase of T_N(Nd) by more than a factor of three was observed by oxygen reduction (see e. g. ref. [42]). This essential change may be influenced by simultaneous copper ordering in the nonsuperconducting samples. Such a 4f-3d coupling resulting in noncollinear magnetic ordering had been recently proposed for Pr1-2-3 by Boothroyd et al. [60], whereas Uma et al. [61] reported only a modest Cu-Pr interaction.

In several cases the magnetic rare-earth ordering was found to be imperfect according to finite correlation length. It is still a question to which extent magnetic chain copper ordering is intrinsic in unsubstituted R1-2-3O_x compounds, i. e. wether it is observed only in single crystals due to better oxygen ordering or caused by rare-earth excess or impurities introduced during crystal growth on mainly chain sites. For R = Y Casalta et al. [49] veryfied the absence of a second antiferromagnetic transition in pure Y1-2-3O_{6+x} single crystals at low temperatures.

A particularly interesting case is antiferromagnetic ordering in Ho1-2-3O_x with singlet crystal-field ground state and magnetic stacking faults along the c-direction for x = 7 [43]. Similar to garnets [147] the magnetic ordering is induced by hyperfine interactions due to nuclear polarisation of ^{165}Ho.

The similarity between the co-ordination polyhedra of rare-earth ions in R1-2-3 and garnets was also emphasised in crystal-field calculations by Nekvasil [2]. However such anaologies should not be taken too far, as may be seen from the differences between calculation and experiment concerning the magnetic rare-earth ordering published for R1-2-3O_x by Misra et al., based on simplified CEF parameters [5]. Crystal-field and magnetic interaction parameters determined by inelastic neutron scattering as in the case of R = Ho [148] would be more appropriate. This has been taken more into account in the new theoretical consideration of possible magnetic ground states of R1-2-3 and R1-2-4 compounds by Misra et al. [10]. According to this work layered antiferromagnetic ordering with magnetic moments parallel to the c-axis is expected for R1-2-3 and R1-2-4 compounds except for R = Sm, Er, and Yb and presumably for Nd1-2-4 where the magnetic moments should orient perpendicular to the [0,0,1]-direction. This prediction does not agree with neutron diffraction results published for Ho1-2-3.

Because of frustration effects associated with the Ammm symmetry of the R1-2-4 structure ideal 2-D antiferromagnetic rare-earth ordering exists in case of Dy1-2-4 with $T_c \approx 80$ K. On the other hand R1-2-4 compounds with R = Pr and Er show 3-D antiferromagnetism, whereas Ho1-2-4 is an intermediate case with finite c-correlation length. Magnetic Cu ordering may be possible in R1-2-4 compounds with partial substitution of Cu by other atoms such as Fe [149], which should be verified by neutron scattering experiments.

In contrast to the 2-D bilayer antiferromagnetic order of Dy ions in $Dy_2Ba_4Cu_7O_{15}$ with $T_c \approx 60$ K quasi 3-D antiferromagnetic Er ordering exists in $Er_2Ba_4Cu_7O_{15}$ with T_c = 89 K below T_N = 0.54 K. Future experiments on other R2-4-7 compounds with maximum T_c will show wether the assumption of an essential influence of T_c and of the related crystal perfection (minimum number of stacking faults) has a direct influence on magnetic ordering. Moreover the dependence of magnetic rare-earth and copper ordering in R2-4-7$O_{14+\delta}$ may provide very interesting results. Such neutron studies are presently in progress at our laboratory in case of R = Dy and Er. For superconducting R1-2-4 and R2-4-7O_{15} compounds T_N < 2.3 K seems to hold too.

Simultaneous magnetic copper and rare earth ordering and complex magnetic moment reorientation phenomena also at copper plane and chain sites as a function of temperature exist in nonsuperconducting systems of the type discussed, e. g. in Pr1-2-

$3O_x$ or doped Y1-2-3 compounds (cf. e.g. ref. [70]). In these fields more neutron scattering experiments on well characterised perfect, pure single crystals are needed where often powder samples turn out to be too intensity limited. In particular temperature dependent reorientations of the magnetic moments as proposed e. g. from μSR experiments for $GdBa_2Cu_3O_{6.3}$ [150] would be of interest. Moreover concerning magnetic ordering phenomena more pressure investigations should be performed in the discussed fascinating classes of compounds.

Acknowledgements

We would like to thank for stimulating discussions with and support by many colleagues, in particular A. Furrer and 'high-T_c' collaborators of LNS, above all P. Allenspach, G. Böttger, M. Guillaume, W. Henggeler, J. Mesot, B. Roessli, U. Staub, A. Dönni and A. W. Hewat, ILL, Grenoble. Many dilution refrigerator measurements would not have been possible without the expert preparation by S. Fischer, LNS and financial support by the Swiss National Science Foundation and ETHZ for our work. Paul Scherrer Institute provided necessary infrastructure such as the operation of the Saphir reactor, of CRG instruments at ILL, Grenoble and now of the spallation neutron source SINQ. We are indebted to E. Kaldis and J. Karpinski et al. of ETH Zurich as well as P. Berastegui from Chalmers University of Technology, Göteborg for the careful preparation and characterisation of the investigated R1-2-3, R1-2-4 and R2-4-7 compounds. Structure plots were produced by programs ATOMS [151] and MolView [152]. Finally we would like to thank J. Rodriguez-Carvajal, LLB, Saclay for the program to calculate dipole energies.

References

1. Johnston, D.C., Sinha, S.K., Jacobson, A.J., and Newsam, J.M. Superconductivity and magnetism in the high T_c copper oxides, *Physica C* **153-155** (1988), 572-577.

2. Nekvasil, V. Crystal field and magnetic moments of rare-earth ions in $REBa_2Cu_3O_{7-x}$ (RE = Ce,...,Yb), *Solid State Commun.* **65** (1988), 1103-1106.

3. Lynn, J.W., Magnetic properties, in J. W. Lynn (ed.) *High Temperature Superconductivity*, Springer, Berlin, (1990), pp. 268-302.

4. Lynn, J.W. Two-dimensional behavior of the rare earth ordering on oxide superconductors, *J. of Alloys and Compounds* **181** (1992), 419-429.

5. Misra, S.K. and Felsteiner, J. Low-temperature ordered states of $RBa_2Cu_3O_{7-\delta}$ (R = rare earth) due to dipole-dipole and exchange interactions, *Phys. Rev. B* **46** (1992), 11033-11039.

6. MacIsaac, A.B., Whitehead, J.P., De'Bell, K., and Narayanan, K.S. Monte Carlo study of two-dimensional Ising dipolar antiferromagnets as a model for rare-earth ordering in the R-Ba-Cu-O compounds (R = rare earth), *Phys. Rev. B* **46** (1992), 6387-6394.

7. Misra, S.K. and Felsteiner, J. Reply to "Comment on 'Low-temperature ordered states of $RBa_2Cu_3O_{7-\delta}$ due to dipole-dipole and exchange interactions'", *Phys. Rev. B* **49** (1994), 12344-12346.

8. Whitehead, J.P., De'Bell, K., and Noakes, D.R. Comment on "Low-temperature ordered states of $RBa_2Cu_3O_{7-\delta}$ due to dipole-dipole and exchange interactions", *Phys. Rev. B* **49** (1994), 12341-12343.

9. Felsteiner, J. and Misra, S.K. Magnetic order of $RBa_2Cu_4O_8$ (R = rare earth) due to dipole-dipole and exchange interactions, *Phys. Rev. B* **50** (1994), 7184-7187.

10. Misra, S.K., Chang, Y.M., and Felsteiner, J. A calculation of effective g-tensor for R^{3+} ions in $RBa_2Cu_3O_{7-\delta}$ and $RBa_2Cu_4O_8$ (R = rare earth): Low-temperature ordering of rare-earth moments, *J. Phys. Chem. Solids* **58** (1997), 1-11.

11. Lynn, J.W. Rare earth magnetic ordering in exchange-coupled superconductors, *J. Alloys Comp.* **250** (1997), 552-558.

12. Park, C. and Snyder, R.L. Structures of high-temperature cuprate superconductors, *J. Am. Ceram. Soc.* **78** (1995), 3171-3194.

13. Thomlinson, W., Shirane, G., Lynn, J.W., and Moncton, D.E., Neutron Scattering Studies of Magnetic Ordering in Ternary Superconductors, in M. B. Maple and O. Fischer (ed.) *Superconductivity in Ternary Compounds (Topics in Current Physics)* Vol. II, Springer, Berlin, (1982), pp. 229-248.

14. Gasser, U., Allenspach, P., Mesot, J., and Furrer, A. Crystal electric field splitting of R^{3+} ions in RNi_2B_2C (R = rare earth), *Physica C* **282** (1997), 1327-1328.

15. Hewat, A.W., Fischer, P., Kaldis, E., Hewat, E.A., Jilek, E., Karpinski, J., and Rusiecki, S. Cu-O bond and T_c changes in 123, 124 & 247-superconductors, *J. Less-Common Met.* **164 &165** (1990), 39-49.

16. Cava, R.J., Hewat, A.W., Hewat, E.A., Batlogg, B., Marezio, M., Rabe, K.M., Krajewski, J.J., Jr., W.F.P., and Jr., L.W.R. Structural anomalies, oxygen ordering and superconductivity in oxygen deficient $Ba_2YCu_3O_x$, *Physica C* **165** (1990), 419-433.

17. Lütgemeier, H., Schmenn, S., Meuffels, P., Storz, O., Schöllhorn, R., Niedermayer, C., Heinmaa, I., and Baikov, Y. A different type of oxygen order in $REBa_2Cu_3O_{(6+x)}$ HT_c superconductors with different ionic radii, *Physica C* **267** (1996), 191-203.

18. Sonntag, R., Hohlwein, D., Brückel, T., and Collin, G. First observation of superstructure reflections by neutron diffraction due to oxygen ordering in $YBa_2Cu_3O_{6.35}$, *Phys. Rev. Lett.* **66** (1991), 1497-1500.

19. Buchgeister, M., Hiller, W., Hosseini, S.M., Kopitzki, K., and Wagener, D., Critical temperature and normal-state resistivity of 1-2-3 HTC-superconductors in dependence of the oxygen stoichiometry, in R. Niclosky (ed.) *Proc. Int. Conf. on Transport Properties of Superconductors, Rio de Janeiro, Brazil (1990)*, World Scientific Publishing, Singapore, (1990), pp. 511-517.

20. Guillaume, M., Allenspach, P., Mesot, J., Roessli, B., Staub, U., Fischer, P., and Furrer, A. A systematic low-temperature neutron diffraction study of the $RBa_2Cu_3O_x$ (R = yttrium and rare earths; x = 6 and 7) compounds, *J. Phys.: Condens. Matter* **6** (1994), 7963-7976.

21. Claus, H., Gebhard, U., Linker, G., Röhberg, K., Riedling, S., Franz, J., Ishida, T., Erb, A., Müller-Vogt, G., and Wühl, H. Phase separation in $YBa_2Cu_3O_{7-\delta}$ single crystals near $\delta = 0$, *Physica C* **200** (1992), 271-276.

22. Kaldis, E., Fischer, P., Hewat, A.W., Hewat, E.A., Karpinski, J., and Rusiecki, S. Low temperature anomalies and pressure effects on the structure and T_c of the superconductor $YBa_2Cu_4O_8$, *Physica C* **159** (1989), 668-680.

23. Yamada, Y., Jorgensen, J.D., Pei, S., Lightfoot, P., Kodama, Y., Matsumoto, T., and Izumi, F. Structural changes of superconducting $YBa_2Cu_4O_8$ under high pressure, *Physica C* **173** (1991), 185-194.

24. Fischer, P., Roessli, B., Mesot, J., Allenspach, P., Staub, U., Kaldis, E., Bucher, B., Karpinski, J., Rusiecki, S., Jilek, E., and Hewat, A.W. Neutron diffraction investigation of structures of 'RE124' (RE = Dy, Ho, Er) and 'Nd247' superconductors; 2-D antiferromagnetism in 'Dy124', *Physica B* **180 &181** (1992), 414-416.

25. Currie, D.B. and Weller, M.T. The crystal structures of $LnBa_2Cu_4O_8$ and $LnBa_2Cu_3O_{7-\delta}$ (Ln = Er, Ho, Dy). A high resolution powder neutron diffraction study, *Physica C* **214** (1993), 204-213.

26. Mori, K., Kawaguchi, Y., Ishigaki, T., Katano, S., Funahashi, S., and Hamaguchi, Y. Crystal structure and critical temperature of $RBa_2Cu_4O_8$ (R = Tm, Er, Ho, Y, Dy and Gd), *Physica C* **219** (1994), 176-182.

27. Hewat, A.W., Fischer, P., Kaldis, E., Karpinski, J., Rusiecki, S., and Jilek, E. High resolution neutron powder diffraction investigation of temperature and pressure effects on the structure of the high-T_c superconductor $Y_2Ba_4Cu_7O_{15}$, *Physica C* **167** (1990), 579-590.

28. Berastegui, P., Fischer, P., Bryntse, I., Johansson, L.-G., and Hewat, A.W. Influence of stacking faults and temperature on the structure of $Y_2Ba_4Cu_7O_{15}$, investigated by high-resolution neutron diffraction and electron microscopy, *J. Solid State Chem.* **127** (1996), 31-39.

29. Kaldis, E. and Karpinski, J. Superconductors in the $Y_2Ba_4Cu_{6+n}O_{14+n}$ family: thermodynamics, structure and physical properties, *Eur. J. Solid State Inorg. Chem.* **27** (1990), 143-190.

30. Karpinski, J., Rusiecki, S., Kaldis, E., and Jilek, E. The high-T_c superconducting phases of the $Y_2Ba_4Cu_{6+n}O_{14+n}$ family, *J. Less-Common Met.* **164 &165** (1990), 3-19.

31. Schwer, H., Kaldis, E., Karpinski, J., and Rossel, C. Effect of structural changes on the transition temperatures in $Y_2Ba_4Cu_7O_{14+x}$ single crystals, *Physica C* **211** (1993), 165-178.

32. Genoud, J.-Y., Graf, T., Junod, A., Triscone, G., and Muller, J. Preparation and magnetic properties of the 95 K superconductor $Y_2Ba_4Cu_7O_{15}$, *Physica C* **185-189** (1991), 597-598.

33. Böttger, G., Fischer, P., Berastegui, P., Dönni, A., Aoki, Y., and Sato, H., Synthesis and characterisation of structure and physical properties of $R_2Ba_4Cu_7O_{15-\delta}$ (R = Er, Dy), in S. R. Leclair T. Chandra, J. A. Meech, B. Verma, M. Smith, and B. Balachandran (ed.) *Proceedings of IPMM'97, Australiasia-Pacific Forum on Intelligent Processing & Manufacturing of Materials*, Vol. **2**: Modelling, Processing and Manufacturing, Watson & Ferguson & Co., Brisbane, (1997), pp. 1452-1458.

34. Currie, D.B., Weller, M.T., Lanchester, P.C., and Walia, R. Superconductivity and crystal structure of $Ln_2Ba_4Cu_7O_{14+\delta}$ (Ln = Er, Y, Ho, Dy, Nd), *Physica C* **224** (1994), 43-50.

35. Böttger, G., Fischer, P., Dönni, A., Berastegui, P., Aoki, Y., Sato, H., and Fauth, F. Long-range magnetic order of the Er ions in $Er_2Ba_4Cu_7O_{14.92}$, *Phys. Rev. B* **55** (1997), R12005-R12007.

36. Zhang, H., Lynn, J.W., and Morris, D.E. Coupled-bilayer two-dimensional magnetic order of the Dy ions in $Dy_2Ba_4Cu_7O_{15}$, *Phys. Rev. B* **45** (1992), 10022-10031.

37. Fischer, P., Kakurai, K., Steiner, M., Clausen, K.N., Lebech, B., Hulliger, F., Ott, H.R., Brüesch, P., and Unternährer, P. Neutron-diffraction evidence for 3-D long-range antiferromagnetic ordering in $DyBa_2Cu_3O_{6.95}$ and for antiferromagnetic correlations in $HoBa_2Cu_3O_{6.8}$, *Physica C* **152** (1988), 145-153.

38. Zhang, H., Lynn, J.W., Li, W.-H., Clinton, T.W., and Morris, D.E. Two- and three-dimensional magnetic order of the rare-earth ions in $RBa_2Cu_4O_8$, *Phys. Rev. B* **41** (1990), 11229-11236.

39. Roessli, B., Fischer, P., Zolliker, M., Allenspach, P., Mesot, J., Staub, U., Furrer, A., Kaldis, E., Bucher, B., Karpinski, J., Jilek, E., and Mutka, H. Crystal-field splitting and temperature dependence of two-dimensional antiferromagnetism in the high-T_c compound $DyBa_2Cu_4O_8$, Z. Phys. B 91 (1993), 149-153.

40. Li, W.-H., Hsieh, W.T., and Lee, K.C. Dependence of powder neutron scattering on the dimensionality of magnetic order, J. Phys.: Condens. Matter 7 (1995), 6513-6522.

41. Clinton, T.W., Lynn, J.W., Lee, B.W., Buchgeister, M., and Maple, M.B. Oxygen dependence of the magnetic order of Nd in $NdBa_2Cu_3O_{6+x}$, J. Appl. Phys. 73 (1993), 6320-6322.

42. Clinton, T.W., Lynn, J.W., Liu, J.Z., Jia, Y.X., Goodwin, T.J., Shelton, R.N., Lee, B.W., Buchgeister, M., Maple, M.B., and Peng, J.L. Effects of oxygen on the magnetic order of the rare-earth ions in $RBa_2Cu_3O_{6+x}$ (R = Dy, Er, Nd), Phys. Rev. B 51 (1995), 15429-15447.

43. Roessli, B., Fischer, P., Staub, U., Zolliker, M., and Furrer, A. Combined electronic-nuclear magnetic ordering of the Ho^{3+}ions and magnetic stacking faults in the high-T_c superconductor $HoBa_2Cu_3O_7$, Europhys. Lett. 23 (1993), 511-515.

44. Rossat-Mignod, J., Regnault, L.P., Bourges, P., Burlet, P., Vettier, C., and Henry, J.Y., Neutron scattering study of the high-T_c superconducting system $YBa_2Cu_3O_{6+x}$, in L. C. Gupta and M. S. Multani (ed.) Frontiers in Solid State Sciences, Selected Topics in Superconductivity Vol. 1, World Scientific Publishing Co., Singapore, (1993), pp. 265-347.

45. Keimer, B., Aksey, I.A., Bossy, J., Bourges, P., Fong, H.F., Milius, D.L., Regnault, L.P., Reznik, D., and Vettier, C. Spin excitations and phonons in $YBa_2Cu_3O_{6+x}$, Physics B 234-236 (1997), 821-829.

46. Tranquada, J.M., Moudden, A.H., Goldman, A.I., Zolliker, P., Cox, D.E., Shirane, G., Shina, S.K., Vaknin, D., Johnston, D.C., Alvarez, M.S., Jacobson, A.J., Lewandowski, J.T., and Newsam, J.M. Antiferromagnetism in $YBa_2Cu_3O_{6+x}$, Phys. Rev. B 38 (1988), 2477-2485.

47. Petitgrand, D. and Collin, G. Antiferromagnetism in $YBa_2Cu_3O_6$, Physica C 153-155 (1988), 192-193.

48. Burlet, P., Vettier, C., Jurgens, M.J.G.M., Henry, J.Y., Rossat-Mignod, J., Noel, H., Potel, M., Gougeon, P., and Levet, J.C. Neutron scattering study of the antiferromagnetic ordering in $YBa_2Cu_3O_{6+x}$ powder and single crystal samples, Physica C 153-155 (1988), 1115-1120.

49. Casalta, H., Schleger, P., Brecht, E., Montfrooij, W., Andersen, N.H., Lebech, B., Schmahl, W.W., Fuess, H., Liang, R., Hardy, W.N., and Wolf, T. Absence of a second antiferromagnetic transition in pure $YBa_2Cu_3O_{6+x}$, Phys. Rev. B 50 (1994), 9688-9691.

50. Kadowaki, H., Nishi, M., Yamada, Y., Takeya, H., Takei, H., Shapiro, S.M., and Shirane, G. Successive magnetic phase transitions in tetragonal $YBa_2Cu_3O_{6+x}$, Phys. Rev. B 37 (1988), 7932-7935.

51. Casalta, H., Schleger, P., Brecht, E., Montfrooij, W., Andersen, N.H., Lebech, B., Schmahl, W.W., Fuess, H., Liang, R., Hardy, W.N., and Wolf, T. No antiferromagnetic reordering at low temperature in pure $YBa_2Cu_3O_{6+x}$, Physica C 235-240 (1994), 1623-1624.

52. Li, W.-H., Lynn, J.W., Mook, H.A., Sales, B.C., and Fisk, Z. Long-range antiferromagnetic order of the Cu in oxygen-deficient $RBa_2Cu_3O_{6+x}$, Phys. Rev. B 37 (1988), 9844-9847.

53. Rossat-Mignod, J., Burlet, P., Jurgens, M.J.G.M., Henry, J.Y., and Vettier, C. Evidence for high temperature antiferromagnetic ordering in $YBa_2Cu_3O_6$, Physica C 152 (1988), 19-24.

54. Sato, M., Shamoto, S., Tranquada, J.M., Shirane, G., and Keimer, B. Two-dimensional antiferromagnetic excitations from a large single crystal of $YBa_2Cu_3O_{6.2}$, Phys. Rev. Lett. 61 (1988), 1317-1320.

55. Korenblit, I.Y. and Aharony, A. Reentrant antiferromagnetism in oxygen-doped cuprates, *Phys. Rev. B* **49** (1994), 13291-13294.

56. Li, W.-H., J. W. Lynn, and Fisk, Z. Magnetic order of the Cu planes and chains in $RBa_2Cu_3O_{6+x}$, *Phys. Rev.* **B41** (1990), 4098-4111.

57. Rosov, N., Lynn, J.W., Cao, G., O'Reilly, J.W., Pernambuco-Wise, P., and Crow, J.E. Magnetic ordering of the Cu spins in $PrBa_2Cu_3O_{6+x}$, *Physica C* **204** (1992), 171-178.

58. Groot, P.A.J.d., Young, T., Rainford, B.D., Weller, M.T., Grasmeder, J.R., and Schärpf, O. Return of antiferromagnetic long range order upon La doping in $YBa_2Cu_3O_{7+y}$, *Physica C* **185-189** (1991), 1169-1170.

59. Brecht, E., Schmahl, W.W., Fuess, H., Casalta, H., Schleger, P., Lebech, B., Andersen, N.H., and Wolf, T. Significance of Al doping for antiferromagnetic AFII ordering in $YBa_2Cu_{3-x}Al_xO_{6+\delta}$ materials: A single-crystal neutron-diffraction study, *Phys. Rev. B* **52** (1995), 9601-9610.

60. Boothroyd, A.T., Longmore, A., Andersen, N.H., Brecht, E., and Wolf, T. Novel Pr-Cu magnetic phase at low temperatures in $PrBa_2Cu_3O_{6+x}$, *Phys. Rev. Lett.* **78** (1997), 130-133.

61. Uma, S., Schnelle, W., Gmelin, E., Rangarajan, G., Skanthakumar, S., Lynn, J.W., Walter, R., Lorenz, T., Büchner, B., Walker, E., and Erb, A. Magnetic ordering in single crystals of $PrBa_2Cu_3O_{7-\delta}$, *J. Phys.: Condens. Matter* **10** (1998), L33-L39.

62. Longmore, A., Boothroyd, A.T., Chankang, C., Yongle, H., Nutley, M.P., Andersen, N.H., Casalta, H., Schleger, P., and Christensen, A.N. Magnetic ordering in $PrBa_2Cu_{3-y}Al_yO_{6+x}$, *Phys. Rev. B* **53** (1996), 9382-9395.

63. Moudden, A.H., Schweiss, P., Hennion, B., Gehring, P.M., Shirane, G., and Hidaka, Y. Neutron diffraction study of the magnetic ordering of the Cu^{++} spins in $Nd_{1.5}Ba_{1.5}Cu_3O_{6+x}$, *Physica C* **185-189** (1991), 1167-1168.

64. Mirebeau, I., Suard, E., Caignaert, V., and Bourée, F. Iron doping in the deoxygenated $YBa_2(Cu_{1-x}Fe_x)_3O_y$, *Phys. Rev. B* **50** (1994), 3230-3238.

65. Suard, E., Mirebaeu, I., Caignaert, V., Imbert, P., and Balagurov, A.M. Influence of a deoxygenation process on the magnetic diagram of iron doped $YBa_2Cu_3O_y$ phases: a neutron diffraction study, *Physica C* **288** (1997), 10-20.

66. Miceli, P.F., Tarascon, J.M., Greene, L.H., Barboux, P., Giroud, M., Neumann, D.A., Rhyne, J.J., Schneemeyer, L.F., and Waszczak, J.V. Antiferromagnetic order in $YBa_2Cu_{3-x}Co_xO_{6+y}$, *Phys. Rev. B* **38** (1988), 9209-9212.

67. Miceli, P.F., Tarascon, J.M., Barboux, P., Greene, L.H., Bagley, B.G., Hull, G.W., Giroud, M., Rhyne, J.J., and Neumann, D.A. Magnetic transitions in the system $YBa_2Cu_{2.8}Co_{0.2}O_{6+y}$, *Phys. Rev. B* **39** (1989), 12375-12378.

68. Garcia-Munoz, J.L., Cywinski, R., Kilcoyne, S.H., and Obradors, X. Magnetic order and disorder in $YBa_2(Cu_{1-x}Fe_x)_3O_{6+y}$, *Physica C* **233** (1994), 85-96.

69. Mirebeau, I., Bellouard, C., Hennion, M., Jehanno, G., Caignaert, V., Dianoux, A.J., Phillips, T.E., and Moorjani, K. Coexistence of antiferromagnetic order and spin glass freezing in $YBa_2(Cu_{0.88}Fe_{0.12})_3O_{6.5}$, *Physica C* **184** (1991), 299-310.

70. Andersen, N.H. and Uimin, G. Model for the low-temperature magnetic phases in doped $YBa_2Cu_3O_{6+x}$, *Phys. Rev. B* **56** (1997), 10840-10843.

71. Gillon, B. Spin density and form factor measurements on HTSC copper oxides, *Physica B* **174** (1991), 340-348.

72. Boucherle, J.X., Henry, J.Y., Papoular, R., Rossat-Mignod, J., Schweizer, J., and Tasset, F. Spin density in the high T_c superconductor $YBa_2Cu_3O_7$, *J. Magn. Magn. Mater.* **104-107** (1992), 630-632.

73. Nutley, M.P., Boothroyd, A.T., and McIntyre, G.J. Magnetisation density in $PrBa_2Cu_3O_7$ by polarised-neutron diffraction, *J. Magn. Magn. Mater.* **104-107** (1992), 623-624.

74. Boucherle, J.X., Henry, J.Y., Papoular, R.J., Rossat-Mignod, J., Schweizer, J., Tasset, F., and Uimin, G. Polarised neutron study of high-T_c superconductors, *Physica B* **192** (1993), 25-38.

75. Lynn, J.W., Li, W.-H., Trevino, S.F., and Fisk, Z. Pressure dependence of the Cu magnetic order in $RBa_2Cu_3O_{6+x}$, *Phys. Rev. B* **40** (1989), 5172-5175.

76. Reeves, M.E., Citrin, D.S., Pazol, B.G., Friedmann, T.A., and Ginsberg, D.M. Specific heat of $GdBa_2Cu_3O_{7-\delta}$ in the normal and superconducting states, *Phys. Rev. B* **36** (1987), 6915-6919.

77. Dunlap, B.D., Slaski, M., Hinks, D.G., Soderholm, L., Beno, M., Zhang, K., Segre, C., Crabtree, G.W., Kwok, W.K., Malik, S.K., Schuller, I.K., Jorgensen, J.D., and Sungaila, Z. Electronic and magnetic properties of rare-earth ions in $REBa_2Cu_3O_{7-x}$ (RE = Dy, Ho, Er), *J. Magn. Magn. Mater.* **68** (1987), L139-L144.

78. Goldman, A.I., Yang, B.X., Tranquada, J., Crow, J.E., and Jee, C.-S. Antiferromagnetic order in $DyBa_2Cu_3O_7$, *Phys. Rev. B* **36** (1987), 7234-7236.

79. Paul, D.M., Mook, H.A., Hewat, A.W., Sales, B.C., Boatner, L.A., Thompson, J.R., and Mostoller, M. Magnetic ordering in the high-temperature superconductor $GdBa_2Cu_3O_7$, *Phys. Rev. B* **37** (1988), 2341-2344.

80. Fischer, P., Schmid, B., Brüesch, P., Stucki, F., and Unternährer, P. Three-dimensional antiferromagnetic ordering in the light rare-earth high-T_c superconductor $NdBa_2Cu_3O_{6.86}$, *Z. Phys. B* **74** (1989), 183-189.

81. Yang, K.N., Ferreira, J.M., Lee, B.W., Maple, M.B., Li, W.-H., Lynn, J.W., and Erwin, R.W. Antiferromagnetic ordering in superconducting and oxygen-deficient nonsuperconducting $RBa_2Cu_3O_{7-\delta}$ compounds (R = Nd and Sm), *Phys. Rev. B* **40** (1989), 10963-10972.

82. Li, W.-H., Lynn, J.W., Skanthakumar, S., Clinton, T.W., Kebede, A., Jee, C.-S., Crow, J.E., and Mihalisin, T. Magnetic order of Pr in $PrBa_2Cu_3O_7$, *Phys. Rev. B* **40** (1989), 5300-5303.

83. Maple, M.B., Ferreira, J.M., Hake, R.R., Lee, B.W., Neumeier, J.J., Seaman, C.L., Yang, K.N., and Zhou, H. 4f electron effects in high T_c $RBa_2Cu_3O_{7-\delta}$ (R = rare earth) superconductors, *J. Less-Common Met.* **149** (1989), 405-425.

84. Liu, S.H. Magnetic interaction between rare-earth moments in high-temperature superconductors $RBa_2Cu_3O_{7-x}$, *Phys. Rev. B* **37** (1988), 7470-7471.

85. Mook, H.A., Paul, D.M., Sales, B.C., Boatner, L.A., and Cussen, L. Magnetic ordering in $GdBa_2Cu_3O_{6.14}$, *Phys. Rev. B* **38** (1988), 12008-12010.

86. Drössler, H., Jostarndt, H.-D., Harnischmacher, J., Kalenborn, J., Walter, U., Severing, A., Schlabitz, W., and Holland-Moritz, E. Magnetic interactions between copper and RE in $REBa_2Cu_3O_{7-\delta}$, *Z. Phys. B* **100** (1996), 1-11.

87. Takenaka, K., Imanaka, Y., Tamasaku, K., Ito, T., and Uchida, S. Anisotropic optical spectrum of untwinned $PrBa_2Cu_3O_7$: Persistence of the charge-transfer insulating state of the CuO_2 plane against hole doping, *Phys. Rev. B* **46** (1992), 5833-5836.

88. Fehrenbacher, R. and Rice, T.M. Unusual electronic structure of $PrBa_2Cu_3O_7$, *Phys. Rev. Lett.* **70** (1993), 3471-3474.

89. Cooke, D.W., Kwok, R.S., Lichti, R.L., Adams, T.R., Boekema, C., Dawson, W.K., Kebede, A., Schwegler, J., Crow, J.E., and Mihalishin, T. Magnetic ordering in $(Y_{1-x}Pr_x)Ba_2Cu_3O_7$ as observed by muon-spin relaxation, *Phys. Rev. B* **41** (1990), 4801-4804.

90. Reyes, A.P., MacLaughlin, D.E., Takigawa, M., Hammel, P.C., Heffner, R.H., Thompson, J.D., Crow, J.E., Kebede, A., Mihalishin, T., and Schwegler, J. Observation of Cu NMR in antiferromagnetic $PrBa_2Cu_3O_7$: Evidence for hole-band filling, *Phys. Rev. B* **42** (1990), 2688-2691.

91. Skanthakumar, S., Lynn, J.W., Rosov, N., Cao, G., and Crow, J.E. Observation of Pr magnetic order in $PrBa_2Cu_3O_7$, *Phys. Rev. B* **55** (1997), R3406-R3409.

92. Guillaume, M., Fischer, P., Roessli, B., Podlesnyak, A., Schefer, J., and Furrer, A. Magnetic Order of Pr Ions in $PrBa_2Cu_3O_6$, *Solid State Commun.* **88** (1993), 57-61.

93. Li, W.-H., Jou, C.J., Shyr, S.T., Lee, K.C., Lynn, J.W., Tsay, H.L., and Yang, H.D. Effects of Ga doping on the magnetic ordering of Pr in $PrBa_2Cu_3O_7$, *J. Appl. Phys.* **76** (1994), 7136-7138.

94. Li, W.-H., Chang, K.J., Hsieh, W.T., Lee, K.C., Lynn, J.W., and Yang, H.D. Magnetic ordering of Pr in $PrBa_2Cu_{2.7}Zn_{0.3}O_{7-y}$, *Phys. Rev. B* **48** (1993), 519-523.

95. Hsieh, W.T., Chang, K.J., Li, W.-H., lee, K.C., Lynn, J.W., Lai, C.C., and Ku, H.C. Magnetic ordering of Pr and Cu in $TlBa_2PrCu_2O_{7-y}$, *Phys. Rev. B* **49** (1994), 12200-12205.

96. Rosov, N., Lynn, J.W., Radousky, H.B., Bennahmias, M., Goodwin, T.J., Klavins, P., and Shelton, R.N. Crystal structure and magnetic ordering of the rare-earth and Cu moments in $RBa_2Cu_2NbO_8$ (R = Nd, Pr), *Phys. Rev. B* **47** (1993), 15256-15264.

97. Hsieh, W.T., Li, W.-H., lee, K.C., Lynn, J.W., Shieh, J.H., and Ku, H.C. Magnetic ordering of Pr in $Pb_2Sr_2Cu_3O_8$, *J. Appl. Phys.* **76** (1994), 7124-7126.

98. Goodwin, T.J., Shelton, R.N., Radousky, H.B., Rosov, N., and Lynn, J.W. Pr and Cu magnetism in $(Pr_{1.5}Ce_{0.5})Sr_2Cu_2MO_{10-\delta}$ (M = Nb, Ta): Correlations with a suppression of superconductivity, *Phys. Rev. B* **55** (1997), 3297-3307.

99. Allenspach, P., Furrer, A., and Hulliger, F. Neutron crystal-field spectroscopy and magnetic properties of $DyBa_2Cu_3O_{7-\delta}$, *Phys. Rev. B* **39** (1989), 2226-2232.

100. Yang, C.N. The spontaneous magnetization of a two-dimensional Ising model, *Phys. Rev.* **85** (1952), 808-816.

101. Onsager, L. Crystal statistics. I. A two-dimensional model with order-disorder transition, *Phys. Rev.* **65** (1944), 117-149.

102. Clinton, T.W., Lynn, J.W., Liu, J.Z., Jia, Y.X., and Shelton, R.N. Magnetic order of Dy in $DyBa_2Cu_3O_7$, *J. Appl. Phys.* **70** (1991), 5751-5753.

103. Clinton, T.W., Lynn, J.W., Liu, J.Z., Jia, Y.X., and Shelton, R.N. Two-dimensional magnetic correlations and magnetic ordering of Dy and Er in $DyBa_2Cu_3O_7$ and $ErBa_2Cu_3O_7$, *J. Magn. Magn. Mater.* **104-107** (1992), 625-626.

104. Lynn, J.W., Li, W.-H., Li, Q., Ku, H.C., Yang, H.D., and Shelton, R.N. Magnetic fluctuations and two-dimensional ordering in $ErBa_2Cu_3O_7$, *Phys. Rev. B* **36** (1987), 2374-2377.

105. Chattopadhyay, T., Brown, P.J., Bonnenberg, D., Ewert, S., and Maletta, H. Evidence for three-dimensional magnetic ordering in the high-T_c superconductor $ErBa_2Cu_3O_7$, *Europhys. Lett.* **6** (1988), 363-368.

106. Paul, D.M., Mook, H.A., Boatner, L.A., Sales, B.C., Ramey, J.O., and Cussen, L. Magnetic ordering in the high-temperature superconductor $ErBa_2Cu_3O_{7-\delta}$, *Phys. Rev. B* **39** (1989), 4291-4294.

107. Chattopadhyay, T., Brown, P.J., Sales, B.C., Boatner, L.A., Mook, H.A., and Maletta, H. Single-crystal neutron-diffraction investigation of the magnetic ordering of the high-temperature superconductor $ErBa_2Cu_3O_{7-\delta}$, *Phys. Rev. B* **40** (1989), 2624-2626.

108. Lynn, J.W., Clinton, T.W., Li, W.-H., Erwin, R.W., Liu, J.Z., Vandervoort, K., and Shelton, R.N. 2D and 3D magnetic behavior of Er in ErBa$_2$Cu$_3$O$_7$, *Phys. Rev. Lett.* **63** (1989), 2606-2609.

109. Maletta, H., Pörschke, E., Chattopadhyay, T., and Brown, P.J. 2-D and 3-D magnetic ordering of Er in ErBa$_2$Cu$_3$O$_x$ (6≤x≤7), *Physica C* **166** (1990), 9-14.

110. Clinton, T.W. and Lynn, J.W. Magnetic ordering of Er in powder and single crystals of ErBa$_2$Cu$_3$O$_7$, *Physica C* **174** (1991), 487-490.

111. Maletta, H., Chattopadhyay, T., and Brown, P.J. Reply to the comment by Clinton and Lynn, *Physica C* **174** (1991), 489-490.

112. Roessli, B., Allenspach, P., Fischer, P., Mesot, J., Staub, U., Maletta, H., Brüesch, P., Ritter, C., and Hewat, A.W. Crystal structures and long-range antiferromagnetic ordering in REBa$_2$Cu$_3$O$_{7-\delta}$ (RE = Yb, Nd), *Physica B* **180 & 181** (1992), 396-398.

113. Lynn, J.W., Li, W.-H., Mook, H.A., Sales, B.C., and Fisk, Z. Nature of the magnetic order of Cu in oxygen-deficient NdBa$_2$Cu$_3$O$_{6+x}$, *Phys. Rev. Lett.* **60** (1988), 2781-2784.

114. Lynn, J.W. and Li, W.-H. Magnetic order in RBa$_2$Cu$_3$O$_{6+x}$ (invited), *J. Appl. Phys.* **64** (1988), 6065-6070.

115. Moudden, A.H., Shirane, G., Tranquada, J.M., Birgeneau, R.J., Endoh, Y., Yamada, K., Hidaka, Y., and Murakami, T. Antiferromagnetic ordering of Cu ions in NdBa$_2$Cu$_3$O$_{6.1}$, *Phys. Rev. B* **38** (1988), 8720-8723.

116. Moudden, A.H., Shirane, G., Tranquada, J.M., Birgeneau, R.J., Endoh, Y., Yamada, K., Hidaka, Y., and Murakami, T. Antiferromagnetism in NdBa$_2$Cu$_3$O$_{6.1}$, *Physica B* **156 & 157** (1989), 861-863.

117. Trounov, V. and Fischer, P. unpublished results (1993).

118. Hodges, J.A., Imbert, P., and Jéhanno, G. Magnetic ordering on Yb^{3+} in YBa$_2$Cu$_3$O$_{7-x}$, *Solid State Commun.* **64** (1987), 1209-1211.

119. Guillaume, M., Allenspach, P., Mesot, J., Staub, U., Furrer, A., Osborn, R., Taylor, A.D., Stucki, F., and Unternährer, P. Neutron spectroscopy of the crystalline electric field in high-T$_c$ YbBa$_2$Cu$_3$O$_7$, *Solid State Commun.* **81** (1992), 999-1002.

120. Birrer, P., Gygax, F.N., Hitti, B., Lippelt, E., Schenck, A., Weber, M., Barth, S., Hulliger, F., and Ott, H.R. Structural and dynamic properties of the magnetic order in the 90-K superconductor HoBa$_2$Cu$_3$O$_7$, *Phys. Rev. B* **39** (1989), 11449-11456.

121. Staub, U. and Ritter, C. Two-dimensional spin fluctuations of Ho^{3+} in HoBa$_2$Cu$_3$O$_7$, *Phys. Rev. B* **54** (1996), 7279-7283.

122. Li, W.-H., Chuang, W.Y., Wu, S.Y., Lee, K.C., Lynn, J.W., Tsay, H.L., and Yang, H.D. Superconductivity, magnetic fluctuations, and magnetic order in TbSr$_2$Cu$_{2.69}$Mo$_{0.31}$O$_7$, *Phys. Rev. B* **56** (1997), 5631-5636.

123. Staub, U., Soderholm, L., Skanthakumar, S., Rosenkranz, S., Ritter, C., and Kagunya, W. Quasi two-dimensional magnetic order of Tb^{3+} spins in Pb$_2$Sr$_2$Tb$_{1-x}$Ca$_x$Cu$_3$O$_8$ (x = 0 and 0.5), *Z. Phys. B* **104** (1997), 37-43.

124. Wu, S.Y., Hsieh, W.T., Li, W.-H., Lee, K.C., Lynn, J.W., and Yang, H.D. Two-dimensional magnetic order in Pb$_2$Sr$_2$TbCu$_3$O$_8$, *J. Appl. Phys.* **75** (1994), 6598-6600.

125. Roessli, B., Fischer, P., Staub, U., Zolliker, M., and Furrer, A. Combined electronic-nuclear magnetic ordering of the Ho^{3+} ions and magnetic stacking faults in HoBa$_2$Cu$_3$O$_x$ (x=7.0, 6.8, 6.3), *J. Appl. Phys.* **75** (1994), 6337-6339.

126. Chattopadhyay, T., Maletta, H., Wirges, W., Fischer, K., and Brown, P.J. Evidence for the dependence of the magnetic ordering on the oxygen occupancy in the high-T$_c$ superconductor GdBa$_2$Cu$_3$O$_{7-\delta}$, *Phys. Rev. B* **38** (1988), 838-840.

127. Guillaume, M., Fischer, P., Roessli, B., Allenspach, P., and Trounov, V. Neutron diffraction investigation of antiferromagnetic rare-earth ordering in $DyBa_2Cu_3O_{6.1}$ and $^{160}GdBa_2Cu_3O_{6.1}$, *Physica C* **235-240** (1994), 1637-1638.

128. Wu, S.Y., Li, W.-H., Lee, K.C., Lynn, J.W., Meen, T.H., and Yang, H.D. Two- and three-dimensional magnetic correlations of Tb in $Pb_2Sr_2TbCu_3O_8$, *Phys. Rev. B* **54** (1996), 10019-10026.

129. Mesot, J., Allenspach, P., Staub, U., Furrer, A., Mutka, H., Osborn, R., and Taylor, A. Neutron-spectroscopic studies of the crystal field in $ErBa_2Cu_3O_x$ ($6 \leq x \leq 7$), *Phys. Rev. B* **47** (1993), 6027-6036.

130. Chattopadhyay, T., Zeilinger, A., Wacenovsky, M., Weber, H.W., Hyun, O.B., and Finnemore, D.K. Search for magnetic ordering of Tm moments in $TmBa_2Cu_3O_{7-\delta}$ down to 90 mK, *Solid State Commun.* **73** (1990), 721-723.

131. Allenspach, P., Maple, M.B., and Furrer, A. Interpretation of the magnetic specific heat of oxygen deficient and doped $NdBa_2Cu_3O_x$ and $DyBa_2Cu_3O_x$ compounds, *J. Alloys Comp.* **207/208** (1994), 213-220.

132. Wu, S.Y., Lin, Y.-C., Li, W.-H., Lee, K.C., Lynn, J.W., and Yang, H.D. Neutron diffraction studies of Pr ordering in $PrBa_2Cu_4O_8$, *Abstract ICNS'97, Toronto, August 17-21* **BA28** (1997), 180.

133. Roessli, B., Fischer, P., Guillaume, M., Mesot, J., Staub, U., Zolliker, M., Furrer, A., Kaldis, E., Karpinski, J., and Jilek, E. Antiferromagnetic ordering and crystal-field splitting of the Ho^{3+} ions in $HoBa_2Cu_4O_8$, *J. Phys.: Condens. Matter* **6** (1994), 4147-4152.

134. Zhang, H., Lynn, J.W., and Morris, D.E. Two-dimensional bilayer magnetic order of Dy ions in $Dy_2Ba_4Cu_7O_{15}$, *J. Magn. Magn. Mater.* **104-107** (1992), 821-822.

135. Böttger, G., Allenspach, P., Dönni, A., Aoki, Y., and Sato, H. Low-temperature specific heat of $Er_2Ba_4Cu_7O_{15-\delta}$, *Z. Phys. B* **104** (1997), 195-198.

136. Abulafia, Y., Barak, J., and Peng, J.L. Susceptibility of $ErBa_2Cu_3O_6$ single crystal: A dipolar antiferromagnet, *J. Appl. Phys.* **76** (1994), 7468-7472.

137. Bertaut, E.F. Electrostatic potentials, fields and field gradients, *J. Phys. Chem. Solids* **39** (1978), 97-102.

138. Bertaut, E.F. The equivalent charge concept and its application to the electrostatic energy of charges and multipoles, *J. de phys.* **39** (1978), 1331-1348.

139. Boothroyd, A.T., Doyle, S.M., and Osborn, R. The magnetic state of Pr in $PrBa_2Cu_3O_7$, *Physica C* **217** (1993), 425-438.

140. Allenspach, P., Mesot, J., Staub, U., Guillaume, M., Furrer, A., Yoo, S.-I., Kramer, M.J., McCallum, R.W., Maletta, H., Blank, H., Mutka, H., Osborn, R., Arai, M., Bowden, Z., and Taylor, A.D. Magnetic properties of Nd^{3+} in Nd-Ba-Cu-O-compounds, *Z. Phys. B* **95** (1994), 301-310.

141. Allenspach, P., Furrer, A., Brüesch, P., Marsolais, R., and Unternährer, P. A neutron spectroscopic comparison of the crystalline electric field in tetragonal $HoBa_2Cu_3O_{6.2}$ and orthorhombic $HoBa_2Cu_3O_{6.8}$, *Physica C* **157** (1989), 58-64.

142. MacIsaac, A.B., Whitehead, J.P., Robinson, M.C., and de'Bell, K. Phase diagram for a two-dimensional uniaxial dipolar antiferromagnet with an exchange interaction, *Physica B* **194-196** (1994), 223-224.

143. MacIsaac, A.B., Whitehead, J.P., Robinson, M.C., and de'Bell, K. Striped phases in two-dimensional dipolar ferromagnets, *Phys. Rev. B* **51** (1995), 16033-16045.

144. Allenspach, P., Magnetic ordering and oxygen deficiency in $R_2Ba_4Cu_{6+n}O_{14+n}$ (R = rare earth; n = 0, 1, 2), to be published in *Handbook on the Physics and Chemistry of Rare Earths* (1998).

145. Sundaresan, A., Chinchure, A.D., Gosh, K., Ramakrishnan, S., Marathe, V.R., Gupta, L.C., Sharon, M., and Shah, S.S. Superconducting and magnetic properties of $TlSr_2Ca_{1-x}R_xCu_2O_7$ [R = Pr, Tb], *Phys. Rev B* **51** (1995), 3893-3898.

146. Zou, Z., Ye, J., Oka, K., and Nishihara, Y. Superconducting $PrBa_2Cu_3O_x$, *Phys. Rev. Lett.* **80** (1998), 1074-1077.

147. Hammann, J. and Manneville, P. Ordre magnétique électronique induit par les interactions hyperfines dans les grenats de gallium-holmium et de gallium-terbium, *J. Physique* **34** (1973), 615-622.

148. Staub, U., Fauth, F., Guillaume, M., Mesot, J., Furrer, A., Dosanjh, P., and Zhou, H. Collective magnetic excitations of Ho^{3+} ions in grain-aligned $HoBa_2Cu_3O_7$, *Europhys. Lett.* **21** (1993), 845-850.

149. Felner, I., Nowik, I., Brosh, B., Hechel, D., and Bauminger, E.R. Superconductivity and magnetic order in the Cu(2) planes in Fe-doped $YBa_2Cu_4O_8$, *Phys. Rev. B* **43** (1991), 8737-8740.

150. Niedermayer, C., Gückler, H., Golnik, A., Binninger, U., Rauer, M., Recknagel, E., Budnick, J.I., and Weidinger, A. Simultaneous magnetic ordering of the Gd and Cu subsystems in oxygen-deficient $GdBa_2Cu_3O_{6+x}$, *Phys. Rev. B* **47** (1993), 3427-3430.

151. Dowty, E. (1997), ATOMS for Windows and Macintosh, computer program, Shape Software.

152. Cense, J.-M. MolDraw and MolView Macintosh computer programs, *Tetrahedron Computer Methodology* **2** (1989), 65-71.

COLLECTIVE MAGNETIC EXCITATIONS OF 4f IONS IN $R_{2-x}Ce_xCuO_4$ (R=Nd, Pr)

W. HENGGELER AND A.FURRER
Laboratory for Neutron Scattering
ETHZ Zürich & Paul Scherrer Institut
CH-5232 Villigen PSI, Switzerland

1. Introduction

The $R_{2-x}Ce_xCuO_4$ (R=Nd,Pr, $0 \leq x \leq 0.2$) compounds have been the subject of intensive investigations since the discovery of superconductivity in some of these substances [1,2]. The superconducting transition is observed for samples in a narrow doping range ($0.13 \leq x \leq 0.18$) after oxygen reduction, with a highest T_c of ≈ 24 K [3]. In contrast to the previously found high-T_c superconductors like $YBa_2Cu_3O_x$ or $La_{2-x}Sr_xCuO_4$, these substances have a simple tetragonal structure (Fig. 1), which makes them attractive for the investigation of superconductivity. There are still many open questions concerning the properties of these compounds, for instance it is still controversial for what doping range superconductivity is observed, see e.g. [4], and whether the charge carriers are electron- or hole-like. Recent experiments suggest the existence of two types of charges carriers [5], but the main part of carriers are nevertheless considered to be electron-like.

Another intriguing property of these compounds has been discovered by specific-heat experiments. $Nd_{2-x}Ce_xCuO_4$ exhibits a heavy-fermion like behaviour, with a large linear specific-heat coefficient $\gamma=4.4$ J/K^2 per mole Nd for x=0.2 [6]. It was commonly assumed that this behaviour is based on the Zeeman effect, arising from the interaction of the Nd-ions with the strongly correlated electrons in the copper oxide planes [7]. With this description, $Nd_{2-x}Ce_xCuO_4$ would represent a novel type of heavy-fermion system, and thereby eliminate the long-standing idea that the Kondo effect is the only route to heavy electrons [8].

Both superconductivity and heavy-fermion behaviour are intimately connected with the magnetism of these systems. While for superconductivity the copper spin fluctuations play a crucial role, the rare-earth magnetism is considered to be the source of the heavy-fermion like behaviour. Therefore, for an understanding if these properties, a detailed knowledge of the magnetism in these compounds is essential.

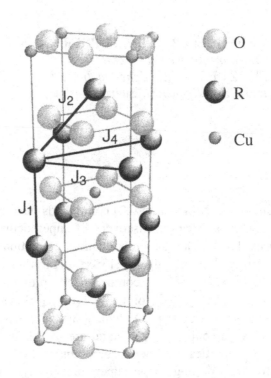

Figure 1.   Crystal structure of R_2CuO_4 with the various R-R exchange constants J_i indicated.

The ordering of the Cu spins has been studied extensively [9-28]. In the undoped substances, the Cu spins order at a temperature of about 280 K in a noncollinear spin-structure with propagation vector $q=(1/2,1/2,0)$. For R=Nd, two Cu spin-reorientations occur at ≈ 70 K and at ≈ 30 K. For R=Pr, no such reorientations have been observed with neutrons. Upon Ce-doping, the Cu-ordering and -spin reorientation temperatures decrease. No Cu-ordering is observed for $x \geq 0.14$ for R=Nd and Pr. The Cu spin dynamics in both compounds seem to follow a simple xy model with in-plane nearest neighbour exchange constants of 130 ± 13 meV for R=Pr [29] and 155 ± 3 meV for R=Nd [30]. The xy-anisotropy and the biquadratic in-plane exchange are expected to lead to spin gaps [31] which have been observed at about 2 meV and 6 meV for R=Pr at 10K and at about 12 meV and 14 meV for R=Nd at 5 K [29,32,33].

In this article, we will concentrate on the rare-earth magnetism in $R_{2-x}Ce_xCuO_4$. It will be shown that the magnetic ground state is mainly determined by the crystalline electric field (CEF) interaction, and that the R-Cu and R-R exchange interactions significantly affect the thermodynamic properties. The results will give a straightforward explanation for the high γ-values observed in $Nd_{2-x}Ce_xCuO_4$.

2. Theory of 4f magnetic excitations

2.1. THE CRYSTALLINE ELECTRIC FIELD (CEF) INTERACTION

For rare-earth ions, the degeneracy of the electronic configuration is lifted by electrostatic and spin-orbit interaction, which leads to the formation of J-multiplets, with energy-spacings of the order of 100-1000meV. The effect of the CEF on the rare-earth ions is to partially or totally remove the (2J+1)-fold degeneracy of these J-multiplets. To parametrise the CEF interaction one expands the CEF potential in terms of tensor operators U_n^m [34]:

$$\mathcal{H}_{CEF} = \sum_{n=0}^{\infty} \sum_{m=-n}^{n} A_n^m U_n^m \qquad (1)$$

The A_n^m denote the CEF parameters, of which only a limited number is non-zero due to symmetry reasons. These parameters define the energies of the crystal-field-levels and the corresponding wave-functions.

2.2. EXCHANGE COUPLED R IONS

In an extended, undiluted system, the spin Hamiltonian is assumed to have the following form:

$$\mathcal{H} = \sum_i \mathcal{H}_{CEF,i} - \sum_{i>j} \hat{\mathbf{J}}_i \bar{\bar{J}}(\mathbf{r}_i - \mathbf{r}_j) \hat{\mathbf{J}}_j \qquad (2)$$

where the exchange tensors $\bar{\bar{J}}(\mathbf{r}_i - \mathbf{r}_j)$ describe the coupling between the ions at positions $\mathbf{r}_i$ and $\mathbf{r}_j$, and $\hat{\mathbf{J}}_i$ are the total angular momentum operators. In general, the $\bar{\bar{J}}(\mathbf{r}_i - \mathbf{r}_j)$ are anisotropic. The many possible causes of two-site exchange anisotropy have been summarised in Ref. [35].

A different source of anisotropy are crystal-field effects. In this case the angular momentum contribution to $\hat{\mathbf{J}}_i$ can lead to single-ion anisotropy (see Sec. 3.1).

A convenient way to calculate directly from Eq. (2) the wave-vector dependent susceptibility is the mean-field random-phase approximation (RPA). A description of this method is given in Ref. [36]. The approximation consists in neglecting longitudinal fluctuations of $\mathbf{J}_i$, therefore it is clearly best justified when the fluctuations are small, i.e. at low temperatures. The susceptibility for a system with n magnetic ions per magnetic unit cell is given by

$$\chi(\mathbf{q},\omega) = 1/2 \sum_{rs} \bar{\bar{\chi}}_{rs}(\mathbf{q},\omega) \,, \ (r,s=1..n), \tag{3}$$

where the $\bar{\bar{\chi}}_{rs}(\mathbf{q},\omega)$ are the building blocks of the 3n×3n tensor

$$\bar{\bar{\chi}}(\mathbf{q},\omega) = \begin{pmatrix} \bar{\bar{\chi}}_{11}(\mathbf{q},\omega) & \cdots & \bar{\bar{\chi}}_{1n}(\mathbf{q},\omega) \\ \vdots & \ddots & \vdots \\ \bar{\bar{\chi}}_{n1}(\mathbf{q},\omega) & \cdots & \bar{\bar{\chi}}_{nn}(\mathbf{q},\omega) \end{pmatrix} \tag{4}$$

which has to be determined by the RPA equation

$$\bar{\bar{\chi}}(\mathbf{q},\omega) = (1 - \bar{\bar{\chi}}^0(\omega)\bar{\bar{J}}(\mathbf{q}))^{-1} \, \bar{\bar{\chi}}^0(\omega) \tag{5}$$

$\bar{\bar{\chi}}^0(\omega)$ is built up of the single-ion susceptibility tensors $\bar{\bar{\chi}}_r^0(\omega)$ (r=1...n):

$$\bar{\bar{\chi}}^0(\omega) = \begin{pmatrix} \bar{\bar{\chi}}_1^0(\omega) & & 0 \\ & \ddots & \\ 0 & & \bar{\bar{\chi}}_n^0(\omega) \end{pmatrix} \tag{6}$$

For $\omega \neq 0$ the $\bar{\bar{\chi}}_r^0(\omega)$ are given by

$$(\overline{\overline{\chi}}{}_r^0)^{\alpha\beta}(\omega) = \sum_{\substack{i,j \\ E_{\Gamma_i} \neq E_{\Gamma_j}}} \frac{\langle \Gamma_i | \hat{j}^\alpha | \Gamma_j \rangle_r \langle \Gamma_j | \hat{j}^\beta | \Gamma_i \rangle_r}{E_{\Gamma_i} - E_{\Gamma_j} - \omega} (p_{\Gamma_j} - p_{\Gamma_i}) \qquad (7)$$

where E_{Γ_i} denotes the energy of the crystal field level Γ_i.

The exchange coupling tensor is constructed from the Fourier-transformed coupling constants $\overline{\overline{J}}(\mathbf{q})_{r,s}$ (r,s=1...n) of the exchange between ions of sublattices r and s:

$$\overline{\overline{J}}(\mathbf{q}) = \begin{pmatrix} \overline{\overline{J}}(\mathbf{q})_{11} & .... & \overline{\overline{J}}(\mathbf{q})_{1n} \\ \vdots & \ddots & \vdots \\ \overline{\overline{J}}(\mathbf{q})_{n1} & ... & \overline{\overline{J}}(\mathbf{q})_{nn} \end{pmatrix} \qquad (8)$$

The magnetic excitation energies are the poles of the wave-vector dependent susceptibility $\chi(\mathbf{q}, \omega)$. In many cases it is possible to analytically calculate these poles and therefore to determine the dispersion $\omega(\mathbf{q})$. For a two-level system with one ion per unit cell, the energies are given by

$$\omega_\alpha(\mathbf{q}) = [\Delta^2 - 2M_\alpha^2 \Delta(J_\alpha(\mathbf{q})]^{1/2} \qquad (9)$$

with $M_\alpha = \langle \Gamma_i | \hat{j}^a | \Gamma_j \rangle$ and $\Delta = E_{\Gamma_i} - E_{\Gamma_j}$.

In the case of two ions per unit cell, but identical single-ion susceptibilities (para- or ferromagnet), the energies of the resulting two excitation branches can be expressed by

$$\omega_\alpha(\mathbf{q}) = [\Delta^2 - 2M_\alpha^2 \Delta(J_\alpha(\mathbf{q}) \pm v|J'_\alpha(\mathbf{q})|)]^{1/2} \qquad (10)$$

v is defined as the sign of $J'(0)$. $J_\alpha(\mathbf{q})$ and $J'_\alpha(\mathbf{q})$ are the Fourier-transformed intra-sublattice and inter-sublattice exchange functions. The + and - sign corresponds to the acoustic and optical branch, respectively.

In Section 5 experiments on Nd_2CuO_4 will be discussed, where a noncollinear magnetic order of Nd is observed, which leads to eight magnetic sublattices. Even in this case it is possible to obtain closed expressions for the excitation energies. They are more complex and given elsewhere [37,38].

2.3. NEUTRON SCATTERING CROSS SECTIONS

The wave-vector dependent susceptibility can directly be determined by inelastic magnetic neutron scattering (see also "Introduction to neutron scattering", this volume). The neutron cross-section can be expressed by

$$\left(\frac{d^2\sigma}{d\Omega d\omega}\right) = [\gamma r_0 F(Q)]^2 \frac{k}{k'} \exp(-2W(Q)) \sum_{\alpha,\beta} (\delta_{\alpha\beta} - \frac{Q_\alpha Q_\beta}{Q^2}) S^{\alpha\beta}(\mathbf{Q},\omega) \quad (11)$$

with the scattering function

$$S^{\alpha\beta}(\mathbf{Q},\omega) = \frac{N\hbar}{\pi}\left(1 - \exp\left(-\frac{\hbar\omega}{k_B T}\right)\right)^{-1} \operatorname{Im}\chi^{\alpha\beta}(\mathbf{Q},\omega)$$

$\mathbf{Q} = \mathbf{k} - \mathbf{k'}$ is the neutron scattering vector, with $\mathbf{k}$ and $\mathbf{k'}$ being the wavevector of the incoming and outgoing neutrons, respectively. γ denotes the gyromagnetic ratio of the neutron, r_0 the classical electron radius, $\exp(-2W(Q))$ the Debye-Waller factor and $F(Q)$ the dimensionless magnetic form factor defined as the Fourier transform of the normalised spin density associated with the magnetic ions.

In the case of single-ion excitations, corresponding to a CEF transition $\Gamma_n \to \Gamma_m$, the scattering function is independent of the momentum transfer $\mathbf{Q}$ and given by

$$S^{\alpha\beta}_{CEF}(\omega) = N p_{\Gamma_n} \langle \Gamma_n | J^\alpha | \Gamma_m \rangle \langle \Gamma_m | J^\beta | \Gamma_n \rangle \delta(\hbar\omega + E_{\Gamma_n} - E_{\Gamma_m}) \quad (12)$$

For interacting ions the scattering function becomes wave-vector dependent. For some cases it is possible to obtain a closed expression for the cross section. For a two-level system with one ion per unit cell, the scattering function is given by

$$S^{\alpha\alpha}(\mathbf{Q},\omega) = \frac{N\hbar}{\pi}\left(1 - \exp\left(-\frac{\hbar\omega}{k_B T}\right)\right)^{-1} M_\alpha^2 \frac{\Delta}{\omega_\alpha(\mathbf{q})} \quad (13)$$

In the case of two ions per unit cell, but identical single-ion susceptibilities (para- or ferromagnet) it can be expressed by

$$S^{\alpha\alpha}(\mathbf{Q},\omega) = \frac{N\hbar}{2\pi}\left(1-\exp\left(-\frac{\hbar\omega}{k_B T}\right)\right)^{-1} M_\alpha^2(1\pm\cos\varphi)\frac{\Delta}{\omega_\alpha(\mathbf{q})} \qquad (14)$$

The + and - sign corresponds to the acoustic and optical branch, respectively, and the phase φ is defined through $J'(\mathbf{Q}) = J'(\mathbf{q})\exp(-i\boldsymbol{\tau}\cdot\mathbf{r}) = \nu|J'(\mathbf{q})|\exp(-i\varphi)$, with $\boldsymbol{\tau}=\mathbf{Q}-\mathbf{q}$: reciprocal lattice vector and $\mathbf{r}$: vector connecting the two sublattices.

In the case of the noncollinear magnetic structure of the Nd ions in Nd_2CuO_4 closed expressions for $S(\mathbf{Q},\omega)$ are given in Ref. [37].

3. Rare-earth magnetic properties in $R_{2-x}Ce_xCuO_4$ (R=Nd,Pr)

3.1 CEF LEVEL SCHEME OF $R_{2-x}Ce_xCuO_4$ (R=Nd,Pr)

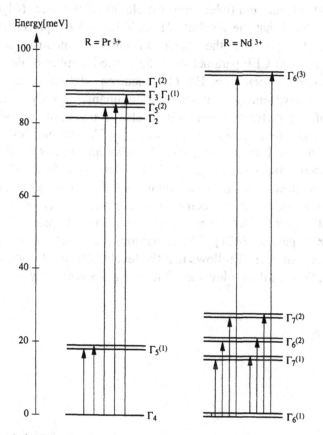

Figure 2. Crystal field level schemes of $R_{2-x}Ce_xCuO_4$ (R=Nd,Pr). The arrows denote the allowed transitions.

The crystalline electric field interaction in the $R_{2-x}Ce_xCuO_4$ (R=Nd,Pr) compounds has been extensively studied [39-45]. For R=Pr, the CEF potential decomposes the J=4 ground-state multiplet into a singlet ground state (Γ_4), a first excited doublet at 18 meV (Γ_5) and various other excited states around 80 meV ($2\times\Gamma_1,\Gamma_2,\Gamma_3,\Gamma_5$) (see Fig. 2). For R=Nd, the 10-fold degeneracy of the J=9/2 ground state multiplet is split by the CEF potential into a magnetic ground state doublet ($\Gamma_6^{(1)}$) and four Kramers doublets at 14 meV ($\Gamma_7^{(1)}$), 21 meV ($\Gamma_6^{(2)}$), 27 meV ($\Gamma_7^{(2)}$), and 93 meV ($\Gamma_6^{(3)}$) (Fig. 2).

For Nd_2CuO_4 there was some controversy in the literature about the symmetry of the CEF levels at 21 meV and 27 meV, respectively. One group claimed that the level at 21 meV has a Γ_6 and the one at 27 meV a Γ_7 representation [42,46]. Another group however recently assigned the level at 21 meV to Γ_7 and the one at 27 meV to a Γ_6 symmetry [41]. Because the ground state doublet has Γ_6 symmetry, the two cases represent different polarisations for the two CEF transitions. This controversy was resolved by a Raman study of intermultiplet crystal-field excitations in Nd_2CuO_4 [45] where it was found that the level at 21 meV has a Γ_6 representation. It was argued in that paper that the major difficulty of neutron scattering in determining a set of CEF parameters is the limited number of detected levels. However, as can be seen from Eq. (11), neutron scattering experiments on single crystals give enough information to unambiguously determine the symmetry of the different excitation levels. A transition between two Γ_6 levels has longitudinal character, i.e. $<\Gamma_n|J_{x,y}|\Gamma_m>=0$ and $<\Gamma_n|J_z|\Gamma_m>\neq0$, whereas a transition between a Γ_6 and a Γ_7 level has transverse character, i.e. $<\Gamma_n|J_{x,y}|\Gamma_m>\neq0$ and $<\Gamma_n|J_z|\Gamma_m>=0$. The polarisation factor $(1-Q_\alpha^2/Q^2)$ in Eq. (11) allows these cases to be distinguished. Fig. 3 shows a measurement at two different positions in reciprocal space [47]. It is obvious that the transition at 21 meV has longitudinal character, because it cannot be observed for $\mathbf{Q}$ parallel [001]. The transition at 27 meV on the other hand has transverse character. It follows that the level at 21 meV corresponds to a Γ_6 representation, while the level at 27 meV has Γ_7 symmetry.

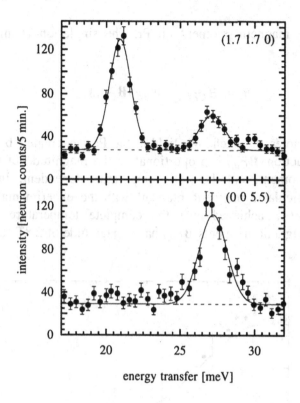

Figure 3. Energy spectra of neutrons scattered from Nd_2CuO_4 at 4 K for two different **Q** positions.

In the Ce-doped compounds, the CEF level scheme is similar. However, significant line-broadening has been observed, presumably due to an inhomogenous charge distribution in the planes (see J. Mesot and A. Furrer, this volume).

As mentioned in Section 2.2, the crystal field effects induce single-ion anisotropies. In the case of R=Nd, where the ground state is a doublet, this anisotropy is given by $|\langle\Phi_0|J_z|\Phi_1\rangle|^2/|\langle\Phi_0|J_{x,y}|\Phi_1\rangle|^2\approx0.2$, where $|\Phi_0\rangle$ and $|\Phi_1\rangle$ denote the wave functions of the ground state $\Gamma_6^{(1)}$ doublet which is split by the exchange field.

3.2 RARE-EARTH MAGNETIC ORDER IN $R_{2-x}Ce_xCuO_4$ (R=Nd,Pr)

From the CEF interaction in Pr_2CuO_4, one would expect that the magnetic moment on the Pr site is quenched due to the nonmagnetic singlet ground state. However, a small moment on the Pr-site has been detected by neutron diffraction experiments [18]. This can be explained by the Pr-Cu exchange interaction which creates a staggered magnetic field at the Pr-site and induces a small magnetic moment on Pr. The single-ion Hamiltonian can then be written as

$$\mathcal{H}_i = \mathcal{H}_{CEF,i} - \mu_B g_J \mathbf{B}_{Cu,i} \hat{\mathbf{J}}_i \qquad (15)$$

where $\mathbf{B}_{Cu,i}$ denotes the exchange field at the Pr site, created by the Pr-Cu exchange interaction. $\mathbf{B}_{Cu,i}$ is proportional to the magnitude of the ordered Cu magnetic moment, and therefore temperature-dependent. In Fig. 4 we compare the calculated magnetic moment with the experimental results. A good agreement is achieved over the complete temperature range. For saturated Cu magnetic moments $\mathbf{B}_{Cu,i}$ has a magnitude of 2.8T for all i.

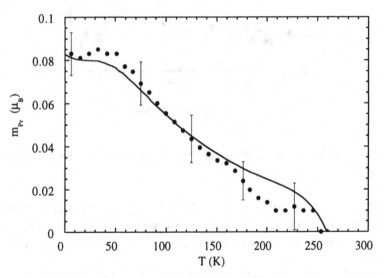

Figure 4. Temperature dependence of the Pr magnetic moment in Pr_2CuO_4 [18]. The line corresponds to a calculation explained in the text.

For R=Nd, the situation is different. Due to the doublet ground state, one expects ordering of the Nd spins. Specific heat measurements indicated indeed ordering of the Nd spins at a temperature of about 1.5 K [48-50]. However, it is not correct to talk about a Nd ordering temperature. The order of the Nd moments gradually builds up over a large temperature range as revealed by neutron diffraction [18,51] and x-ray magnetic scattering [52]. As in the case of R=Pr, this is explained by a Cu-Nd exchange interaction which creates a staggered magnetic field at the Nd site. In the case of R=Nd, the Nd-Nd exchange interaction is however not negligible. Fig. 5 compares the experimental and calculated values of the Nd magnetic moment. It is obvious that the ordering of the Nd spins can not fully be explained by the presence of $\mathbf{B}_{Cu,i}$. We can include the Nd-Nd exchange interaction in a simple mean-field model, where the magnetic moment $m_{Nd}(T)$ of Nd has to be determined self-consistently by

$$m_{Nd}(T) = m_{Nd}(T=0) \cdot \tanh\{(h_{Cu} + h_{Nd} \cdot \frac{m_{Nd}(T)}{m_{Nd}(T=0)})/k_B T\} \qquad (16)$$

h_{Cu} denotes the energy-splitting of the Nd-ground state doublet, induced by $\mathbf{B}_{Cu,i}$. h_{Nd} is the splitting due to the mean Nd-Nd exchange field at the Nd-site for saturated Nd moments. In this expression, we neglect crystal field effects. These effects are however negligible in this temperature range. For h_{Cu} we obtain a value of 0.5meV, which corresponds to $\mathbf{B}_{Cu,i}$ =3.1T, and for h_{Nd} we obtain -0.25meV. This indicates that the Nd-Nd exchange interaction is opposite to the Nd-Cu exchange. However, because Eq. (16) completely neglects any fluctuations, the size of h_{Nd} can not be taken too seriously.

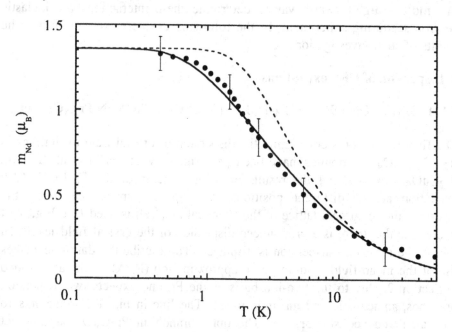

Figure 5. Temperature dependence of the Nd magnetic moment in Nd$_2$CuO$_4$ [18]. The line corresponds to a calculation explained in the text;---: h_{Cu}=0.5 meV and h_{Nd}=0; —: h_{Cu}=0.5 meV and h_{Nd}=-0.25meV

In the case of R=Pr, it is obvious why the R-R exchange interactions does not affect the magnetic order. Because the Pr-moment is more than one order of magnitude smaller than the Nd moment, one expects that the Pr-Pr exchange field at the Pr site is negligible. This exchange interaction will however lead to a dispersion of the crystal field levels, as demonstrated in section 2.2.

Concerning the Ce-doped sample, the doping dependence of the Pr magnetic moment in $Pr_{2-x}Ce_xCuO_4$, can be deduced from measurements performed on samples with x=0, 0.08 and 0.125 [17]. As expected from the decrease of $B_{Cu,i}$, the Pr-moment decreases with increasing Ce-concentration. In Ce-doped $Nd_{2-x}Ce_xCuO_4$, long-range magnetic order of Nd is observed up to x=0.13. Short range order persists up to x=0.17 [53]. The magnitude of the saturated Nd-moment as a function of doping has not been established so far. One would however not expect a doping dependence, because the magnitude is only defined by the wave-functions of the Nd ground-state doublet.

As shown above, the R-R exchange interactions seem to play an important role in the magnetic behaviour of the system, at least for R=Nd. The most straightforward way to determine these interactions are inelastic neutron scattering experiment. In the following chapters, we will discuss the results of such investigations.

4. Dispersion of CEF excitations

4.1. DISPERSION OF THE Γ_4-Γ_5 Pr CEF EXCITATION IN $Pr_{2-x}Ce_xCuO_4$

The first experiments concerning the dispersion of crystal field excitations in $R_{2-x}Ce_xCuO_4$ compounds have been performed by Sumarlin et al. [29] on Pr_2CuO_4. Fig. 6 shows the results of measurements of the Γ_4-Γ_5 Pr CEF excitation at two different positions in reciprocal space. To index the positions the reciprocal lattice of the chemical unit cell is used (see Fig.7). It is obvious that there is a pronounced dispersion of the crystal field levels. In Fig. 8, the measured dispersion is displayed. To describe the data one makes use of the mean-field random phase approximation (RPA) model as outlined in Section 2. Due to the two-ion basis of the Pr, one expects two dispersion branches, an acoustic and an optic mode. The line in Fig. 8 corresponds to the calculated acoustic branch. The optic branch in Pr_2CuO_4 has not yet been unambiguously determined. Nevertheless it was possible to obtain reliable values for the Pr-Pr exchange constants. The following coupling constants indicated in Fig.1 were derived from a fitting procedure with $M=|<\Gamma_4|J_{x,y}|\Gamma_5>|=2.63$ [40]:

$J_1=-52\pm3\mu eV$, $J_2=-17\pm2\mu eV$, $J_3=-15\pm2\mu eV$

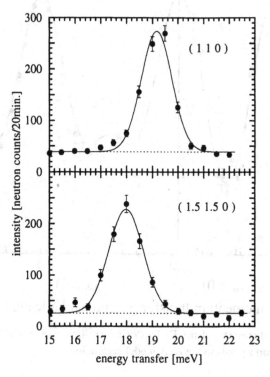

Figure 6. Energy spectra of neutrons scattered from Pr_2CuO_4 at 10 K for two different **Q** positions, taken from [29].

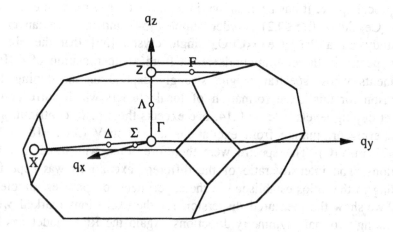

Figure 7. Brillouin zone of the body centred tetragonal lattice.

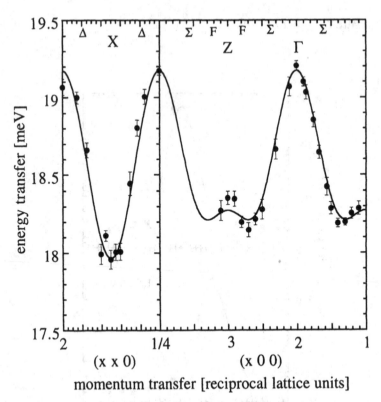

Figure 8. Measured dispersion of the Γ_4-Γ_5 Pr CEF excitation in Pr_2CuO_4 at 10 K, taken from [29]. The line corresponds to the RPA model calculation.

To examine the change of the coupling constants upon Ce-doping, experiments on a $Pr_{1.86}Ce_{0.14}CuO_4$ single crystal have been performed [47]. Fig. 9 shows the results of the measurements at two different positions in reciprocal space. It has been shown in an inelastic neutron scattering study on $Pr_{2-x}Ce_xCuO_4$ ($0 \leq x \leq 0.2$) powder samples [54] and in a Raman crystal-field study on a $Pr_{1.85}Ce_{0.15}CuO_4$ single crystal [55] that the observed energy spectra in these compounds result from a superposition of different components whose spectral weights strongly depend on the doping level. The reason for this is the formation of local clusters which correspond to different doping levels. For $x \approx 0.14$, one expects three main contributions to the scattering originating from excitations at ≈ 15 meV (A) , ≈ 18 meV(B) and ≈ 21 meV(C). The spectra were therefore fitted with three individual excitations. The intensity ratio of the different excitations was kept fixed according to the ratios established in the experiments on powder samples. In Fig. 10 we show the measured dispersion for the excitations marked with A and B along two main symmetry directions. Again the RPA model has been used to describe the data. In this case, because of the dilution of the Pr spin systems, the coupling corresponds only to an effective spin exchange interaction which is reduced in comparison with the actual exchange coupling. The coupling constants derived from a least-square fitting procedure were:

$$J_1 = -27 \pm 5 \mu eV, \quad J_2 = -10 \pm 3 \mu eV, \quad J_3 = -16 \pm 3 \mu eV$$

From these measurements one can conclude that the doping of Ce ions into Pr_2CuO_4 only slightly reduces the exchange couplings. In a first approximation one would expect a reduction of 7% according to the fraction of substituted Pr ions in $Pr_{1.86}Ce_{0.14}CuO_4$. The coupling that is mostly affected by doping is the exchange J_1 mediated by the copper-oxide planes. Because of the big errors it is however not possible to conclude that the doped electrons in the copper oxide planes affect the exchange interaction. Nevertheless, our results indicate that we would not expect a drastic change of the exchange coupling upon Ce doping in the related $Nd_{2-x}Ce_xCuO_4$ compounds.

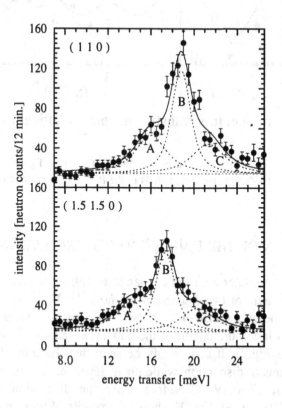

Figure 9. Energy spectra of neutrons scattered from $Pr_{1.86}Ce_{0.14}CuO_4$ at 10 K at the two different **Q** positions (1 1 0) and (1.5 1.5 0).

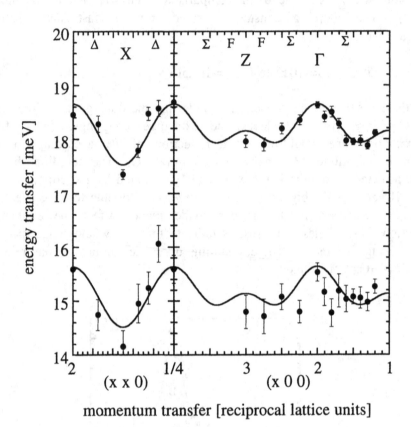

Figure 10. Measured dispersion of the acoustic branch of the Γ_4 - Γ_5 - Pr CEF excitation in $Pr_{1.86}Ce_{0.14}CuO_4$ at 10 K. The line corresponds to the RPA model calculation as explained in the text.

4.2. DISPERSION OF THE $\Gamma_6^{(1)}$-$\Gamma_6^{(2)}$ Nd CEF EXCITATION IN Nd_2CuO_4

The first attempt to determine the exchange coupling constants in Nd_2CuO_4 was the measurement of the dispersion of the $\Gamma_6^{(1)}$-$\Gamma_6^{(2)}$ Nd CEF excitation [56]. Fig. 11 shows experiments at two different positions in reciprocal space. Again, to index the positions the reciprocal lattice of the chemical unit cell is used (see Fig.7). It can clearly be seen that the transition at 21 meV shows a pronounced dispersion while no q-dependence can be detected for the transition at 27 meV. Therefore, only the dispersion of the former transition was determined. Fig. 12 shows the results of these measurements at four different Q vectors. The intensities of the acoustic and the optic branch continuously change when going along the [001] direction, as predicted by the RPA model calculation (see Eq. (14)). Fig. 13 shows the measured dispersion along the three main symmetry directions. The line drawn corresponds to the RPA model calculation and perfectly agrees with the measurements. Also the calculated intensity ratio between the two branches along the [001] direction agrees well with the measurements, as shown in Fig. 14. The three coupling constants J_1, J_2, and J_3 indicated in Fig. 1 were included in the calculation. The following coupling constants for this CEF excitation, with $M=|<\Gamma_6^{(1)}|J_z|\Gamma_6^{(2)}>|=2.60$, have been derived:

$J_1= -7\pm2\mu eV$, $J_2= -19\pm1\mu eV$, $J_3= -2.5\pm1\mu eV$.

It is also obvious from Eq. (10) why it was not possible to observe a $\mathbf{q}$-dependence for the $\Gamma_6^{(1)}$-$\Gamma_7^{(2)}$ CEF transition at 27 meV. Using the above exchange parameters and $M=|<\Gamma_6^{(1)}|J_{x,y}|\Gamma_7^{(2)}>|=1.74$, the resulting dispersion of less than 0.4 meV could not be determined within the experimental resolution.

In the discussion of the dispersion of the crystal field excitations, we neglected the influence of the Nd-Cu exchange interaction on the dispersion relation. As we will discuss later, for the low-energy Nd-spin excitations it is justified to assume that the only consequence of the Cu-Nd exchange interaction is the creation of a staggered magnetic field at the Nd site. However, in principle the case is different for the CEF excitation at energies where Cu spin-waves are present. Therefore, numerical calculations of the wave-vector dependent susceptibility including Cu-Cu, Cu-Nd and Nd-Nd exchange interaction were performed [57]. They show that the dispersion of the CEF excitation is indeed affected at the X-point, where the Cu spin-wave excitations cross the CEF level. The anomaly at this point is however only small, but nevertheless indications for its presence have been found experimentally [58]. Apart from that anomaly, the results of the calculations correspond to Eq. (10).

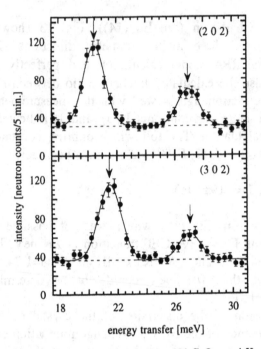

Figure 11. Energy spectra of neutrons scattered from Nd_2CuO_4 at 4 K for two different **Q** positions.

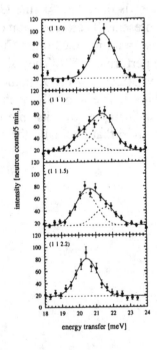

Figure 12. Energy spectra of neutrons scattered from Nd_2CuO_4 at 4 K for different **Q**=(11*l*).

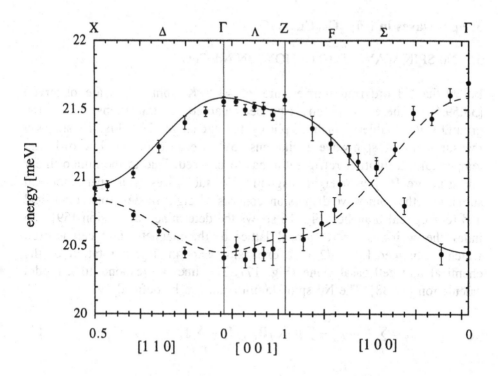

Figure 13. Measured dispersion of the $\Gamma_6^{(1)}$ - $\Gamma_6^{(2)}$ - Nd CEF excitation in Nd_2CuO_4 at 4 K. The lines correspond to the RPA model calculation.

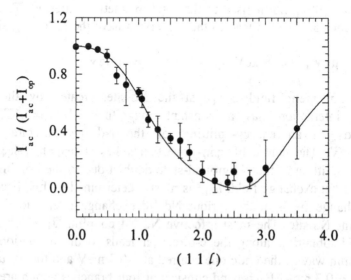

Figure 14. Measured intensity of the acoustic branch at (11 *l*) positions, normalised to the total magnetic scattering. The line corresponds to the RPA model calculation.

5. Spin waves in $Nd_{2-x}Ce_xCuO_4$

5.1. Nd SPIN WAVE EXCITATIONS IN Nd_2CuO_4

Below the Nd ordering temperature of ≈ 1.5 K, spin waves are observed [38,59,60]. These excitations can be regarded as transitions within the ground-state doublet, which is split by the exchange field. Fig. 15 shows a measurement of spin-wave excitations [59]. Because of the low ordering temperature, a dilution refrigerator has to be used. Due to the noncollinear AF structure [61] with eight magnetic Nd sublattices (four per chemical sublattice) the spin wave dispersion consists of eight modes, four acoustical and four optical branches. Fig. 16 shows the determined dispersion [59]. To index the positions the reciprocal lattice of the magnetic unit cell is used which is obtained by a $\sqrt{2} \times \sqrt{2}$ expansion and 45 degree rotation of the chemical unit cell basal plane (Fig. 17). The lines correspond to a model calculation [37,38]. The Nd-spin Hamiltonian can be defined by

$$\mathcal{H} = \sum_i \mathcal{H}_{CEF,i} - \sum_i \mu_B g_J \mathbf{B}_{Cu,i}\hat{\mathbf{J}}_i - \sum_{i>j} \hat{\mathbf{J}}_i \bar{\bar{J}}(\mathbf{r}_i - \mathbf{r}_j)\hat{\mathbf{J}}_j \qquad (17)$$

The model gives a good description of the energies as well as of the intensities. Fig. 18 shows the measured and calculated intensity of the optic branch along [001], normalised to the total magnetic scattering. The Nd-Nd exchange constants which lead to the dispersion have the following size:

$J_1 = -33$ μeV , $J_2 = 18$ μeV, $J_3 = -8$ μeV, $J_4 = -4$ μeV

The mean magnetic field $\mathbf{B}_{cu,i}$ at the Nd-site, created by the Nd-Cu exchange interaction has a constant magnitude $|\mathbf{B}_{Cu,i}| = 3.1$T, which corresponds to an energy-splitting of the Nd ground-state doublet $h_{Cu} \approx 0.5$ meV. The centre of spin-wave bands lies at $h_{MF} = h_{Cu} + h_{Nd}$, where h_{MF} is the splitting of the Nd ground-state doublet due to the combined Nd-Cu and Nd-Nd exchange field. h_{MF} is mostly determined by h_{Cu} because the mean exchange fields of the various Nd-Nd exchange interactions roughly cancel at the Nd-site. The most effective Nd-Nd coupling J_1 which connects the two Nd sublattices along the c-direction leads to the formation of two bands of spin waves, the optic one centred at ≈ 0.3 meV and the acoustic one centred at ≈ 0.7 meV. Each band consists of four branches which are split by ≈ 0.2 meV by the additional exchange couplings J_2, J_3, and J_4. The optic band is lower than the acoustic one because the antiferromagnetic coupling

J_1 is opposed to the ferromagnetic ordering along the c direction induced by the Nd-Cu exchange interaction.

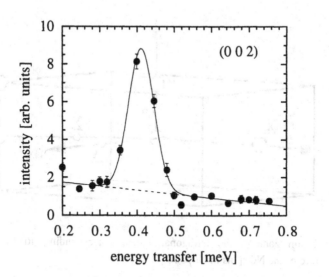

Figure 15. Energy spectra of neutrons scattered from Nd_2CuO_4 at T=50 mK at (0 0 2).

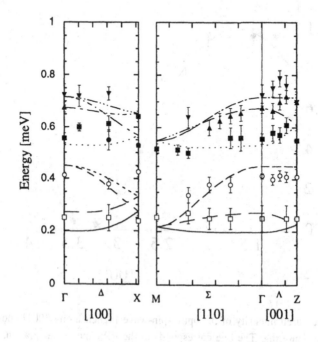

Figure 16. Measured dispersion of the Nd spin waves in Nd_2CuO_4 at T=50 mK. The lines correspond to the model calculation with the exchange constants given in the text.

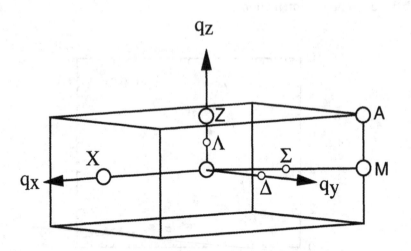

Figure 17. Brillouin zone of the tetragonal lattice corresponding to the noncollinear magnetic structure of the Nd spins in Nd_2CuO_4.

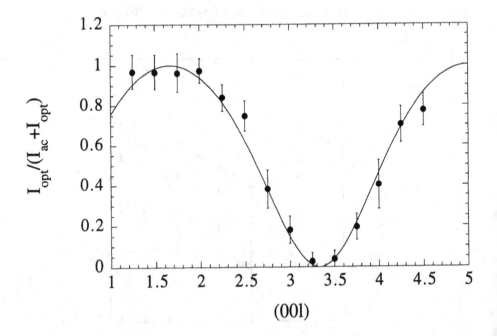

Figure 18. Measured intensity of the optic spin-wave branch along [001], normalised to the total magnetic scattering. The line corresponds to the RPA model calculation.

Note that these exchange constants, especially J_2, differ from the ones obtained previously [47] with another model [62]. Here a description is used where the difference of the single-ion susceptibilities in the eight magnetic sublattices is explicitly taken into account. Recently a model has been presented where the interactions of the Cu spins have been included [63]. The two dispersions deviate however only close to the Gamma point. Moreover, four-site exchange interactions may lead to Cu-spin gaps [31] which have indeed been observed at 12 meV and 14 meV at 5 K [29,32,33]. This justifies the assumption that the only consequence of the Cu-Nd exchange interaction is the creation of a staggered magnetic field at the Nd site.

A broadening of lines observed at some reciprocal lattice positions indicates that the degeneracy of the excitations may be further lifted than predicted by the model calculation. Possible reasons for this could be a small tilting ($\approx 1°$) of the Nd- with respect to Cu-moments [37], or higher-order exchange interactions.

The comparison of the exchange couplings for the $\Gamma_6^{(1)}$-$\Gamma_6^{(2)}$ CEF excitation with the coupling parameters derived from the measurements of the spin wave excitations shows that the exchange interaction is dependent on the initial and the final state of the CEF excitation, i.e. the exchange interaction has to be regarded as a tensor $J_{ij}(\sigma_i, \sigma_j, \sigma_i', \sigma_j')$ where σ denotes the initial and σ' the final state of a CEF excitation. Such a behaviour has so far only been observed in the dimer compound $Cs_3Ho_2Br_9$ [64].

5.2. INFLUENCE OF Ce-DOPING ON THE MAGNETIC EXCITATIONS IN $Nd_{2-x}Ce_xCuO_4$

A study of the doping dependence of the magnetic excitations of Nd in $Nd_{2-x}Ce_xCuO_4$ has been presented in Ref. [65]. Fig. 19 shows inelastic neutron scans measured at the reciprocal lattice vector $\mathbf{Q}=(0\ 0\ 1.5)$ for the differently doped crystals. At this $\mathbf{Q}$-value the spin-wave model calculation [37] predicts a dominant scattering of an optic excitation. Additional scattering with less than 5% of the total magnetic cross section is expected to originate from an acoustic branch. The spectra were therefore fitted with one inelastic line, using a damped harmonic oscillator convoluted with the resolution function of the spectrometer. The additional scattering at higher energies originates from the weak acoustic excitation. It can be seen that the energy of the optic spin-wave excitation is well defined and the line widths are only slightly broader than the experimental resolution.

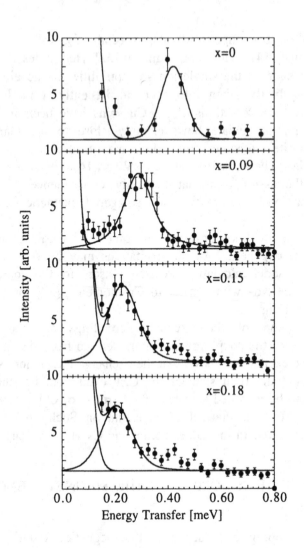

Figure 19. Spectra of neutrons scattered from $Nd_{2-x}Ce_xCuO_4$ (x=0,0.09,0.15,0.18) at T=50mK and **Q**=(0 0 1.5).

Measurements at many **Q**-positions, varying between fifteen (x=0.18) and sixty-five (x=0) were performed. Fig. 20 shows the excitation energies as a function of doping for a selection of four different momentum transfers. It is obvious that there is a pronounced softening of the spin waves upon doping. Almost the same amount of softening is observed for all the measured excitations. This shows that the doping process leads to an overall shift of the spin-wave dispersion to lower energies, without significantly altering the energy splitting between the different excitation branches. Because this energy splitting is solely given by the Nd-Nd exchange interactions, this demonstrates that the Nd-Nd exchange couplings are nearly unaffected by the doping process. This is not surprising, because even for the x=0.18 compound there are only 9% Ce-ions present, which is not enough to drastically change the effective Nd-Nd interaction. The only explanation for the softening is therefore a decrease of h_{Cu}, the energy-splitting of the Nd-ground state doublet caused by the Nd-Cu exchange interaction. This is easily explained by the decrease of the Cu magnetic moments induced by the doping of electrons into the copper-oxide planes.

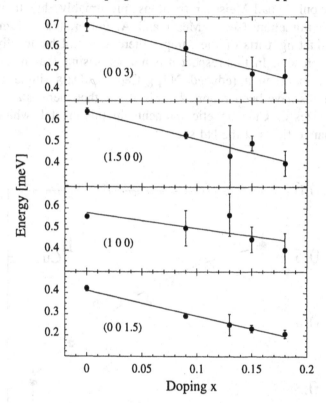

Figure 20. Energies of the spin-wave excitations of Nd in $Nd_{2-x}Ce_xCuO_4$ as a function of Ce-doping at different positions in reciprocal space. The **Q**-values are given in units of the reciprocal lattice of the magnetic unit cell. Lines are guides to the eye.

The reduction of h_{Cu} can directly be calculated by fitting the measured dispersions of the differently doped samples with a spin-wave model. As mentioned before, the Nd-Nd exchange parameters are almost independent of the doping. The calculated reduction of h_{Cu} is then nearly equal to the amount of softening of the spin waves, regardless which model is used. The values obtained for h_{Cu} are shown in Fig. 21. One can see that for x=0.18 the Nd-Cu exchange field h_{Cu} has a size of 0.3 meV. This seems to be rather high, because no long-range magnetic order of the Cu-ions is observed in the x=0.18 sample. One possible explanation is the presence of locally ordered Cu moments, which has indeed been deduced from μSR experiments on an x=0.2 sample [66], The situation may be similar for reduced, superconducting samples, where no long-range antiferromagnetic order of Cu is observed for x≥0.14 [53]. There one would expect a more drastic softening of the spin waves. However, the superconducting samples usually show only small Meissner fractions, presumably due to the phase-separation phenomenon (see J. Mesot and A. Furrer, this volume) In the nonsuperconducting parts of the sample there will still be locally ordered Cu-moments present. In this respect it is not surprising that in an inelastic neutron experiment on a reduced $Nd_{1.85}Ce_{0.15}CuO_{4-\delta}$ single crystal[67] large line widths have been reported. This shows that there is a distribution of different sizes of Cu magnetic moments in this crystal, which leads to varying exchange fields at the Nd site.

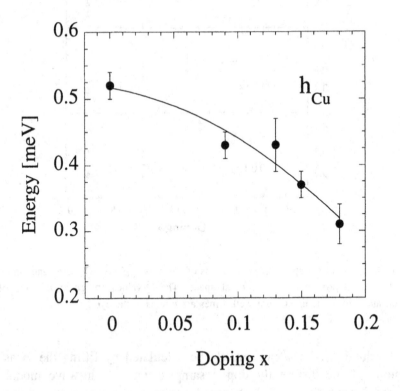

Figure 21. Doping dependence of the exchange field h_{Cu} at the Nd site created by the Nd-Cu exchange interaction. The line is a guide to the eye.

5.3. INFLUENCE OF THE Nd MAGNETIC CORRELATIONS ON THE SPECIFIC HEAT IN $Nd_{2-x}Ce_xCuO_4$

The behaviour of the specific heat C can easily be calculated from the density of states $g(\omega)$ of the spin-wave excitations:

$$C = k_B \int_0^{\omega_{max}} \left(\frac{\hbar\omega}{2k_BT} \right)^2 \sinh^{-2}\left(\frac{\hbar\omega}{2k_BT} \right) g(\omega)\, d\omega \qquad (18)$$

The softening shifts the spin-wave branches and accordingly the peaks in the density of states to lower values. At a critical Nd-Cu exchange field, corresponding to $h_{Cu}=0.38$ meV in the calculation, one spin-wave branch shows a complete softening. In Fig. 22 the calculated $\gamma=C/T$-values for three different h_{Cu} are shown and compared with the experimental results of Brugger et al [6]. Note that the spin-wave model is only valid for $T<<h_{MF}/k_B\approx5K$, where h_{MF} is the combined Nd-Cu and Nd-Nd exchange field. For values of h_{Cu} lower than 0.38 meV, the complete softening of one spin-wave branch is expected to lead to an instability of the Nd spin system. The Nd moments will tend to align antiferromagnetically along the c-axis, which follows from the fact that the soft mode is optic. Indications for such an instability have indeed been found in a diffraction experiment on an x=0.17 sample [53]. These measurements show that even in a highly doped sample (x=0.17) the Nd-moment is gradually building up over a large temperature range, as reflected in the temperature dependence of the (1/2 1/2 3) reflection (see Fig. 23). When the Nd moment is large enough so that the Nd-Nd exchange interaction becomes important, one observes broad magnetic scattering at (1/2 1/2 5/2) positions, which corresponds to a doubling of the unit cell along the c direction. But presumably due to the effect of disorder, no new long-range spin order is established. For $h_{Cu}<0.38$ meV it is therefore not appropriate to calculate the specific heat as in a truly spin-ordered state. The observed continuation of the spin-wave softening upon further doping will however lead to an even higher density of states at low energies, and consequently to the observed higher γ-values.

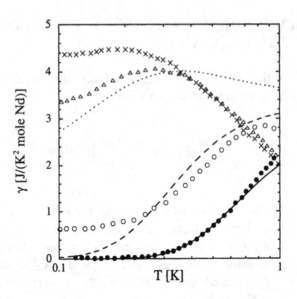

Figure 22. Specific-heat coefficient $\gamma = C/T$. Symbols correspond to experimental data of Brugger et al [6]: ●: x=0, ○: x=0.1, △: x=0.15, ✕: x=0.2. The lines show calculations described in the text: —: h_{Cu}=0.52 meV, — —: h_{Cu}=0.42 meV, ----: h_{Cu}=0.38 meV.

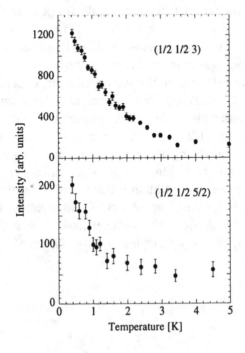

Figure 23. Temperature dependence of the (1/2 1/2 3) and the (1/2 1/2 5/2) magnetic Bragg peak for a $Nd_{2-x}Ce_xCuO_4$ crystal with x=0.17 [53].

6. Conclusions

We have shown how the electronic ground state of the R^{3+} ions in $R_{2-x}Ce_xCuO_4$ can be determined by neutron scattering. The magnetic coupling between the R^{3+} ions gives rise to a dispersion behaviour of the CEF- and spin-wave excitations. The RPA model was used to analyse these dispersion relations, which yields direct information on the exchange couplings constants. It was shown that the exchange interaction between the R ions is only slightly affected by the Ce-doping. For R=Nd, there is however a strong decrease of the exchange field at the Nd site created by the Nd-Cu exchange interactions. This leads to a strong softening of the spin-wave branches upon Ce-doping. Consequently a high spin-wave density of states is created at low energies, which leads to giant $\gamma=C/T$-values at low temperatures.

Acknowledgements

Financial support by the Swiss National Science Foundation is gratefully acknowledged. For their participation in the experimental and theoretical work as well as for numerous stimulating discussions we wish to thank our colleagues P. Vorderwisch, B. Roessli, P. Thalmeier and Tapan Chatterji. We are grateful to A. Metz for important suggestions.

References

1. Y. Tokura, H. Takagi and S. Uchida, Nature **337** (1989) 345.
2. H. Takagi, S. Uchida and Y. Tokura, Phys. Rev. Lett. **62** (1989) 1197.
3. O.K. Singh, B.D. Padalia, O. Prakash, K. Suba, A.V. Narlikar and L.C. Gupta, Physica C **219** (1994) 156.
4. J.L. Peng, E. Maiser, T. Venkatesan, R.L. Greene and G. Czjzek, Phys. Rev. B **55** (1997) R6145.
5. J. Fournier, X. Jiang, W. Jiang, S.N. Mao, T. Venkatesan, C.J. Lobb and R.L. Greene, Phys. Rev. B **56** (1997) 14149.
6. T. Brugger, T. Schreiner, G. Roth, P. Adelmann and G. Czjzek, Phys. Rev. Lett. **71** (1993) 2481.
7. P. Fulde, V. Zevin and G. Zwicknagl, Z. Phys. B **92** (1993) 133.
8. P. Fulde, Physica B **230** (1997) 1.
9. R. SaezPuche, M. Norton, T.R. White and W.S. Glaunsinger, J. Solid State Chem. **50** (1983) 281.
10. C.L. Saeman, N.Y. Ayoub, T. Bjornholm, E.A. Early, S. Ghamaty, B.W. Lee, J.T. Markert, J.J. Neumeier, P.K. Tsai and M.B. Maple, Physica C **159** (1989) 391.

11. M.J. Rosseinsky and K. Prassides, Physica C **162-164** (1989) 522.

12. D.E. Cox, A.I. Goldman, M.A. Subramanian, J. Gopalakrishnan and A.W. Sleight, Phys. Rev. B **40** (1989) 6998.

13. P. Allenspach, S.-W. Cheong, A. Dommann, P. Fischer, Z. Fisk, A. Furrer, H.R. Ott and B. Rupp, Z. Phys. B **77** (1989) 185.

14. S. Skanthakumar, H. Zhang, T.W. Clinton, W.-H. Li, J.W. Lynn, Z. Fisk and S.-W. Cheong, Physica C **160** (1989) 124.

15. J. Akimitsu, H. Sawa, T. Kobayashi, H. Fujiki and Y. Yamada, J. Phys. Soc. Jpn. **58** (1989) 2646.

16. Y. Endoh, M. Matsuda, K. Yamada, K. Kakurai, Y. Hidaka, G. Shirane and R.J. Birgeneau, Phys. Rev. B **40** (1989) 7023.

17. T.R. Thurston, M. Matsuda, K. Kakurai, K. Yamada, Y. Endoh, R.J. Birgeneau, P.M. Gehring, Y. Hidaka, M.A. Kastner, T. Murakami and G. Shirane, Phys. Rev. Lett. **65** (1990) 263.

18. M. Matsuda, K. Yamada, K. Kakurai, H. Kadowaki, T.R. Thurston, Y. Endoh, Y. Hidaka, R.J. Birgenau, M.A. Kastner, P.M. Gehring, A.H. Moudden and G. Shirane, Phys. Rev. B **42** (1990) 10098.

19. H. Yoshizawa, S. Mitsuda, H. Mori, Y. Yamada, T. Kobayashi, H. Sawa and J. Akimitsu, J. Phys. Soc. Jpn. **59** (1990) 428.

20. S. Skanthakumar and J.W. Lynn, Physica C **170** (1990) 175.

21. S. Skanthakumar, H. Zhang, T.W. Clinton, I.W. Sumarlin, W.-H. Li, J.W. Lynn, Z. Fisk and S.-W. Cheong, J. Appl. Phys. **67** (1990) 4530.

22. T. Chattopadhyay, P.J. Brown and U. Köbler, Physica C **177** (1991) 294.

23. M.J. Rosseinsky, K. Prassides and P. Day, Inorg. Chem. **30** (1991) 2680.

24. A.G. Gukasov, S.Y. Kokovin, V.P. Plakhty, I.A. Zobkalo, S.N. Barilo and D.I. Zhigunov, Physica B **180 & 181** (1992) 455.

25. S. Skanthakumar, J.W. Lynn, J.L. Peng and Z.Y. Li, J. Magn. Mag. Mat. **104-107** (1992) 519.

26. D. Petitgrand, L. Boudarène, P. Bourges and P. Galez, J. Magn. Magn. Mat. **104-107** (1992) 585.

27. G.M. Luke, L.P. Le, B.J. Sternlieb, Y.J. Uemura, J.H. Brewer, R. Kadono, R.F. Kiefl, S.R. Kreitzman, T.M. Riseman, C.E. Stronach, M.R. Davis, S. Uchida, H. Takagi, Y. Tokura, Y. Hidaka, T. Murakami, J. Gopalakrishnan, A.W. Sleight, M.A. Subramanian, E.A. Early, J.T. Markert, M.B. Maple and C.L. Saeman, Phys. Rev. B **42** (1990) 7981.

28. J. Akimitsu, J. Amano, M. Yoshinari and M. Kokubun, Hyperfine Interactions **85** (1994) 187.

29. I.W. Sumarlin, J.W. Lynn, T. Chattopadhyay, S.N. Barilo and D.I. Zhigunov, Phys. Rev. B **51** (1995) 5824.

30. P. Bourges, H. Casalta, A.S. Ivanov and D. Petitgrand, Phys. Rev. Lett. **79** (1997) 4906.

31. V.L. Sobolev, H.L. Huang, Y.G. Pashkevich, M.M. Larionov, I.M. Vitebsky and V.A. Blinkin, Phys. Rev. B **49** (1994) 1170.

32. P. Bourges, A.S. Ivanov, D. Petitgrand, J. Rossat-Mignod and L. Boudarene, Physica B **186** (1993) 925.
33. A.S. Ivanov, P. Bourges, D. Petitgrand and J. Rossat-Mignod, Physica B **213** (1995) 60.
34. B.G. Wybourne: *Spectroscopic Properties of Rare Earths*, Interscience, New York, 1965.
35. J. Jensen, J.G. Houmann and H.B. Møller, Phys. Rev. B **12** (1975) 303.
36. J. Jensen and A.R. Mackintosh: *Rare Earth Magnetism, Structures and Excitations*, Oxford Science Publications, Clarendon Press, Oxford, 1991.
37. P. Thalmeier, preprint
38. A. Metz, unpublished
39. P. Allenspach, A. Furrer, R. Osborn and A.D. Taylor, Z. Phys. B **85** (1991) 301.
40. A.T. Boothroyd, S.M. Doyle, D.M. Paul and R. Osborn, Phys. Rev. B **45** (1992) 10075.
41. C.K. Loong and L. Soderholm, Phys. Rev. B **48** (1993) 14001.
42. U. Staub, P. Allenspach, A. Furrer, H.R. Ott, S.-W. Cheong and Z. Fisk, Solid State Commun. **75** (1990) 431.
43. P. Hoffmann, M. Loewenhaupt, S. Horn, P.v. Aken and H.-D. Jostarndt, Physica B **163** (1992) 10075.
44. M. Loewenhaupt, P. Fabi, S. Horn, P. v.Aken and A. Severing, J. Magn. Magn. Mat. **140-144** (1995) 1293.
45. S. Jandl, P. Dufour, T. Strach, T. Ruf, M. Cardona, V. Nekvasil, C. Chen and B.M. Wanklyn, Phys. Rev. B **52** (1995) 15558.
46. A. Furrer, P. Allenspach, J. Mesot and U. Staub, Physica C **168** (1990) 609.
47. W. Henggeler, T. Chattopadhyay, B. Roessli, P. Vorderwisch, P. Thalmeier, D.I. Zhigunov, S.N. Barilo and A. Furrer, Phys. Rev. B **55** (1997) 1269.
48. M.F. Hundley, J.D. Thompson, S.-W. Cheong, Z. Fisk and S.B. Oseroff, Physica C **158** (1989) 102.
49. S. Ghamaty, B.W. Lee, J.T. Markert, E.A. Early, T. Bjornholm, C.L. Saeman and M.B. Maple, Physica C **160** (1989) 217.
50. M.B. Maple, N.Y. Ayoub, T. Bjornholm, E.A. Early, S. Ghamaty, B.W. Lee, J.T. Markert, J.J. Neumeier and C.L. Saeman, Physica C **162-164** (1989) 296.
51. J.W. Lynn, I.W. Sumarlin, S. Skanthakumar, W.-H. Li, R.N. Shelton, J.L. Peng, Z. Fisk and S.-W. Cheong, Phys. Rev. B **41** (1990) 2569.
52. J.P. Hill, A. Vigliante, D. Gibbs, J.L. Peng and R.L. Greene, Phys. Rev. B **52** (1995) 6575.
53. S. Skanthakumar, PhD Thesis University of Maryland (1993).
54. W. Henggeler, G. Guntze, M. Klauda, J. Mesot, A. Furrer and G. Saemann-Ischenko, Europhys. Lett. **29** (1995) 233.
55. J.A. Sanjurjo, G.B. Martins, P.G. Pagliuso, E. Granado, I. Torriani, C. Rettori, S. Oseroff and Z. Fisk, Phys. Rev. B **51** (1995) 1185.

56. W. Henggeler, T. Chattopadhyay, B. Roessli, D.I. Zhigunov, S.N. Barilo and A. Furrer, Z. Phys. B **99** (1996) 465.
57. W. Henggeler, unpublished.
58. A.S. Ivanov, private communication
59. W. Henggeler, T. Chattopadhyay, B. Roessli, D.I. Zhigunov, S.N. Barilo and A. Furrer, Europhys. Lett. **34** (1996) 537.
60. H. Casalta, P. Bourges, D. Petitgrand and A. Ivanov, Solid State Comm. **100** (1996) 683.
61. S. Skanthakumar, J.W. Lynn, J.L. Peng and Z.Y. Li, Phys. Rev. B **47** (1993) 6173.
62. P. Thalmeier, Physica C **266** (1996) 89.
63. R. Sachidanandam, T. Yildirim, A.B. Harris, A. Aharony and O. Entin-Wohlman, Phys. Rev. B **56** (1997) 260.
64. A. Furrer, H.U. Güdel, E.R. Krausz and H. Blank, Phys. Rev. Lett. **64** (1990) 68.
65. W. Henggeler, B. Roessli, A. Furrer, P. Vorderwisch and T. Chatterji, Phys. Rev. Lett. **80** (1998) 1300.
66. M. Hillberg, M.A.C.de Melo, H.H. Klauß, W. Wagener, F.J. Litterst, P. Adelmann and G. Czjzek, Hyperfine Interactions **104** (1997) 221.
67. P.M. Pyka, M. Loewenhaupt, A. Metz and N.T. Hien, Physica B **234-236** (1997) 808.

THE CRYSTAL FIELD AS A LOCAL PROBE IN RARE EARTH BASED HIGH-TEMPERATURE SUPERCONDUCTORS

JOEL MESOT AND ALBERT FURRER

Laboratory for Neutron Scattering
ETH Zürich & Paul Scherrer Institut
CH-5232 Villigen PSI, Switzerland

1. Introduction

The discovery of high-temperature superconductivity in the perovskites La_2CuO_4 [1] and $YBa_2Cu_3O_7$ [2] has given rise to a large amount of materials research. Of particular importance in determining the nature of the superconductivity is the effect of substituents at various sites in these compounds. It has been realized that the superconducting transition temperature T_c is essentially unchanged upon replacing the Y and La ions by paramagnetic rare-earth (R) ions. This surprising observation is in contrast to conventional superconductors, for which paramagnetic ions usually have a large detrimental effect on superconductivity. It is therefore important to achieve a detailed understanding of the low-energy electronic properties which define the magnetic ground state of the R ions. In particular, information on the crystal-field (CF) interaction at the R site is highly desirable, for the following reasons:

(i) For many high-T_c compounds superconductivity and long-range three-dimensional magnetic ordering of the R ion sublattice coexist at low temperatures. An understanding of both the nature of the magnetic ordering and its apparent lack of influence on T_c requires a detailed knowledge of the CF states of the R ions.

(ii) There are a few exceptions to the above statement, e.g., Ce and Pr substitution for Y in $YBa_2Cu_3O_7$ are known to have a drastic detrimental effect on superconductivity. Again, information on the CF states of these particular R ions will help to elucidate the observed loss of superconductivity.

iii) In most high-T_c compounds the R ions are situated close to the superconducting copper-oxide planes (see Fig. 1 and Chapter 2 of this book) where it is widely believed that the superconducting carriers are located, thus the CF interaction at the R sites constitutes an ideal probe of the local symmetry as well as the local charge distribution of the superconducting CuO_2 planes and thereby monitors directly changes of the carrier concentration induced, e.g., by oxygen nonstoichiometry, pressure, doping and disordering effects. This property results from the short-range nature of the leading terms in the CF potential.

(iv) The temperature dependence of the intrinsic linewidths of CF transitions is found to vary with temperature which is essentially a reflection of the density-of-states associated with the charge carriers at the Fermi energy. Linewidth studies can therefore reveal information about the opening as well as the symmetry of the energy gap.

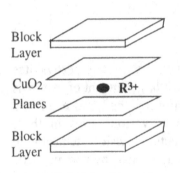

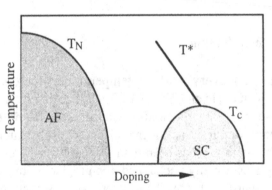

Figure 1: Schematic layered structure of $RBa_2Cu_3O_{6+x}$ (R=rare earth or Y)

Figure 2: Schematic phase diagram of high-T_c superconductors versus doping.

The methods of choice for studying the CF interaction are spectroscopic techniques which provide direct information on the CF energy levels. For the high-T_c compounds most results have been obtained by inelastic neutron scattering in the past, but Raman scattering is now increasingly being used. Both techniques have their merits and disadvantages and should be considered as complementary methods. Raman scattering, on the one hand, can be applied to very small samples of the order of $(10 \ \mu m)^3$, it provides highly resolved spectra so that small line shifts and splittings can be detected [3, 4] and it covers a large energy range so that intermultiplet transitions can be observed [5]. Neutron scattering, on the other hand, is not restricted to particular points in reciprocal space, i.e., interactions between the rare-earth ions can be observed through the wavevector dependence [6], the intensities of CF transitions can easily be interpreted, and data can be taken over a wide temperature range which is of importance when studying linewidths of CF transitions.

In the present work we summarize the basic principles and provide representative examples of CF excitations observed by neutron spectroscopy in high-T_C superconducting materials. Section 2 summarizes the theoretical aspects of the CF interaction. In Section 3 we exemplify how to arrive at a unique set of CF parameters and thereby to a detailed understanding of the electronic ground state of the R ions. The results are used to calculate various thermodynamic magnetic properties which are compared to experimental data as a consistency check. Then there follow Sections 4 and 5 which are dealing with the aspects (iii) and (iv) mentioned above, and we believe that they are most crucial for our current understanding of the phenomenon of high-T_c superconductivity. High-temperature superconductors are basically different from classical superconducting compounds in the sense that superconductivity is usually realized through doping. The undoped system is antiferromagnetic (AF) and insulating (Fig. 2). At small doping, the Néel temperature T_N decreases rapidly and vanishes. The system enters then a spin-glass (SG) phase. Upon further doping superconductivity (SC) appears below a critical temperature T_c, but the system behaves like a strange metal. The homogeneous parent compound is structurally distorted by the doping elements and the question then arises whether the doping process results in an extended electronic structure or whether the injected charge carriers are locally trapped. In the latter case the system becomes electronically inhomogeneous, and the superconducting state is reached by a percolation process [7]. Such inhomogeneities have actually been observed by a variety of local techniques [8] and most prominently by neutron CF spectroscopy as described in Section 4. Of similar importance is the relaxation behavior of CF excitations discussed in Section 5. The lifetime of CF states is finite because of the interaction with the charge carriers, thus the temperature dependence of the intrinsic linewidth of CF splittings reflects the density of states at the Fermi energy similar to NMR and NQR experiments and thereby reveals information on both the opening and the symmetry of the energy gap. There is evidence that the energy gap opens at $T^*>T_c$ for underdoped high-T_c compounds [9]. CF spectroscopy has the advantage to probe the static susceptibility in zero magnetic field (NMR needs a static external field) as well as at THz frequencies (NMR and NQR work at MHz frequencies). The latter point is crucial for the understanding of the data, since it yields new information not available by NMR or NQR. Some final conclusions are given in Section 6.

2. The crystal field (CF) interaction

2.1 BASIC FORMALISM

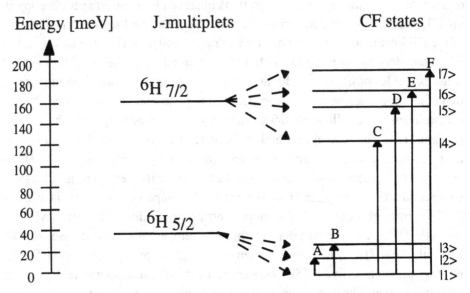

Figure 3: Energy-level scheme of Sm^{3+} ions in $SmBa_2Cu_3O_7$ for the lowest two J-multiplets as derived by neutron spectroscopy [10]. The arrows denote the observed CF transitions.

The electrostatic and spin-orbit interactions lift the degeneracy of the unfilled $4f^n$ configuration of the R ions and give rise to the J-multiplets (Fig. 3). The degeneracy of a J-multiplet of a magnetic ion embedded in a crystal lattice is partly removed by the crystal-field potential V_{CF} created by the surrounding ions. V_{CF} has to remain invariant under the operations of the point group which characterizes the symmetry of the R site and can be expressed by spherical harmonics, which are the basis vectors of the representations of the related rotation group. Using the tensor operator method and due to the fact that for 4f-shells terms up to *6th* order are relevant, the related Hamiltonian can be constructed in a very simple way [11]:

$$H_{CF} = \sum_{n=1}^{6} \sum_{m=0}^{n} A_n^m \left(Y_n^m + Y_n^{-m} \right). \tag{1}$$

Y_n^m are tensor operators of rank n and A_n^m are complex CF parameters. The number of non-zero CF parameters depends on the point symmetry at the R site.

The wave functions Γ_i and energy levels E_i are obtained by diagonalisation of the CF Hamiltonian (1).

2.2. AB INITIO DETERMINATION OF THE CF PARAMETERS

An *ab initio* calculation of the CF parameters is difficult. The simplest approach is to consider that the electrostatic potential felt by the R ions arises from point electric charges $Z_j|e|$ located at the center of the ions j at the position $\mathbf{R}_j$ relative to the R ion:

$$V(r) = \sum_j \frac{e^2 Z_j}{|r - R_j|} = \sum_S e^2 Z_S \sum_{j \in S} \frac{1}{|r - R_j|}, \tag{2}$$

where S represents a coordination shell of a particular atomic species with charge $Z_S|e|$. In the frame of this point-charge model the CF parameters are given by:

$$A_n^m = \sum_S Z_S \frac{<r^n>}{R_S^{n+1}} \frac{4\pi}{2n+1} f_n^m(S), \tag{3a}$$

with

$$f_n^m(S) = \sum_{j \in S} (-1)^m Y_n^{-m}(\theta_j, \varphi_j), \tag{3b}$$

where $<r^n>$ is the *nth* moment of the radial distribution of the 4f-electrons and the Y_n^{-m} are spherical harmonics. The $f_n^m(S)$ are geometrical coordination factors. In general this model is not able to reproduce the observed energies and intensities. Indeed, the second and sixth order CF parameters are usually found to be an order of magnitude too large and too small, respectively. Some corrections to this model can be considered:

1) Sternheimer [12] showed that due to the screening of the 4f electrons by the outer shells, the $<r^n>$ terms should be replaced by $<r^n>(1-\sigma_n)$. The shielding factors σ_n are given as a function of the number of electrons N of the R ion as [13]:

$$\sigma_2 = 0.6846 - 0.00854 \cdot N$$
$$\sigma_4 = 0.02356 - 0.00182 \cdot N \tag{4}$$
$$\sigma_6 = -0.04238 - 0.00014 \cdot N$$

2) Morrison [13] showed that the n*th* moment of the radial wave functions $<r^n>$ for the free ion have to be replaced by $<r^n>/\tau^n$ when it is placed into a crystal. τ as a function of N is given for the system $R:CaWO_4$ as:

$$\tau = 0.75(1.0387 - 0.0129 \cdot N) \tag{5}$$

3) Although in high-T_c compounds the carrier concentration resulting from doping is considerably smaller than in ordinary metals, the screening effect is enhanced by the fact that the carriers are located in the CuO_2 planes and therefore have approximately two-dimensional character. The screening effects can be taken into account by a Yukawa-type potential:

$$V(\mathbf{r}) = \sum_j \frac{e^2 Z_j}{|\mathbf{r} - \mathbf{R}_j|} \exp\left(-\kappa |\mathbf{r} - \mathbf{R}_j|\right), \tag{6}$$

where κ is the inverse screening length. As a consequence, $1/R_S^{n+1}$ in eq. (3a) has to be replaced by [14]:

$$K_{n+1/2}\left(\kappa |\mathbf{r} - \mathbf{R}_j|\right)/R_S^{n+1}. \tag{7}$$

The $K_{n+1/2}(x)$ are the modified Bessel functions, normalised to obtain $K_{n+1/2}(0)=1$. Using Eqs 3-7 and assuming nominal charges for the neighboring ions, we are left with a model containing only one parameter κ. As we will see in Section 3, this model works well for the perovskites [15, 16].

2.3. EXTRAPOLATION SCHEMES

The determination of the CF parameters in high-T_c superconductors is not an easy task:

-Firstly, the splitting produced by the CF ($\approx$ 100 meV) is comparable to the intermultiplet splitting energies (e.g. 250 meV for Nd^{3+}). This leads to an admixture of different J-multiplets through the CF interaction (J-mixing). Furthermore, due to the spin-orbit coupling S and L are no longer good quantum numbers, and every J-multiplet is contaminated by states of different S and L but same J (intermediate coupling approximation). For some particular high-T_c compounds it has been essential to include these effects in order to achieve a quantitative understanding of the CF interaction [17].

-Secondly, for low symmetry sites such as orthorhombic symmetry D_{2h} realized for $RBa_2Cu_3O_7$ there are nine independent CF parameters, so that a correct set of starting parameters is difficult to find. As shown in Ref. [18], the number of parameters can be drastically reduced by using geometrical constraints. By considering in a first step only the nearest-neighbour oxygen shell Eq. (3) reduces to

$$A_n^m = A_n^0 f_n^m(O(2,3))/f_n^0(O(2,3)), \tag{8}$$

and we are left with only three independent parameters A_n^0, n=2,4,6. Following Lea, Leask, and Wolf [19] we introduce a parametrization which covers all possible ratios of A_2^0/A_4^0 and A_6^0/A_4^0:

$$B_2^0 F_2 = A_2^0 \chi_2 F_2 = W(1 - |y|),$$

$$B_4^0 F_4 = A_4^0 \chi_4 F_4 = Wxy,$$

$$\tag{9}$$

$$B_6^0 F_6 = A_6^0 \chi_6 F_6 = W(1 - |x|)y.$$

The F_n are numerical factors tabulated by Lea, Leask, and Wolf [19], $-1 \leq x,y \leq 1$, and W is a scale factor. The χ_n are reduced matrix elements tabulated by Hutchings [11] and relate our notation to the CF parameters B_n^m used in Ref. [19]. Eq. (9) corresponds to the most general combination of second-, fourth-, and sixth-order CF parameters. In high-T_c compounds only one set of x,y parameters gives a reasonable agreement between the observed and calculated energies and intensities. In a second step of the least-squares fitting procedure, after having determined the most reasonable start values of the diagonal CF parameters, all the leading "tetragonal" parameters (m=0,4) and later also the "orthorhombic" parameters (m=2,6) are allowed to vary independently.

An alternative way to derive a starting set of CF parameters is to consider the model presented in Section 2.2, Eqs 3-7. Since the averaged structure is known from neutron diffraction experiments, it is possible to obtain a set of CF parameters which allows a reasonable description of the observed spectra as will be shown is Section 3.1. Once a good starting set of CF parameters has been found, either the charges of the neighbouring ions or the CF parameters directly are adjusted to improve the fit.

2.4. CALCULATION OF THERMODYNAMIC PROPERTIES

In order to check the reliability of the CF parameters it is important to calculate various thermodynamic magnetic properties and to compare the results with experimental data. These properties depend explicitely upon both energies E_i and wavefunctions $|\Gamma_i\rangle$. Based on general expressions of statistical mechanics for the Gibbs free energy $F = -k_B T \ln Z$ and the internal energy $U = F - T(\partial F/\partial T)_V$ where Z is the partition function we obtain for the magnetization:

$$M_\alpha = \frac{1}{k_B T} \frac{\partial \ln Z}{\partial H_\alpha} = g\mu_B \sum_i p_i \langle \Gamma_i | J_\alpha | \Gamma_i \rangle, \qquad (10)$$

the single-ion susceptibility:

$$\chi_{\alpha\alpha} = \frac{\partial M_\alpha}{\partial H_\alpha} = g^2 \mu_B^2 \left[\sum_i \frac{\left|\langle \Gamma_i | J_\alpha | \Gamma_i \rangle\right|^2}{k_B T} p_i + \sum_{i \neq j} \frac{\left|\langle \Gamma_j | J_\alpha | \Gamma_i \rangle\right|^2}{E_i - E_j} (p_j - p_i) \right], \quad (11)$$

and the Schottky heat capacity:

$$C_v = \left(\frac{\partial U}{\partial T}\right)_V = k_B \left[\sum_i \left(\frac{E_i}{k_B T}\right)^2 p_i - \sum_i \left(\frac{E_i}{k_B T} p_i\right)^2 \right], \qquad (12)$$

with the Boltzmann population factor

$$p_i = \frac{1}{Z} \exp\left(-\frac{E_i}{k_B T}\right). \qquad (13)$$

2.5. RELAXATION EFFECTS

The linewidth of CF excitations in metallic systems is mainly governed by relaxation effects with the charge carriers [20], similar to the most prominent relaxation methods NMR and NQR. However, CF spectroscopy has the advantage to probe the static susceptibility in zero magnetic field (NMR needs a static external field) as well as at THz frequencies (NMR and NQR work at

MHz frequencies). As shown in Section 5, the latter point is crucial for the understanding of the data, since it yields new information not available by NMR or NQR.

Based on the theory developed by Becker, Fulde and Keller [21], in a metal the linewidth of a transition (i,j) with energy $\hbar\omega_{ij} = \hbar\omega_i - \hbar\omega_j$ between the CF states $|\Gamma_i\rangle$ and $|\Gamma_j\rangle$ is given by [22]:

$$\Gamma^{ij} = 2J_{ex}^2\left[M_{ij}^2 \coth\left(\beta\hbar\omega_{ij}/2\right)\chi''\left(\hbar\omega_{ij}\right)\right.$$

$$\left. + \sum_{n\neq i}M_{in}^2\frac{\chi''\left(\hbar\omega_{in}\right)}{(e^{\beta\hbar\omega_{in}}-1)} + \sum_{n\neq j}M_{nj}^2\frac{\chi''\left(\hbar\omega_{nj}\right)}{(e^{\beta\hbar\omega_{nj}}-1)}\right], \qquad (14)$$

where M_{ij} is the transition matrix element of the CF transition, J_{ex} the exchange integral between the f-electrons of the R ions and the charge carriers, $\beta=1/k_BT$, and $\chi''(\hbar\omega)$ is the imaginary part of the susceptibility summed over the Brillouin zone. Eq. (14) is also known as the Korringa law in the high-temperature limit. In the normal (N) state and for a non-interacting Fermi liquid $\chi''(\hbar\omega) = \chi_N''(\hbar\omega) = \pi N^2(E_F)\hbar\omega$, where $N(E_F)$ is the electronic density-of-states at the Fermi energy. The use of a Fermi liquid model at the R sites in high-T_c materials is justified by [89]Y NMR observations [23] which show that the [89]Y relaxation rate is close to the non-interacting limit in the normal state and obeys the Korringa law. Due to the opening of an electronic gap Δ, the relaxation rate of the CF transitions is strongly affected in the superconducting (SC) state. If $\hbar\omega < 2\Delta$ the relaxation mechanism cannot occur, and the susceptibility has to be modified in order to account for the deviation of the linewidth in the SC state from the linewidth in the normal state (see Section 5).

3. Examples

3.1. OPTIMALLY DOPED $ErBa_2Cu_3O_7$

The ground-state of the Er^{3+} ions has a total angular momentum J=15/2. The symmetry of the R ions is orthorhombic for x > 6.4 and tetragonal for x ≤ 6.4, and the CF will split the 16-fold degeneracy of the ground-state multiplet into eight Kramers doublets. All seven CF transitions out of the ground-state |0> and in addition some excited-state CF transitions have been measured on time-of-flight spectrometers [17]. As shown in Fig. 4, at low temperature (T=10 K) we observe three inelastic lines A, B, C in a low-energy window ($\Delta E < 12$ meV) and four inelastic lines D, E, F, G in a high-energy window (65<ΔE<82 meV). By using the model described in Section 2.2, together with the known structure [24] and by adjusting $\kappa = 1.25$ Å^{-1}, we derive a starting set of CF parameters (Table 1) which were then subject to a least-squares fitting procedure to reproduce the observed spectra. The starting CF parameters (except the orthorhombic ones with m=2) are found to be extremely close to the fitted ones, which provide a very good description of both the energies and intensities of the observed CF transitions displayed in Fig. 4.

TABLE 1. Calculated ($\kappa = 1.25$ Å^{-1}) and Fitted CF A_n^m parameters (in meV) for $ErBa_2Cu_3O_7$.

(n, m)	(2, 0)	(2, 2)	(4, 0)	(4, 2)	(4, 4)	(6, 0)	(6, 2)	(6, 4)	(6, 6)
Calc.	13.7	0.2	-32.0	8.0	149	3.58	1.0	102.6	-0.9
Fit	13.9(7)	12(3)	-32.3(2)	10(3)	157(1)	3.67(6)	-0.6(1)	104.6(3)	0.6(1)

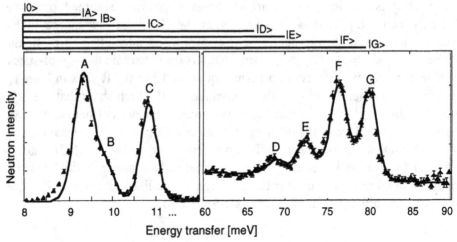

Figure 4: Energy spectra of neutrons scattered from $ErBa_2Cu_3O_{6.98}$ at T=10 K taken at the time-of-flight spectrometers IN4 (left part, E_i=17.2 meV) and HET (right part, E_i=100 meV) [17]. The lines represent the calculated spectra.

In order to check the reliability of the CF parameters it is important to calculate various thermodynamic properties with the help of Eqs. 10-13 and to compare the results with experimental data. Fig. 5 shows the good agreement between high-field magnetization [25] and specific heat measurements [26, 27] of a powder sample and calculations using the CF parameters listed in Table 1. The zero-field moment of the R ion at saturation is drastically reduced below the free ion value of 9 μ_B. For $ErBa_2Cu_3O_7$ a mean-field calculation (Eq. 10) yields $4.2\mu_B$ which agrees well with the value derived from Mössbauer experiments [28].

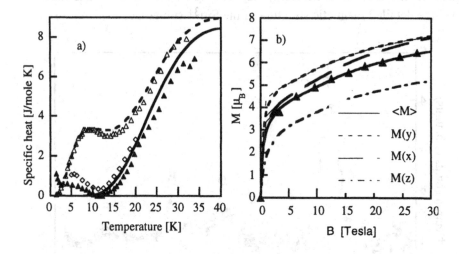

Figure 5: Observed (symbols) and calculated (lines) thermodynamic magnetic properties of $ErBa_2Cu_3O_7$. a) Schottky anomaly of the heat capacity, the symbols are taken from [26, 27]. b) high-field magnetisation, the symbols are taken from [25].

3.2. INTERMULTIPLET CF TRANSITIONS

Often the CF splittings observed in the ground-state J-multiplet are not sufficient to unambiguously determine the CF parameters. This is the case, e.g., for $SmBa_2Cu_3O_7$, since there are only three doublets within the ground-state J-multiplet $^6H_{5/2}$ associated with Sm^{3+}, which does not allow to determine the nine non-vanishing CF parameters. In principle, the number of observables can be increased by measurements of intermultiplet transitions (Fig. 3) which have become possible due to the copious flux of epithermal neutrons produced at spallation neutron sources. In spite of this established potential, the INS technique has so far made only a limited contribution in the field of intermultiplet spectroscopy [29].

The energy spectrum displayed in Fig. 6 shows CF ground-state transitions to the four doublet states within the first-excited J-multiplet $^6H_{7/2}$ of the Sm^{3+} ions in $SmBa_2Cu_3O_7$. The corresponding energy level scheme is shown in Fig. 3. This information, together with the observed CF transitions within the ground state has allowed a precise determination of the CF parameters [10]. Magnetic susceptibility and heat capacity measurements on polycrystalline $SmBa_2Cu_3O_7$ [30] indicate long-range antiferromagnetic ordering of the Sm^{3+} sublattice below T_N=0.61 K. The CF parameters predict the c axis to be the easy axis of magnetisation with a drastically quenched moment of 0.11 μ_B at saturation. Because of this very small moment, magnetic neutron diffraction has failed to unravel the detailed magnetic structure of the Sm^{3+} sublattice.

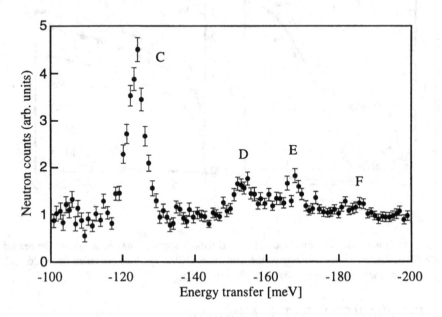

Figure 6: Energy spectrum of neutrons scattered from $SmBa_2Cu_3O_7$ at T=10 K taken at MARI [10].

3.3. OTHER $RBa_2Cu_3O_7$ COMPOUNDS

Table 2 lists the CF parameters for other $RBa_2Cu_3O_7$ compounds as determined by inelastic neutron scattering. We do not recognize any systematics for the CF parameters A_n^m with n=2 due to the long-range nature of the *2nd* order CF potential as well as for those with m=2,6 due to their insensitivity in the least-squares fitting procedure. The values of the leading CF parameters A_4^0, A_4^4, A_6^0, A_6^4, on the other hand, are found to decrease continuously when going from the light to the heavy rare-earth ions, and the question arises whether this decrease can be quantitatively accounted for. As long as the charge distribution around the R^{3+} sites is not changed throughout the rare-earth series, the leading CF parameters are expected to scale according to Eq. (3) with constant charges Z_S|e|. We find that the charges Z_S|e| exhibit a continuous increase, typically by 50% when going from Pr to Yb, similar to CF results for metallic as well as insulating compounds [38, 39]. This effect has been thoroughly investigated and discussed in terms of dipolar polarizability and covalency [40, 41] which seem to have an increased relative importance for the heavy members of the rare-earth compounds. This means that any extrapolation of CF parameters from one rare-earth compound to another has to be considered with extreme caution.

4. Doping effects

The critical temperature T_c can be influenced by various techniques, such as oxygen reduction, copper substitution or external pressure. What is the origin of the observed changes of T_c in each of these cases ? Are these changes only due to modifications of the electronic structure, or must some more delicate effects be involved ?

In order to answer these questions, CF spectroscopy is certainly a method of choice, since the short-range nature of the 4*th* and 6*th* order CF parameters enables us to directly monitor the charge distribution of the CuO_2 planes.

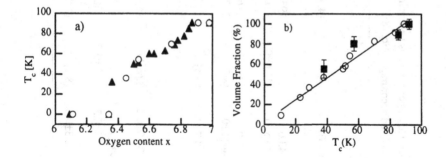

Figure 7: a) Superconducting transition temperature T_c versus oxygen content x for $ErBa_2Cu_3O_x$ [17, 24]. b) Superconducting volume fraction (Meissner fraction) versus T_c determined for $ErBa_2Cu_3O_x$. The circles refer to SQUID magnetometry results [42]. The squares result from the analysis of the CF spectra as explained in the text.

4.1. p-TYPE SUPERCONDUCTORS

4.1.1. *ErBa₂Cu₃Oₓ (6<x<7) : charge transfer, inhomogeneous properties*
One of the most interesting aspects of the superconductivity in the $RBa_2Cu_3O_x$ compounds is the relation between T_c and the oxygen stoichiometry (see Fig. 2). Annealed systems show the well-known two-plateau structure of T_c (Fig. 7a).

TABLE 2. CF parameters A_n^m (in meV) of RBa2Cu3O7 compounds determined by inelastic neutron scattering. The numbers in parantheses correspond to CF parameters extrapolated according to Eq. (8).

R	Ref.	(2, 0)	(2, 2)	(4, 0)	(4, 2)	(n, m) (4, 4)	(6, 0)	(6, 0)	(6, 4)	(6, 6)
Pr	[31]	28.0	24.4	-43.0	1.4	193.3	6.1	-27.0	227.2	12.3
	[32]	18.0±0.5	7.4±1.3	-46.5±0.2	-31.6±0.8	178.9±1.5	8.2±0.4	3.2±1.3	210.5±1.5	17.1±1.9
Nd	[31]	25.8	22.5	-42.0	1.2	216.4	4.8	-21.3	172.0±9.7	9.7
	[33]	36.7±1.3	(4.4)	-43.7±0.2	(8.3)	207.9±1.2	4.2±0.1	(-0.4)	174.3±0.8	(0.8)
Sm	[10]	22.8±1.2	(3.2)	-39.7±0.8	(6.4)	163.8±4.2	5.4±0.3	(1.1)	173.3±2.8	(1.2)
Eu	[34]	49.3±1.4	4.9±0.7							
Ho	[35]	17.6±0.8	12.0±1.0	-34.0±0.2	3.6±0.8	161.3±1.7	4.0±0.1	-2.5±0.8	116.5±0.2	-0.4±0.2
Er	[36]	20.1	22.8	-30.1	-0.4	173.4	3.6	-4.4	107.6	0.
	[17]	13.9±0.8	11.6±3.1	-32.2±0.2	10.3±3.2	156.8±1.5	3.7±0.1	-0.6±0.2	104.6±0.3	0.6±0.2
Yb	[37]	5.7±2.3	(2.0)	-32.3±0.6	(8.1)	(149.2)				

Since charge transfer from the CuO chains to the CuO$_2$ planes is certainly playing a crucial role in the oxygen-vacancy induced suppression of superconductivity in YBa$_2$Cu$_3$O$_x$, we have investigated the oxygen stoichiometry dependence of the observed CF energy spectra in ErBa$_2$Cu$_3$O$_x$ (Fig. 8) [17]. When going from x=6 to x=7, the transitions B, D, E shift slightly to lower energies and the transitions A,F,G move up to higher energies, whereas the energy of the transition C remains unchanged.

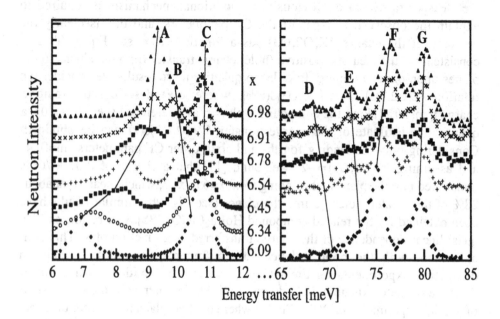

Figure 8: Energy spectra of neutrons scattered from ErBa$_2$Cu$_3$O$_x$ at T=10 K taken at IN4 (left part, E$_i$=17.2 meV) and HET (right part, E$_i$=100 meV) [17]. The lines indicate the x dependence of the observed CF transitions.

In the analysis of the energy spectra displayed in Fig. 8, we only consider the center of gravity and the integrated intensity of each of the seven CF transitions. Once a set of CF parameters has been derived for each sample [17], we can estimate the charge being transferred from the chains to the planes upon oxygen doping. In the following, and for clarity, we restrict our considerations to the nearest-neighbour oxygen shell. If we consider relative changes of the CF parameters, we obtain from Eq. (3):

$$\frac{{}^x A_n^m}{{}^7 A_n^m} = \frac{{}^x Z(O2,O3)}{{}^7 Z(O2,O3)} \frac{{}^x f_n^m(O2,O3)}{{}^7 f_n^m(O2,O3)} \cdot \left(\frac{{}^7 R(O2,O3)}{{}^x R(O2,O3)} \right)^{n+1}, \qquad (15)$$

where x is the oxygen content. Since the ${}^x f_n^m(O2,O3)$ and ${}^x R(O2,O3)$ are known from crystallographic measurements [24] we can use Eq. 15 to estimate the relative changes of the CF parameters as a function of x due to structural modifications alone. As shown in Fig. 9a the structural deformations upon oxygen doping have almost no influence on A_4^0, while the values derived from the CF spectra decrease. Obviously, an additional mechanism is required to explain the observed changes of the CF parameters, and this is clearly the decrease of the charge ${}^x Z(O2,O3)$ as a function of x, see Eq. 3. This is consistent we the idea of a positive (hole) charge transfer from the chains to the planes upon oxygen doping. In order to quantify these results we consider the relation ${}^x Z(O2,O3) = {}^6 Z(O2,O3)\delta(x)|e|$, where $\delta(x)$ describes the relative charge transfer. Taking the compound with x=6 as a reference (i.e., $\delta(6) = 0$), a quantitative estimate of the charge transfer ${}^6 Z(O2,O3)\delta(x)$ can be obtained. Considering only the leading fourth- and sixth-order CF parameters (m=0, 4) and assuming ${}^6 Z(O2,O3) = -2$, we found [17] that a charge of 0.07 |e|/O is transferred into the planes when going from x=6 to x=7, which means that about 28% of the created holes go into the planes (see Fig. 9b). Similar results have been obtained for the related compound $HoBa_2Cu_3O_x$ [35]. Our results show a quasi linear dependence of the charge transferred as a function of x. This is in agreement with the bond valence sum arguments derived from neutron diffraction experiments on $ErBa_2Cu_3O_x$ which give evidence for a linear decrease of the c axis upon hole doping [42], but in contrast to the conclusions of similar experiments on $YBa_2Cu_3O_x$ where the two-plateau structure of T_c has been suggested to be due to the non-linearity of the hole transfer into the planes [43].

Upon increased resolution conditions we observe (Fig. 10) that the lowest CF transition A in $ErBa_2Cu_3O_x$ is actually built up by three different components A_1, A_2 and A_3 whose main characteristics can be summarised as follows:

1) Intensities: The individual components A_i have maximum weight close to x=7.0, x=6.5, and x=6.0, respectively (Figs 11a-c). With the CF interaction being a local probe, there is no doubt that these substructures originate from different local environments of the Er^{3+} ions which obviously coexist in the compound $ErBa_2Cu_3O_x$.

2) Energies: Whereas all the CF transitions are independent of energy for oxygen content x≥6.5 within experimental error, they shift slightly when going

from x≈6.5 to x≈6.0. This may be due to the structural discontinuities at the orthorhombic to tetragonal phase transition at x≈6.4.

3) Linewidth: as visualized in Fig. 11d, the intrinsic linewidths of the transitions A_i are much smaller for oxygen contents where these transitions individually reach their maximum weight, namely, for x≈6.0, 6.5, and 7.0.

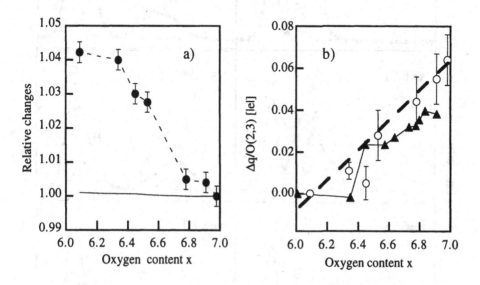

Figure 9: a) Relative variation of the CF parameter A_4^0 of $ErBa_2Cu_3O_x$ versus oxygen content x [17]. The dashed line is a guide to the eye. The solid line corresponds to A_4^0 values extrapolated from the structural changes alone. b) Charge transfer versus oxygen content x derived from CF spectroscopic experiments on $ErBa_2Cu_3O_x$ (circles and dashed line) [17], in comparison with the results obtained for $YBa_2Cu_3O_x$ from bond valence sum considerations (triangles and solid line) [43].

Our data and their interpretation provide clear experimental evidence for cluster formation. It is tempting to identify the three clusters associated with the transitions A_1, A_2 and A_3 by two local regions of metallic ($T_c ≈ 90$ K, $T_c ≈ 60$ K) and a local region of semiconducting character, respectively [44]. Figs 11a-c show the fractional proportions of the three cluster types which exhibit a continuous behaviour versus the oxygen content x, consistent with our earlier findings (see Section 4.1.1) that the transfer of holes into the CuO_2 planes is linearly related to the oxygenation process [17]. Furthermore, the continuous increase of the metallic states A_1 and A_2 can explain the increase of the superconducting volume fraction as observed by magnetic susceptibility measurements [42] when the oxygen content is raised from x=6 to x=7 (Fig.7b).

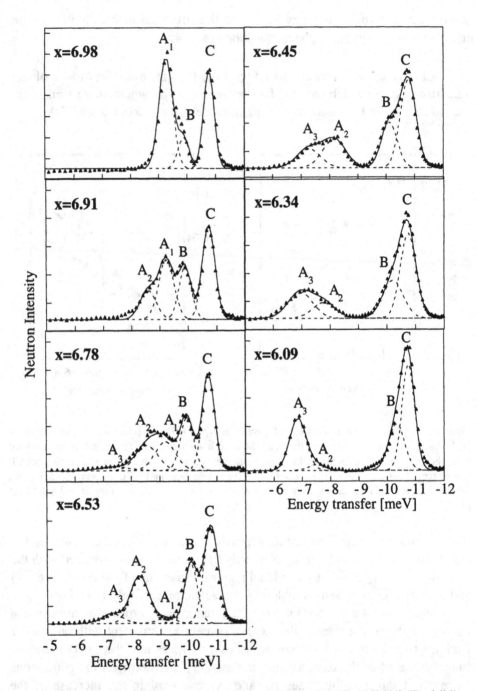

Figure 10: Energy spectra of neutrons scattered from $ErBa_2Cu_3O_x$ at T=10 K [44]. The full lines are the result of a least-squares fitting procedure. The broken lines indicate the subdivision into individual CF transitions.

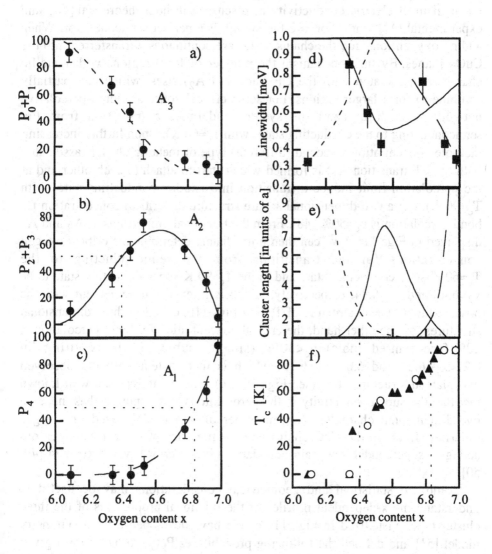

Figure 11: a-c) Proportions (expressed as a percentage) of the lowest-lying CF transitions A_i of ErBa$_2$Cu$_3$O$_x$ as a function of the oxygen content x [44]. The lines refer to geometrical probability functions as explained in the text. The dotted lines mark the critical concentration for bond percolation. d) Intrinsic linewidths of the CF transitions A_i of ErBa$_2$Cu$_3$O$_x$ at 10 K. e) Mean cluster length versus x [45]. The lines are the result of a Monte Carlo simulation as explained in the text. The long dashed, full and short dashed lines correspond to A_1, A_2 and A_3, respectively. f) Superconducting transition temperature T_c versus oxygen content x for ErBa$_2$Cu$_3$O$_x$.

Our current understanding of the superconducting properties of $ErBa_2Cu_3O_x$ (and more generally all the $RBa_2Cu_3O_x$ compounds) involves a percolation mechanism of electric conductivity as discussed in both theoretical [46] and experimental [47] work. For x=6 the system is a perfect semiconductor. When adding oxygen ions into the chains, holes are continuously transferred into the CuO_2 planes. By this mechanism the number of local regions with metallic character (associated with the CF transition A_2) rises, which can partially combine to form larger regions. For some critical concentration a percolative network is built up, and the system undergoes a transition from the semiconducting to the conducting state (with $T_c \approx 60$ K). Upon further increasing the hole concentration a second (different) type of metallic cluster (associated with the CF transition A_1) is formed which start to attach to each other and at the percolation limit induce a transition into another conducting state (with $T_c \approx 90$ K). For a two-dimensional square structure the critical concentration for bond percolation is p_c=50% [48]. From the fractional proportions of A_2 and A_1 displayed in Figs 11a-c we can then immediately determine the critical oxygen concentrations for the transitions from the semiconducting to the $T_c \approx 60$ K-superconducting state and to the $T_c \approx 90$ K-superconducting state to be x_2=6.40 and x_1=6.84, respectively, which is in excellent agreement with the observed two-plateau structure of T_c (see Figs 11b, c, f). For three-dimensional structures, on the other hand, the critical concentration for bond percolation is 20%(face–centred cubic)<p_c<30% (simple cubic) [48], resulting in 6.21<x_2<6.31 and 6.64<x_1<6.73, which is inconsistent with the observed two-plateau structure of T_c (see Figs 11b, c). This reinforces the well known fact that the superconductivity in the perovskite-type compounds has indeed a two-dimensional character. A similar percolation model based on oxygen ordering effects in the CuO chains and the presence of oxygen-poor regions acting as superconducting grain boundaries gives a critical value x_1=6.74 [49, 50].

Combined statistical and geometrical considerations may be useful to understand the x-dependent profiles of the fractional proportions of the three cluster types visualized in Figs 11a-c. We have developed a local symmetry model [51] and defined the following probabilities $P_k(y)$ to have, for a given oxygen content x=6+y, k of the four oxygen chain sites (0,1/2,0), (1,1/2,0), (0,1/2,1) and (1,1/2,1) nearest to the R^{3+} ion occupied (see Fig. 12):

$$P_k(y) = \binom{4}{k} y^k (1-y)^{4-k}, \quad (0 \le k \le 4). \tag{16}$$

The fractional proportion of the cluster type A_1 exhibits the behavior predicted by the probability function $P_4(y)$ (i.e., all the oxygen chain sites being occupied). Similarly, the fractional proportions of the cluster types A_2 and A_3 follow the sum of the probability functions $P_3(y)+P_2(y)$ (i.e., one or two empty oxygen chain sites) and $P_1(y)+P_0(y)$ (i.e., one or none oxygen chain site being occupied), respectively. The above probability functions are shown in Figs 11a-c by lines which excellently reproduce the experimental data.

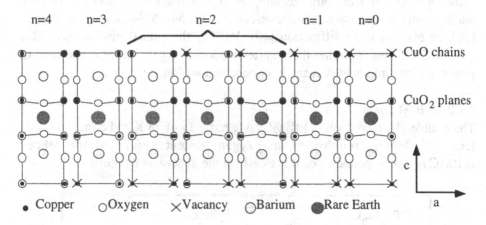

Figure 12: Schematic visualisation of the different chain-oxygen configurations in $RBa_2Cu_3O_x$.

The x-dependent linewidth of the CF transitions allows to estimate the size of the clusters, since a line broadening is due to structural inhomogeneities and gradients in the charge distribution which occur predominantly at the border lines between different cluster types. For a cluster of a mean length L (in units of the lattice parameter a) in the (a,b)-plane the fraction of unit cells at the border of the cluster and inside the cluster is $4(L-1)/L^2$ and $\{(L-2)/L\}^2$, respectively. In a first approximation the total linewidth is then given by [45]

$$\Gamma = \frac{1}{L}\sqrt{(L-2)^2\Gamma_0^2 + 4(L-1)\left[\left(\frac{\Delta E}{2}\right)^2 + \Gamma_0^2\right]},\qquad(17)$$

where Γ_0 is the intrinsic linewidth for infinite cluster size (i.e., for x=6 and x=7 exactly realized) and ΔE corresponds to the energetic separation of the CF transitions A_i for two coexisting cluster types. More specifically, from Fig. 10 we derive $\Delta E=E(A_1)-E(A_2)\approx1.0$ meV and $\Delta E=E(A_2)-E(A_3)\approx1.5$ meV at high and low oxygen concentrations x, respectively, and $0.3\leq\Gamma_0\leq0.4$ meV.

The lines of Fig. 11d are the results of Monte-Carlo simulations of the doping mechanism [52]. These simulations start from the ortho-II structure at x=6.5 with some degree of disorder and show that the sizes of the clusters at percolation (Fig. 11e) are of the order of L=2-3 (7-10Å), which is of the order of the superconducting coherence length in these materials. These numbers compare favourably with theoretical estimates by Hiznyakov et al. [46].

Evidence for the strongly inhomogeneous distribution of holes in the CuO_2 planes also results from other techniques such as magnetic susceptibility [47], nuclear and electron paramagnetic resonance [53, 54], Mössbauer [55], Raman [56], or even neutron diffraction [42]. Various theoretical models have also predicted that under doping these systems become highly unstable and tend to phase separate into hole-rich and hole-poor regions [46, 57, 58].

4.1.2. $ErBa_2Cu_4O_8$

The double-chain compound $ErBa_2Cu_4O_8$ has a T_c of 78 K and corresponds, in terms of doping, roughly to an oxygen content $x \approx 6.78$ of the related $ErBa_2Cu_3O_x$ compounds, i.e., it is clearly in the underdoped regime.

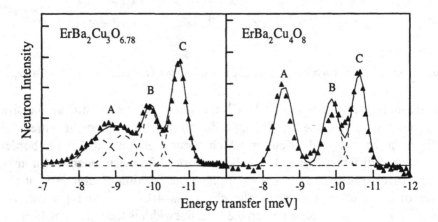

Figure 13: Energy spectra of neutrons scattered from $ErBa_2Cu_4O_8$ and $ErBa_2Cu_3O_{6.78}$ [44] at T=10 K taken at IN4 (E_i=17.2 meV). The lines are as in Fig. 10.

Fig.13 shows that the measured averaged peak positions and intensities of the CF transitions in $ErBa_2Cu_4O_8$ are very close to those measured for the $ErBa_2Cu_3O_{6.78}$ compound. This observation is important since it is an additional proof of the short range nature of the CF interaction in high-T_c materials. In other words, the CF interaction is reflecting predominantly the charge distribution of the CuO_2 planes and is not sensitive to the details of the CuO chains. It should also be noticed that the stoichiometric $ErBa_2Cu_4O_8$

compound exhibits narrower linewidths than the corresponding non-stoichiometric $ErBa_2Cu_3O_{6.78}$ compound, see Fig. 13.

4.1.3. *Cu substitution by Ni and Zn*

Substitution of Cu by both Ni and Zn has a detrimental effect on T_c, but the suppression rates of T_c are very different: for the samples $ErBa_2(Cu_{1-y}M_y)_3O_7$ (M=Ni, Zn) we determined $\partial T_c/\partial y$ to be -2.5 K/% and -7 K/% for M=Ni and M=Zn, respectively. From neutron diffraction experiments [59] we found that Zn occupies the Cu(2) sites in the planes, while Ni occupies the Cu(1) sites in the chains, thus it is obvious that different mechanisms have to be invoked for the suppression of T_c in the Ni and Zn substituted samples.

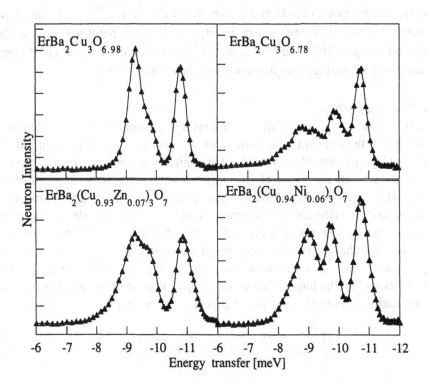

Figure 14: Energy spectra of neutrons scattered from $ErBa_2(Cu_{1-y}M_y)_3O_x$ (M=Zn,Ni) at T=10 K taken at IN4 (E_i=17.2 meV) [44, 59].

Fig. 14 shows the CF measurements obtained for $ErBa_2(Cu_{0.95}Zn_{0.05})_3O_7$ and $ErBa_2(Cu_{0.94}Ni_{0.06})_3O_7$. Although the CF transitions are broadened with respect to the non-substituted analog due to disorder, we clearly recognize the following interesting features: the substitution of Ni atoms at the Cu sites in the chains results in a change of the CF interaction similar to the effect of decreasing the oxygen content in $ErBa_2Cu_3O_x$ from x=7.0 to x=6.7. Ni substitution may therefore be considered as a negative doping, consistent with the observed decrease in the Hall carrier concentration [59]. The CF measurements obtained for the Zn doped sample are similar to those obtained for the optimally doped sample $ErBa_2Cu_3O_7$, i.e., almost no charge-transfer effect occurs upon Zn doping. Moreover, since the decrease of T_c is about three times larger for Zn than for Ni doping, our CF experiments clearly indicate that another mechanism must be invoked to explain the dramatic decrease of T_c upon Cu substitution by Zn. According to NMR data, the Zn substitution creates a net magnetic moment in the CuO_2 planes [23, 60-62] which can lead to magnetic pair-breaking effects and a subsequent decrease of T_c [63]. Based on this scenario, the observed upturn in H_{c_2} at low temperature for the Zn substituted samples [64] and the observed increase of the isotope coefficient [65] for the Pr substituted samples could also be explained [66].

4.1.4. *External pressure*

Another important property of the $RBa_2Cu_3O_x$ compounds is the strong dependence of the critical temperature upon external pressure. As shown in Fig. 15 $\partial T_c(x)/\partial p$ increases by an order of magnitude around x=6.7 [67]. Is this anomaly due to an enhanced charge transfer into the planes upon pressure $(\partial n(x)/\partial p)$ around x=6.7 ? In order to answer this question we have performed both neutron diffraction and CF experiments on $ErBa_2Cu_3O_x$ under pressure for several oxygen contents x (see Fig. 15). From the derived CF parameters we found that the charge transfer into the planes under pressure, $\partial n(x)/\partial p$, is independent of the oxygen content x and equals 0.12 lel/O/10 kbar [68]. This result indicates that the large value of $\partial T_c(x)/\partial p$ measured for x≈6.7 is not due to an anomalously high charge transfer process. Let us rewrite

$$\frac{\partial T_c}{\partial p}(x) = \frac{\partial T_c}{\partial n}\frac{\partial n}{\partial p}(x) = \frac{\partial T_c}{\partial x}\frac{\partial x}{\partial n}\frac{\partial n}{\partial p}(x). \tag{18}$$

From the above mentioned result, $\partial n(x)/\partial p$=constant, and from the linear charge-transfer process upon oxygen doping derived in Section 4.1.1, $\partial n/\partial x$=constant, we obtain:

$$\frac{\partial T_c}{\partial p}(x) \propto \frac{\partial T_c}{\partial x}(x), \tag{19}$$

In other words, Eq. (19) clearly tells us that the same mechanism is responsible for the large enhancement of both $\partial T_c/\partial x$ and $\partial T_c/\partial p$ in the vicinity of $x \approx 6.7$, i.e. the proximity of the percolation limit plays an important role and has to be considered.

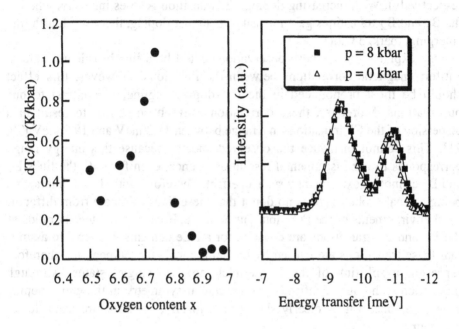

Figure 15: Left: $\partial T_c/\partial p$ versus oxygen content x observed for $YBa_2Cu_3O_x$ [67]. Right: pressure dependence of the CF spectra of $ErBa_2Cu_3O_{6.97}$ at T=10 K taken at IN4 (E_i=17.2 meV) [68].

4.2. n-TYPE SUPERCONDUCTORS

The R_2CuO_4 (R=La, Pr, Nd) family is extremely interesting, since these compounds can either become p- or n-type conductors upon doping. Thus, doping La_2CuO_4 with the divalent Ba or Sr ions oxidizes the CuO_2 planes leaving holes as carriers. In contrast doping Nd_2CuO_4 and Pr_2CuO_4 with tetravalent Ce or Th ions reduces the CuO_2 planes, giving rise to electron carriers. It should be noticed that both p- and n-type compounds are found to crystallize in a tetragonal structure.

In order to detect the effect of doping onto the CuO_2 planes in n-type superconductors we have performed a very detailed neutron spectroscopic study of the lowest ground-state CF excitation of Pr^{3+} in $Pr_{2-x}Ce_xCuO_4$ [69]. This compound is superconducting in a narrow doping range ($0.14 \leq x \leq 0.17$) with maximum T_c=25 K. As for the $RBa_2Cu_3O_x$ compounds (see Section 4.1.1), the observed energy spectra are found to separate into different stable states whose spectral weights distinctly depend on the doping level (see Fig. 16). For the undoped compound we observe an intense CF transition at 18 meV (A) with some minor shoulders on both the low- and high-energy side (B_1 and B_2, respectively). With increasing doping, the transition A loses intensity, whereas the B_1 and B_2 transitions gain intensity. At higher doping, there is a fourth line emerging, marked with C.

One explanation for these observations could be a line broadening due to antiferromagnetic correlations between the Pr^{3+} ions. However, this effect should be most pronounced in the low-doping regime, in contrast to our observations. Moreover, these correlations have been shown to result in a dispersion of the CF transitions in a range between 18.2 meV and 19.2 meV [70, 71]. This cannot influence the observed spectra because this energy range corresponds to our instrumental resolution. Hence, even for x=0, the lines B_1 and B_2 cannot be explained by such an effect. Therefore, with the CF interaction being a local probe, there is no doubt that these lines originate from different local environments of the Pr^{3+} ion. Furthermore, because the intensity ratios of the B_1 and B_2 transitions are constant for all Ce contents, we have to assume that these transitions are caused by the same local environment and therefore represent a splitting of the Γ_5 doublet into two singlet states. Although diffraction techniques confirmed the tetragonal symmetry in the entire doping range, our observation clearly shows that, locally, the structure has a lower symmetry.

Our data and their interpretation provide clear experimental evidence for cluster formation. We tentatively identified the clusters associated with the transitions A, B (= B_1+ B_2) and C by local regions of undoped, intermediately doped and highly doped character, respectively. Upon doping with Ce^{4+} ions and reducing oxygen, electrons are continuously transferred into the CuO_2 planes. By this mechanism the number of local electron-rich regions (associated with the transitions B and C) rises, which can partially combine to form larger regions. Finally, we could show [69] that for a critical volume fraction of 50% a two-dimensional percolative network is built up [48], and the system undergoes a transition form the insulating to the metallic state.

The change of the relative intensities of the transitions A, B and C could be well reproduced [69] by assuming a similar statistical model as the one presented in Section 4.1.1 for $ErBa_2Cu_3O_x$.

It should be pointed out that different local environments of the R ions in these systems have also been inferred from Raman scattering [3, 4, 72].

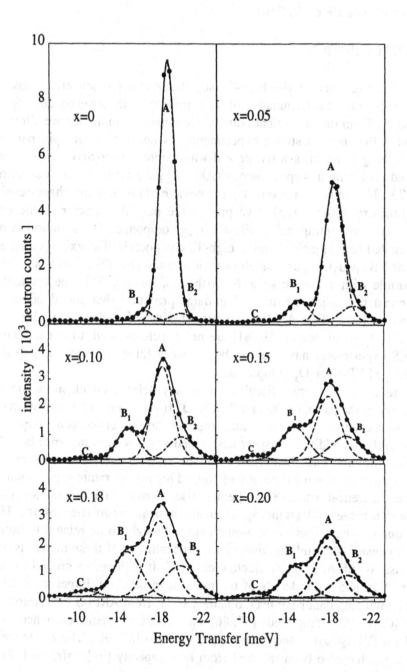

Figure 16: Energy spectra of neutrons scattered from $Pr_{2-x}Ce_xCuO_4$ measured at T=10 K and Q=1.8 Å^{-1}. The lines are the result of a least-squares fitting procedure as explained in Ref. [69].

5. Linewidth of CF excitations

5.1. INTRODUCTION

Since the discovery of the high-T_c superconductors much effort has been concentrated on the determination of the symmetry of the superconducting order parameter. From the theoretical point of view many symmetries are allowed in principle, but there is strong experimental evidence that the spin pairing is singlet, suggesting an s-wave or a d-wave state. In $Bi_2Sr_1CaCu_2O_8$, angle resolved photoemission spectroscopy (ARPES) gave evidence for an anisotropic gap [73, 74], thus ruling out a pure s-wave state. Recent phase-sensitive tunneling experiments [75] have proven the pure d character of the order-parameter in the tetragonal $Tl_2Ba_2CuO_{6+\delta}$ compound. The situation is more complicated for the orthorhombic high-T_c compounds. The existing tunneling data for $YBa_2Cu_3O_{7-\delta}$ are somehow controversial [76, 77] but to a large extent compatible with a d-wave state. Nevertheless, an s+id [78] type symmetry is also consistent with the tunneling data, provided that the d component dominates [79]. The ARPES results obtained for the optimally doped $YBa_2Cu_3O_{7-\delta}$ compound [80, 81] are not conclusive and, to our knowledge, ARPES experiments have not yet been successful to measure the gap in the underdoped $YBa_2Cu_3O_x$ compounds.

In general, it is very difficult to draw any definite conclusions about the symmetry of the pairing state in $YBa_2Cu_3O_x$ from a single experiment. In order to obtain a coherent picture, several complementary experiments are required.

Recently, an additional aspect gained very much importance both from the experimental and theoretical point of view, namely the existence of a pseudogap in the normal state of underdoped high-temperature superconductors. Several theoretical works [82, 83] emphasise the role of the pseudogap as a key element for the understanding of many unusual properties of the HTSC compounds. The origin of the pseudogap is thought to be related to both the low-dimensionality and the low-carrier concentration of these materials which give rise to enhanced phase fluctuations [84]. It is therefore crucial to obtain information about the details of the pseudogap function. Experimentally, the most convincing data have been obtained so far from ARPES measurements in $Bi_2Sr_2Ca_1Cu_2O_8$ compounds [85, 86] and nuclear magnetic resonance (NMR) results in $YBa_2Cu_3O_x$ and $YBa_2Cu_4O_8$ systems [87- 90]. The existence of a pseudogap has also been inferred from heat capacity [91], infrared [92], and transport [93] data.

We propose a complementary method to study both the symmetry and the opening of the gap in the $YBa_2Cu_3O_x$ family. This method is based on the temperature dependence of the relaxation rate of CF transitions of rare-earth containing systems.

5.2. STATIC SUSCEPTIBILITY IN THE SUPERCONDUCTING STATE

The CF linewidth is directly related to the static susceptibility $\chi''(\hbar\omega)$ via Eq. 14. Since an exact calculation of $\chi''(\hbar\omega)$ in the superconducting state is rather difficult, we have used the following simple expression which allows us to estimate the temperature dependence of the CF relaxation rate, both below and above a critical temperature T* (not necessarily coinciding with T_c as will be shown below), for several types of gap functions:

$$\chi''(\hbar\omega, T) = \chi''_N \oint_{k_F} dk \, G_k,$$

with   (20)

$$G_k = \begin{cases} 1 & \text{if } \hbar\omega \geq 2|\Delta_k(T)|, \\ \exp\left(-|\Delta_k(T)| / k_B T\right) & \text{if } \hbar\omega < 2|\Delta_k(T)|. \end{cases}$$

The following gap functions have been considered:

Isotropic   s $\Delta_s(k) = \Delta_s$

Anisotropic
$$\begin{cases} d & \Delta_{x^2-y^2}(k) = \Delta_d\left(\cos(k_x) - \cos(k_y)\right) \\ s^* & \Delta_{s^*}(k) = \Delta_{s^*}\left(\cos(k_x) + \cos(k_y)\right) \\ s + id^n & \Delta_{s+id^n}(k) = \Delta_s + i\Delta_d\left(\cos(k_x) - \cos(k_y)\right)^n \end{cases}$$

(21)

The integration in the (k_x,k_y) plane was performed both for squared and t' altered Fermi surfaces as determined from ARPES measurements in $Bi_2Sr_2CaCu_2O_8$ [94]. It is crucial to notice that the extended s*-wave gap function for a Fermi surface of a system with a second-nearest-neighbor term t' can have nodes [95]. This has important consequences since, as already mentioned above, CF relaxation measurements are only sensitive to the presence of nodes and not to the phase, i.e., we cannot distinguish between extended s* and d-wave gap functions. We assumed the following temperature dependence for the gap amplitude:

$$\Delta(T) = \Delta(T = 0)\left(1 - \left(\frac{T}{T^*}\right)^m\right)$$

(22)

A BCS type temperature dependence of the gap [96] can be well approximated by assuming m=4. Actually, it is found that our results are weakly affected by the value of m, and in the following we assume m=4. The CF linewidths Γ_N and Γ are then calculated using Eqs. 14, 20-22. Although this simple model may have several shortcomings, particularly in the vicinity of T* where more involved calculations would be necessary, it is able to explain, at least qualitatively, most of the important observations.

Isotropic   Anisotropic

s   d   s+id³

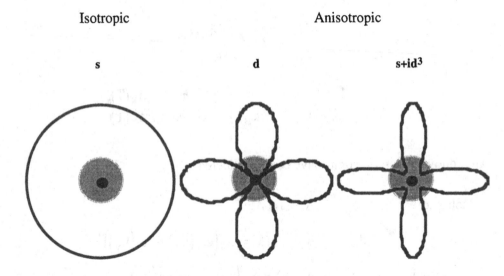

Figure 17: Schematic plots of $|\Delta_k|$ for isotropic s-, and anisotropic d- and s+id³-wave gap functions (full lines). The black and gray circles denote low-energy and high-energy CF transitions, respectively.

Fig. 17 shows schematically isotropic and anisotropic gap functions together with two CF energies. In the isotropic case the temperature dependence of the linewidth of CF transitions will be almost energy independent, while in the anisotropic case two major differences can be recognised. Firstly, the CF energy may cross the gap function, which gives rise to possible relaxation channels even at the lowest temperatures. Secondly, by tuning the energy $\hbar\omega$ of the CF transition, it is possible to investigate different regions of the reciprocal space. In the examples shown in Fig. 17 the low-energy CF transition (black circle) is probing the gap averaged over the entire Brillouin zone, while the high-energy

CF transition (gray circle) is probing the gap predominantly around the k_x- and k_y-directions.

Note that, using Eq.(20), we can also estimate the NMR relaxation rate T_1 above and below T*, since [97]:

$$\frac{1}{T_1 T} \propto \lim_{\hbar\omega \to 0} \frac{\chi''(\hbar\omega)}{\hbar\omega} \tag{23}$$

Eq. 23 can also be used to calculate the NMR Knight shift K, since it has been shown experimentally that in the superconducting state K is proportional to $1/T_1 T$ [98].

5.3. OPTIMALLY DOPED SYSTEMS

Recently a detailed study of the linewidth of the lowest CF transition at $\hbar\omega \approx$ 0.5 meV was performed for the optimally doped high-T_c compound $Ho_{0.1}Y_{0.9}Ba_2Cu_3O_{7-\delta}$ (T_c=92 K) [22]. Figure 18a shows the temperature behaviour of the reduced linewidth Γ/Γ_N, where Γ_N is the normal state linewidth calculated from Eq. 14 with $\chi''(\hbar\omega) = \chi_N''(\hbar\omega)$. Γ/Γ_N was observed to be zero only at the lowest temperatures, to increase linearly far below T_c, and to converge into the Korringa behavior above T_c (in that case T_c=T*). The unusual linewidth behavior below T_c was ascribed to a high degree of gap anisotropy. Figs. 18b and 18c show the calculated reduced linewidth in $Ho_{0.1}Y_{0.9}Ba_2Cu_3O_{7-\delta}$ for both isotropic s and anisotropic d gap functions. Γ/Γ_N depends strongly on the value of the gap maximum Δ_{max}. In the isotropic case and for large ratios $2\Delta_{max}/k_B T_c$, the linewidth rises exponentially below T_c and follows the normal metal behaviouir above T_c. This behaviour has been observed in the classical superconductor $La_{1-x}Tb_xAl_2$ (T*=T_c) [99]. However, although the relaxation behaviour observed for $Ho_{0.1}Y_{0.9}Ba_2Cu_3O_{7-\delta}$ drastically deviates from the classical case, an anisotropic gap function cannot be inferred necessarily. As shown in Figs. 18b and 18c, an isotropic gap function with a smaller value of Δ_{max} produces a similar behaviour as an anisotropic gap with a larger value of Δ_{max}. Identical conclusions result [9] from the analysis of the NMR relaxation rates obtained for O nuclei [88].

As will be shown below, definitive conclusions about the gap symmetry can only be made through the additional energy variable $\hbar\omega$, exclusively available from neutron CF experiments.

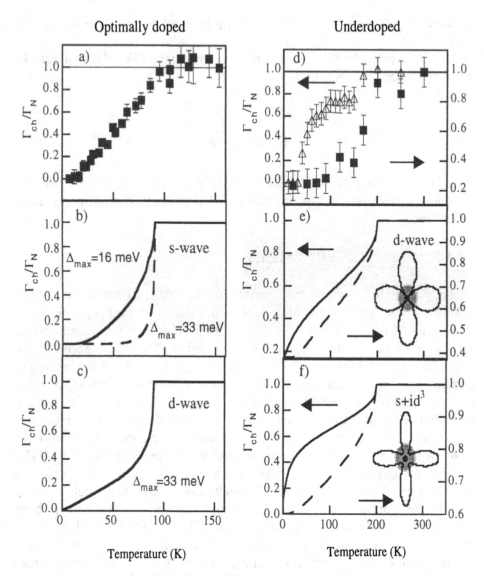

Figure 18: Temperature variation of the reduced linewidth of the lowest CF transition in (a-c) optimally doped and (d-f) underdoped high-T_c superconductors. Experimental data taken from (a) $Ho_{0.1}Y_{0.9}Ba_2Cu_3O_7$ [22] and (d) $HoBa_2Cu_4O_8$ and $Er_2Ba_4Cu_7O_{14.92}$ [9]. (d) The open triangles and filled squares refer to the low-energy ($\hbar\omega=1$ meV) and high-energy ($\hbar\omega=10$ meV) CF transitions, respectively. (b, c and e, f) Results of model calculations as explained in the text. (e and f) Temperature variation of the reduced linewidth Γ_{ch}/Γ_N calculated for $\hbar\omega\approx1$ meV (solid line) and $\hbar\omega\approx10$ meV (dashed line) CF transitions, in the case of anisotropic gap functions. The maximum of the gap was fixed to 33 meV, T* to 200 K and in (f) $\Delta_s/\Delta_d = 0.2$. The inserts represent schematic plots of the related $|\Delta_k|$ (solid lines), together with the energy of two CF transitions shown by the gray ($\hbar\omega=10$ meV) and black ($\hbar\omega=1$ meV) circles.

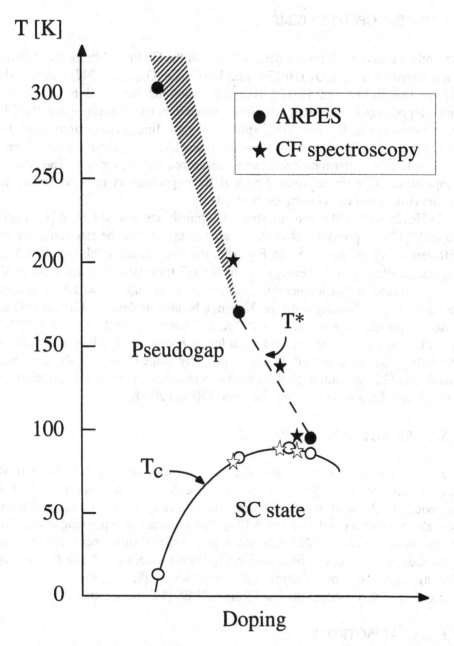

Figure 19: Schematic phase diagram of high-T_c superconductors as a function of doping. Full and open symbols denote T_c and T^*, respectively. Circles stand for ARPES values obtained for $Bi_2Sr_2CaCu_2O_8$ [86] and stars for CF values obtained for (from left to right) $HoBa_2Cu_4O_8$ & $Er_2Ba_4Cu_7O_{14.92}$ [102], $Tm_{0.1}Y_{0.9}Ba_2Cu_3O_{6.9}$ [29] and $Ho_{0.1}Y_{0.9}Ba_2Cu_3O_7$ [22], respectively. The lines and shaded areas are guides to the eye.

5.4. UNDERDOPED SYSTEMS

Recently we have performed a detailed study of the CF linewidth for the slightly underdoped $HoBa_2Cu_4O_8$ (Ho124) and $Er_2Ba_4Cu_7O_{14.92}$ (Er247) compounds [100]. The Er247 and Ho124 HTSC compounds are ideal to study the underdoped region since they are almost stoichiometric. Consequently, the CF transitions do not show line splittings and line broadenings due to inhomogeneities present in all non-stoichiometric compounds (see Sections 4.1.1, 4.1.2), and lifetime effects can be measured rather precisely. The critical temperatures T_c were determined by SQUID magnetometry to be 81 and 89 K for the Ho124 and Er247 samples, respectively.

In Ho124 and Er247 there are strong CF transitions around 1 meV [101] and 10 meV [17], respectively, thus we are able to investigate the electronic gap at different energies. As shown in Fig. 18d, the temperature evolution of Γ/Γ_N depends distinctly on the energy $\hbar\omega$ of the CF transition. For $\hbar\omega \approx 10$ meV Γ/Γ_N is finite at low temperature, remains constant up to T=70 K, and then increases almost linearly until the Korringa behaviour described by Eq.(14) is observed for temperatures $T>T^* \approx 200$ K, far above T_c. For $\hbar\omega \approx 1$ meV Γ/Γ_N is, within the experimental error, zero at low temperature, then increases already far below T_c, as observed for the optimally doped $Ho_{0.1}Y_{0.9}Ba_2Cu_3O_{7-\delta}$ compound [22], and although the increase is much more rapid, the normal-state behaviour is fully recovered only between 150 and 200 K.

5.5. NORMAL STATE PSEUDOGAP

Our measurements clearly support the existence of a pseudogap in the normal state of underdoped high-T_c compounds. Previous measurements of the CF linewidth in the slightly underdoped $Tm_{0.1}Y_{0.9}Ba_2Cu_3O_{6.9}$ compound have revealed a T* of the order of 140 K [29]. The existence of a pseudogap has also been inferred from NMR relaxation and Knight shift measurements in $YBa_2Cu_4O_8$ [88]. As can be shown in Fig. 19 the values of T* as a function of doping obtained from neutron CF spectroscopy [9, 22, 29] are in good agreement with the values obtained from ARPES [86] measurements.

5.6. GAP ANISOTROPY

Figures 18e and 18f show the calculated reduced linewidth Γ/Γ_N of both Er247 and Ho124 samples [9] for anisotropic d- and $s+id^3$ wave functions, respectively. We have fixed T* to 200 K, and the maximum of the gap function was taken to be 33 meV as determined from ARPES measurements in

$Bi_2Sr_2CaCu_2O_8$ [86]. The isotropic s-wave case can be ruled out, since from Eqs 20-21 we find no energy dependence of Γ/Γ_N, in contrast to the anisotropic cases (Figs. 18e). Within our simple model, we are able to reproduce the main features of the temperature dependence of Γ/Γ_N observed for the Ho124 and Er247 samples. Firstly, we find that the relaxation rate at T=0 K depends strongly on the energy of the CF transition. Secondly, for $\hbar\omega=10$ meV Γ/Γ_N remains constant up to 30 K and then increases quasi-linearly up to 200 K, while for $\hbar\omega=1$ meV the increase of Γ/Γ_N is more rapid at the lowest temperatures and slows down around T≈50 K.

More interesting is the quantitative comparison between the calculation performed for both d (Fig. 18e) and $s+id^3$ gap functions (Fig. 18f). It turns out that an $s+id^3$ wave gap can best reproduce the salient features of the temperature dependence of the observed linewidths. The reasons lie both in the power of the cosine term (narrowing the gap along the main axes) and in the isotropic component Δ_0 (no nodes) appearing in the $s+id^3$ gap function (see insert in Fig. 18f). The best agreement between measurement and calculation is obtained for $\Delta_s/\Delta_d \approx 0.2$, i.e., a mixed gap function of s-wave and dominating d-wave symmetries, consistent with the tunnelling data [79].

Note that a natural explanation for the narrowing of the gap function in the underdoped regime could be the increased long-range character of the e-e interaction due to the reduced screening factor [103]. Other possible scenarios are the presence of Van Hove singularities [104] or intra-bilayer interactions [105].

6. Conclusions

We have shown that neutron spectroscopy is a powerful tool to determine unambiguously the CF potential in rare-earth based high-T_c superconducting materials. This provides detailed information on the electronic ground state of the R ions which is important to understand the observed coexistence between superconductivity and long-range magnetic ordering of the R ion sublattice at low temperatures. Moreover, the inhomogeneous decay of the antiferromagnetic state of the parent compound as well as the inhomogeneous evolution of the superconducting state upon doping can be directly and quantitatively monitored, resulting in the onset of bulk superconductivity by a percolation mechanism. For underdoped systems, the origin of the inhomogeneities is believed to lie in the small hole density which gives rise to a strong enhancement of phase fluctuations [84]. A direct consequence of the phase fluctuations is the appearance of two crossovers associated with the superconducting gap and the pseudogap which merge at the optimum doping level. We have observed these effects through studying the temperature dependence of the linewidth of CF splittings, which - in addition - provides information about the symmetry of the gap function. In particular, CF relaxation experiments can be performed at different energies $\hbar\omega$, which is crucial to distinguish between several types of gap functions. A semi-quantitative analysis carried out for the R124 and R247 compounds shows that the neutron CF-data can be best reproduced by considering an anisotropic gap of predominant d character with a small s component.

Acknowledgments

Financial support by the Swiss National Science Foundation is gratefully acknowledged. For their participation in the experimental work as well as for numerous stimulating discussions, we are grateful to our colleagues P. Allenspach, B. Braun, G. Böttger, P. Fischer, M. Guillaume, W. Henggeler, B. Roessli, U. Staub (PSI, Villigen, Switzerland), E. Kaldis, J. Karpinski (ETH, Zurich, Switzerland), S. M. Bennington, Z. Bowden, R. J. Eccleston, R. Osborn, A. D. Taylor (RAL, Didcot, UK), H. Büttner, H. Mutka, C. Vettier (ILL, Grenoble, France), A. Mirmelstein, A. Podlesnyak (IMP, Ekaterinburg, Russia), V. Kresin (UC, Berkeley, USA), V. Trounov (PNPI, Gatchina, Russia), and K. A. Müller (IBM & Univ. of Zurich, Switzerland).

References

1. J. G. Bednorz and K. A. Müller, Z. Phys. B **64** (1986) 189.
2. M. K. Wu, J. R. Ashburn, C. J. Torng, P. H. Hor, R. L. Meng, L. Gao, Z. J. Huang, Y. Q. Wang, and C. W. Chu, Phys. Rev. Lett. **58** (1987) 908.
3. S. Jandl, P. Dufour, T. Strach, T. Ruf, M. Cardona, V. Nekvasil, C. Chen, B. M. Wanklyn, and S. Piñol, Phys. Rev. B **53** (1996) 8632.
4. J. A. Sanjurio, G. B. Martins, P. G. Pagliuso, E. Granado, I. Torriani, C. Rettori, S. Oseroff, and Z. Fisk, Phys. Rev. B **51** (1995) 1185.
5. T. Strach, T. Ruf, M. Cardona, C. T. Lin, S. Jandl, V. Nekvasil, D. I. Zhigunov, S. N. Barilo, and S. V. Shiryaev, Phys. Rev. B **54** (1996) 4276.
6. W. Henggeler, T. Chattopadhyay, B. Roessli, P. Vorderwisch, P. Thalmeier, D. I. Zhigunov, S. N. Barilo, and A. Furrer, Phys. Rev B **55** (1997) 1269.
7. V. Hizhnyakov and E. Sigmund, Physica C **156** (1988) 655.
8. *Phase separation in Cuprate Superconductors,* E. Sigmund and K. A. Müller (eds.), Springer, Berlin, 1994.

9. J. Mesot and A. Furrer, J. Supercond. **10** (1997) 623.
10. M. Guillaume, W. Henggeler, A. Furrer, R. S. Eccleston, and V. Trounov, Phys. Rev. Lett. **74** (1995) 3423.
11. M. T. Hutchings: *Solid State Physics*, Vol. 16, F. Seitz and D. Turnbull (eds), Academic Press, New York, 1964, p. 227.
12. R. M. Sternheimer, Phys. Rev. **146** (1966) 140.
13. C. A. Morrison: *Angular Momentum Theory Applied to Interactions in Solids*, Vol. 47, G. Berthier (ed.), Springer, Berlin, 1988.
14. U. Tellenbach: *Report AF-SSP-75*, ETH Zurich, 1974.
15. J. Mesot, P. Allenspach, U. Gasser, and A. Furrer, J. Alloys Comp. **250** (1997) 559.
16. J. Mesot: *Magnetic Neutron Scattering*, A. Furrer (ed.), World Scientific, Singapore, 1995, p. 178.
17. J. Mesot, P. Allenspach, U. Staub, A. Furrer, H. Mutka, R. Osborn, and A. Taylor, Phys. Rev. B **47** (1993) 6027.
18. A. Furrer, P. Brüesch, and P. Unternährer, Phys. Rev. B **38** (1988) 4616.
19. K. R. Lea, M. J. M. Leask, and W. P. Wolf, J. Phys. Chem. Solids **23** (1962) 1381.
20. P. Fulde and M. Löwenhaupt, Adv. Phys. **34** (1986) 589.
21. K. W. Becker, P. Fulde, and J. Keller, Z. Phys. B **28** (1977) 9.
22. A. T. Boothroyd, A. Mukherjee, and A. P. Murani, Phys. Rev. Lett. **77** (1996) 1600.
23. H. Alloul, T. Ohno, and P. Mendels, Phys. Rev. Lett. **63** (1989) 1700.
24. B. Rupp, E. Pörschke, P. Meuffels, P. Fischer, and P. Allenspach, Phys. Rev. B **40** (1989) 4472.
25. L. W. Roeland, F. R. De Boer, Y. K. Huang, A. A. Menovsky, and K. Kadowaki, Physica C **152** (1988) 72.
26. H. P. Van Der Meulen, J. J. M. Franse, Z. Tarnawski, K. Kadowaki, J. C. P. Klasse, and A. A. Menovski, Physica C **152** (1988) 65.
27. B. D. Dunlap, M. Slaski, D. G. Hinks, L. Soderholm, M. Beno, K. Zhang, C. Segre, G. W. Crabtree, W. K. Kwok, S. K. Malik, I. K. Schuller, J. D. Jorgensen, and Z. Sungaila, J. Magn. Magn. Mater. **68** (1987) L139.
28. J. A. Hodges, P. Imbert, J. B. Marimon de Cunha, and J. P. Sanchez, Physica C **160** (1989) 49.
29. R. Osborn and E. A. Goremychkin, Physica C **185-189** (1991) 1179.
30. K. N. Yang, J. M. Ferreira, B. W. Lee, M. B. Maple, W. H. Li, J. W. Lynn, and R. W. Erwin, Phys. Rev. B **40** (1989) 10963.
31. G. L. Goodman, C.-K. Loong, and L. Soderholm, J. Phys.:Condens. Matter **3** (1991) 49.
32. A. T. Boothroyd, S. M. Doyle, and R. Osborn, Physica C **217** (1993) 425.
33. P. Allenspach, J. Mesot, U. Staub, M. Guillaume, A. Furrer, S.-Y. Yoo, M. J. Kramer, R.W. McCallum, H. Maletta, H. Blank, H. Mutka, R. Osborn, M. Arai, Z. Bowden, and A. D. Taylor, Z. Phys. B **95** (1994) 301.
34. U. Staub, L. Soderholm, R. Osborn, M. Guillaume, A. Furrer, and V. Trounov, J. Alloys Comp. **225** (1995) 591.
35. U. Staub, J. Mesot, M. Guillaume, P. Allenspach, A. Furrer, H. Mutka, Z. Bowden, and A. Taylor, Phys. Rev. B **50** (1994) 4068.
36. L. Soderholm, C.-K. Loong, and S. Kern, Phys. Rev. B **45** (1992) 10062.

37. M. Guillaume, P. Allenspach, J. Mesot, U. Staub, A. Furrer, R. Osborn, A. D. Taylor, F. Stucki, and P. Unternährer, Solid State Commun. **82** (1992) 999.
38. R. J. Birgeneau, E. Bucher, J. P. Maita, L. Passel, and K. C. Turberfield, Phys. Rev. B **8** (1973) 5345.
39. A. Furrer and H. U. Güdel, Phys. Rev. B **56** (1997) 15062.
40. T. R. Faulkner, J. P. Morley, F. S. Richardson, and R. W. Schwartz, Mol. Phys. **40** (1980) 1481.
41. M. F. Reid and F. S. Richardson, J. Chem. Phys. **83** (1985) 3831.
42. P. G. Radaelli, C. U. Segre, D. G. Hinks, and J. D. Jorgensen, Phys. Rev. B **45** (1992) 4923.
43. R. J. Cava, A. W. Hewat, E. A. Hewat, B. Batlogg, M. Marezio, K. M. Rabe, J. J. Krajewski, W. F. Peck Jr., and L. W. Rupp Jr., Physica C **165** (1990) 419.
44. J. Mesot, P. Allenspach, U. Staub, A. Furrer, and H. Mutka, Phys. Rev. Lett. **70** (1993) 865.
45. A. Furrer, J. Mesot, P. Allenspach, U. Staub, F. Fauth, and M. Guillaume: *Phase separation in Cuprate Superconductors,* E. Sigmund and K. A. Müller (eds.), Springer, Berlin, 1994, p. 101.
46. V. Hizhnyakov and E. Sigmund, Physica C **156** (1988) 655.
47. A. K. Kremer, E. Sigmund, V. Hizhnyakov, F. Hentsch, A. Simon, K. A. Müller, and M. Mehring, Z. Phys. B-Condensed Matter **86** (1992) 319.
48. S. Kirkpatrick, Rev. Mod. Phys. **45** (1973) 574.
49. Y. Kubo and H. Igarashi, Phys. Rev. B. **39** (1989) 725.
50. M. S. Osofsky, J. L. Cohn, E. F. Skelto, M. M. Miller, J. R. J. Soulen, S. A. Wolf, and T. A. Vanderah, Phys. Rev. B **45** (1992) 4917.
51. P. Allenspach, A. Furrer, and B. Rupp: *Progress in High-Temperature Superconductivity,* V. L. Aksenov, N. N. Bogolubov, and N. M. Plakida (eds.), World Scientific, Singapore, 1990, p. 318.
52. A. Furrer, P. Allenspach, F. Fauth, M. Guillaume, W. Henggeler, J. Mesot, and S. Rosenkranz, Physica C **235-240** (1994) 261.
53. P. C. Hammel, A. P. Reyes, Z. Fisk, M. Takigawa, J. D. Thompson, K. H. Heffner, and S.-W. Cheong, Phys. Rev. B **42** (1990) 6781.
54. G. Wübbeler and O. F. Schirmer, Phys. stat. sol. (b) **174** (1990) K21.
55. J. A. Hodges, P. Bonville, P. Imbert, G. Jéhanno, and P. Debray, Physica C **184** (1991) 270.
56. N. Poulakis, D. Palles, E. Liarokapis, K. Konder, and E. Kaldis, Phys. Rev. B **53** (1996) R534.
57. V. J. Emery, S. A. Kivelson, and H. Q. Lin, Phys. Rev. Lett. **64** (1990) 475.
58. M. Grilli, R. Raimondi, C. Castellani, C. Di Castro, and G. Kotliar, Phys. Rev. Lett. **67** (1991) 259.
59. A. Podlesnyak, V. Kozhevnikov, A. Mirmelstein, P. Allenspach, J. Mesot, U. Staub, A. Furrer, R. Osborn, S. M. Bennington, and A. D. Taylor, Physica C **175** (1991) 587.
60. T. Miyatake, K. Yamaguchi, T. Takata, N. Koshizuka, and S. Tanaka, Phys. Rev. B **44** (1991) 10139.
61. A. V. Mahajan, H. Alloul, G. Collin, and J. F. Marucco, Phys. Rev. Lett. **72** (1994) 3100.
62. S. Zagoulaev, P. Monod, and J. Jégoudez, Phys. Rev. B **52** (1995) 10474.
63. V. Z. Kresin and S. A. Wolf, Phys. Rev. B **46** (1992) 6458.

64. D. J. C. Walker, O. Laborde, A. P. Mackenzie, S. R. Julian, A. Carrington, J. W. Loram, and J. R. Cooper, Phys. Rev. B **51** (1995) 9375.

65. J. P. Franck, J. Jung, M. A.-K. Mohamed, S. Gygax, and G. I. Sproule, Phys. Rev. B **44** (1991) 5318.

66. Yu. N. Ovchinnikov and V. Z. Kresin, Phys. Rev. B **54** (1996) 1251; V. Z. Kresin, A. Bill, S. A. Wolf, and Yu. N. Ovchinnikov, Phys. Rev. B **56** (1997) 107.

67. B. Bucher, J. Karpinski, E. Kaldis, and P. Wachter, J. Less Common Metals **164 &165** (1990) 20.

68. J. Mesot, P. Allenspach, U. Staub, A. Furrer, H. Blank, H. Mutka, C. Vettier, E. Kaldis, J. Karpinski, and S. Rusiecki, J. Less Common Met.als **164 & 165** (1990) 59.

69. W. Henggeler, G. Cuntze, J. Mesot, M. Klauda, G. Saemann-Ischenko, and A. Furrer, Europhys. Lett. **29** (1995) 233.

70. I. W. Sumarlin, J. W. Lynn, T. Chattopadhyay, S. N. Barilo, and D. I. Zhigunov, Physica C **219** (1994) 195.

71. W. Henggeler, T. Chattopadhyay, B. Roessli, P. Vorderwisch, P. Thalmeier, D. I. Zhigunov, S. N. Barilo, and A. Furrer, Phys. Rev. B **55** (1997) 1269.

72. T. Strach, T. Ruf, M. Cardona, S. Jandl, and V. Nekvasil, Physica B **234-236** (1997) 810.

73. Z. X. Shen, W. E. Spicer, D. M. King, D. S. Dessau, and B. O. Wells, Science **267** (1995) 343.

74. H. Ding, M. R. Norman, J. C. Campuzano, M. Randeria, A. F. Bellman, T. Yokoya, T. Takahashi, T. Mochiku and, K. Kadowaki, Phys. Rev. B **54** (1996) R9678.

75. C. C. Tsuei, J. R. Kirtley, Z. F. Ren, J. H. Wang, H. Raffy, and Z. Z. Li, Nature **387** (1997) 481.

76. K. A. Müller, Nature **377** (1995) 133.

77. K. A. Müller and H. Keller: *High-T_c Superconductivity: Ten Years after the Discovery*, E. Kaldis (ed.), Kluwer, Dodrecht, 1997, p. 7.

78. G. Kotliar, Phys. Rev. B **37** (1988) 3664.

79. D. A. Brawner and H. R. Ott, Phys. Rev. B **53** (1996) 8249.

80. H. Ding, M. R. Norman, J. Giapintzakis, J. C. Campuzano, H. Claus, H. Wühl, M. Randeria, A. Bellman, T. Yokoya, T. Takahashi, T. Mochiku, K. Kadowaki, and D. M. Ginsberg, SPIE Symposium Series **2699** (1996) 496.

81. M. C. Schabel, C.-H. Park, A. Matsuura, Z. X. Shen, D. A. Bonn, R. Liang, and W. N. Hardy, Phys. Rev. B **55** (1997) 2796.

82. V. J. Emery and S. A. Kivelson, Nature **374** (1995) 434.

83. C. Castellani, C. D. Castro, and M. Grilli, Physica C **282-287** (1997) 260.

84. V. Emery and S. A. Kivelson, Phys. Rev. Lett. **74** (1995) 3253.

85. A. G. Loeser, Z.-X. Shen, D. S. Dessau, D. S. Marshall, C. H. Park, P. Fournier, and A. Kapitulnik, Science **273** (1996) 325.

86. H. Ding, T. Yokoya, J. C. Campuzano, T. Takahashi, M. Randeria, M. R. Norman, T. Mochiku, K. Kadowaki, and J. Giapintzakis, Nature **382** (1996) 51.

87. M. Takigawa, A. P. Reyes, P. C. Hammel, J. D. Thompson, R. H. Heffner, Z. Fisk, and K. C. Ott, Phys. Rev. B **43** (1991) 247.

88. I. Tomeno, T. Machi, K. Tai, and N. Koshizuka, Phys. Rev. B **49** (1994) 15327.

89. G. V. M. Williams, J. L. Tallon, R. Michalak, and R. Dupree, Phys. Rev. B **54** (1996) R6909.
90. M. Bankai, M. Mali, J. Roos, and D. Brinkmann, Phys. Rev. B **50** (1994) 6416.
91. J. W. Loram, K. A. Mirza, J. R. Cooper, and J. R. Liang, Phys. Rev. Lett. **71** (1993) 1740.
92. D. N. Basov, T. Timusk, B. Dabrowski, and J. D. Jorgensen, Phys. Rev. B **50** (1994) 3511.
93. J. L. Tallon, J. R. Cooper, P. S. I. P. N. de Silva, G. V. M. Williams, and J. W. Loram, Phys. Rev. Lett. **75** (1995) 4414.
94. H. Ding, M. R. Norman, T. Yokoya, T. Takeuchi, M. Randeria, J. C. Campuzano, T. Takahashi, T. Mochiku and, K. Kadowaki, Phys. Rev. Lett. **78** (1997) 2628.
95. D. J. Scalapino, Physics Reports **250** (1995) 329.
96. B. Mühlschlegel, Z. Phys. **155** (1959) 313.
97. R. Stern, M. Mali, J. Roos, and D. Brinkmann, Phys. Rev. B **51** (1995) 15478.
98. R. E. Walstedt and J. W. W. Warren, Science **248** (1996) 1082.
99. R. Feile, M. Loewenhaupt, J. K. Kjems, and H. E. Hoenig, Phys. Rev. Lett. **47** (1981) 610.
100. A. Furrer, J. Mesot, W. Henggeler, and G. Böttger, J. Supercond. **10** (1997) 273.
101. B. Roessli, P. Fischer, M. Guillaume, J. Mesot, U. Staub, M. Zolliker, A. Furrer, E. Kaldis, J. Karpinski, and E. Jilek, J. Phys.: Condens. Matter **6** (1994) 4147.
102. J. Mesot, G. Boettger, A. Furrer, and H. Mutka, submitted to Phys. Rev. Lett. (1997).
103. M. R. Norman, private communication.
104. J. Maly, D. Z. Liu, and K. Levin, Phys. Rev. B **53** (1996) 6786.
105. S. Chakravarty, A. Sudbø, P. W. Anderson, and S. Strong, Science **261** (1993) 337.

SMALL ANGLE NEUTRON SCATTERING EXPERIMENTS ON VORTICES IN COPPER OXIDE SUPERCONDUCTORS

E.M. FORGAN
School of Physics and Astronomy
University of Birmingham
Birmingham B15 2TT, U.K.

1. History and Introduction

When the concept of magnetic flux lines in superconductors was first proposed by Abrikosov, it was sufficiently revolutionary to require microscopic experimental confirmation. This was eventually supplied by the pioneering experiment of Cribier et *al.* [1], which showed that flux lines could be detected by their ability to scatter neutrons. This resulted in many detailed investigations, using SANS and other techniques, of the structure of flux lines in superconductors. This was essentially a "crystallography" of the flux line lattice (FLL) that forms inside a Type-II superconductor when the applied magnetic field H is sufficient to cause vortices to enter ($>H_{c1}$) but insufficient to destroy superconductivity altogether ($<H_{c2}$). The SANS technique was re-invigorated by the discovery of High-T_c Superconductors, which are all Type-II, and the subsequent neutron observations of flux lines in YBCO [2]. Since then, the interest has spread to flux structures in other materials such as Heavy Fermion [3] and Borocarbide superconductors [4,5], as well as the direct observation of *Flux Lattice Melting* in High-T_c materials [6,7]. In this article, we shall review the fundamentals of the SANS technique applied to the observation of flux lines, the nature of flux structures that one would expect to observe, and recent results in High-T_c materials. We shall conclude with an indication of future possibilities for this and related techniques.

2. The Interaction between Flux Lines and Neutrons

2.1. DIFFRACTION

The diffraction of neutrons by flux lines occurs because of the neutron magnetic moment, which gives it a potential energy in a magnetic field B. Thus a neutron of fixed *total* energy in a spatially varying magnetic field has a *kinetic* energy which is a function of position. Hence the neutron is effectively moving through a medium with a spatially varying refractive index and diffraction occurs. The neutron does not need to be polarised: the interaction occurs for both orientations of the neutron spin relative to the magnetic field, and since the magnetic potential energy is much smaller than the neutron kinetic energy, the magnitude of the effect is independent of spin orientation.

The Bragg d-spacing of a FLL is given by the condition that there is one flux quantum, Φ_0, per unit cell, and hence:

$$d = \sqrt{(\Phi_0 / B)} \quad , \tag{1}$$

for a square lattice (and 7% smaller for the more likely triangular lattice). For a field of 0.2 T, this gives the d-spacing ~ 1000 Å, and with an incident neutron wavelength λ of 10 Å, the Bragg 2Θ is $\sim 0.5°$. Hence FLL diffraction is a SANS experiment.

The expression for the intensity I_{hk} of a single (h,k) reflection (integrated over the rocking curve of the FLL) is given, e.g. in [8] by:

$$I_{hk} = 2\pi\phi\left(\frac{\gamma}{4}\right)^2 \frac{V\lambda^2}{\Phi_0^2\tau_{hk}}\left|F_{hk}\right|^2 \quad , \tag{2}$$

where ϕ is the incident neutron flux, γ is the neutron magnetic moment in nuclear magnetons, V is the sample volume, and F_{hk} is the τ_{hk} Fourier coefficient of the magnetic field in the mixed state. The London theory gives:

$$F_{hk} = \frac{B}{1+(\tau_{hk}\lambda_L)^2} \quad , \tag{3}$$

where λ_L is the (temperature-dependent) magnetic penetration depth. For all inductions larger than B_{c1}, the second term in the denominator of Eq. 3 is dominant, so that the intensity given by Eq. 2 varies with wavevector as $(\tau_{hk})^{-5}$. Higher-order reflections are *much* weaker than low-order ones, because flux lines are finite objects with their field extending over a distance $\sim \lambda_L$. This also means that the intensity falls off as $1/\lambda_L^4$: this rapid variation makes the SANS signals *very* weak for High-T_c (and even more, Heavy Fermion) superconductors, which have long magnetic penetration depths. The field-dependence of intensity for given h,k is however only an effect of reciprocal-space geometry: $I_{hk} \propto 1/\tau_{hk} \propto B^{-0.5}$ (F_{hk} is field-*in*dependent for inductions larger than B_{c1}). One might think that as the flux lines get closer together, their magnetic fields would overlap and reduce the spatial variation of B that gives rise to diffraction. This does not occur because in the London theory the magnetic field near the axis of a vortex line rises logarithmically and this sharp variation is not removed by increasing the flux line density. However, the London theory is a high-κ approximation (κ is λ_L divided by ξ, the superconducting coherence length), in which the vortex cores of radius $\sim \xi$ are ignored. When core effects are included, the intensity falls off much more quickly with increasing

B, going to zero at B_{c2}, where the vortex cores strongly overlap and superconductivity disappears.

There have been some recent attempts to use the deviations from the London B-dependence of F_{hk} to derive values for the coherence length (see e.g. 3,4,5). However, these relied on an estimated correction for the effect of cores, which had no secure theoretical foundation. If microscopic calculations are too unwieldy, then an exact solution of the Ginzburg-Landau equations is the next best thing. Recently, this has been done numerically throughout the mixed state [9], and these numerical results have given confidence in an approximate variational expression for core effects [10], which has the advantage of being available in algebraic form. In High-T_c materials, B_{c2} at low temperatures is so large that one might expect that core effects would be negligible. However, there is the added complication that the superconducting wavefunction in these materials almost certainly has a d-wave character. This has the consequence that the superconducting energy gap is zero at certain lines on the Fermi surface. In these regions, the very large supercurrent velocity near a vortex core can cause depairing and a reduction in the superfluid density. Such d-wave "core" effects are much larger than those in conventional s-wave superconductors; they may well have been observed by μSR [11], and should be taken into account when analysing high-field data.

2.2. NEUTRON SPIN FLIP

In an *isotropic* superconductor, all supercurrents flow in planes which are perpendicular to the direction of the average induction., and the spatially varying part of the magnetisation due to these currents is parallel (or antiparallel) to the average induction. Hence, neutron scattering occurs without flipping of the spins of the neutrons [12]. However, in an anisotropic superconductor - and in particular in the layered High-T_c materials, this may not be the case, since the supercurrents will tend to flow in the CuO_2 planes, which will not be perpendicular to the average induction if the field is applied at an angle to the c-direction. This gives rise to spatially varying local fields which have components perpendicular to the direction of the average induction (which, by flux quantisation, is also the direction of the flux line axes). Transverse components of the local field will - if they have a component perpendicular to the scattering vector [12] - flip the neutron spins from parallel to antiparallel to the field - or vice versa. All components of the spatially varying fields in the mixed state have been calculated within anisotropic London theory [13], and the transverse components are comparable in magnitude to the longitudinal one, unless the field is applied along a principal axis of the crystal. Hence, the effect is quite measurable by *polarisation-analysed* SANS experiments, and the detailed predictions of the theory may be tested.

2.3. NEUTRON DEPOLARISATION

Another way to investigate magnetic field distributions is to investigate the change in polarisation of neutrons *transmitted* through the sample. If the field in the sample is uniform, then all transmitted neutrons travelling at a given speed will have their *degree* of polarisation *un*changed (although the *direction* of polarisation may have altered, due to the Larmor precession of the neutron spins during the time they are in the field). On the other hand, if the field is non-uniform, then the beam may be depolarised. This technique, relying on *transmission*, rather than scattering, is clearly sensitive to field variations over distances rather larger than the flux lattice spacing, and can therefore be used to investigate flux gradients, pinning etc. In the general case, the incident neutron polarisation can be in any direction relative to the incident neutron beam, so the technique is referred to as "three-dimensional neutron depolarisation" (3DND), and has application to magnetic materials [14] as well as superconductors [15].

The theory may be simplified as follows [16]. We suppose that there is a small region in the sample, of dimension ζ, in which the field deviates from the average field and has an extra component B_n perpendicular to the neutron spin direction. This region can cause depolarisation if the apparent turning frequency of the field (= v/ζ, where v is the speed of the neutron) is larger than the Larmor frequency, γB. This condition is usually satisfied inside a sample (e.g. up to $\zeta = 0.2$ mm for $\lambda = 1.5$ Å neutrons at $B = 100$ mT), so the neutron spin cannot adiabatically follow the field direction and in the time (ζ/v) precesses by a small angle $\phi = \gamma B_n(\zeta/v)$ about the local field. As a result the neutron polarisation is reduced by the factor $\cos\phi \sim 1 - \phi^2/2$. If there are n such independent regions along the neutron path, then the depolarisations add in a random walk fashion and the net polarisation becomes:

$$P \approx \exp(-n\gamma^2\langle B_n^2\rangle\zeta^2/2v^2) = \exp(-\gamma^2\langle B_n^2\rangle\zeta t/2v^2) \quad , \tag{4}$$

where $t = n\zeta$ is the thickness of the sample.

One possible ND setup would be with the initial neutron spin and average magnetic field parallel to the neutron beam: in that case, if bundles of flux lines of dimension ζ are *tilted* with respect to the average field direction, then depolarisation will occur [16]. Alternatively, with longitudinally polarised neutrons, the field may be aligned perpendicular to the neutron beam. In this case, the average field causes an overall *rotation* of the polarisation but in addition, either tilts of flux bundles or variations of the *magnitude* of the field will cause depolarisation. It should be mentioned that variations of magnetic field around an *individual* flux line occur on the length scale of the flux line spacing, d. These variations will give rise to SANS if the flux lines are oriented correctly for diffraction. They will *not* give rise to detectable depolarisation unless the detector has sufficient aperture to accept the diffracted beams, and even in this case the effects are often small [17]. This may be regarded as arising because the length scale $\zeta \sim d$ is too short. Thus SANS and neutron depolarisation may be regarded as complementary: one is sensitive to flux lattice structure and the other to longer-range variations of magnetic field.

3.   Experimental Geometry for SANS

3.1.   DIFFRACTION GEOMETRY

The diffraction geometry used to observe flux lines is important: there are two obvious possibilities, illustrated in Fig. 1.

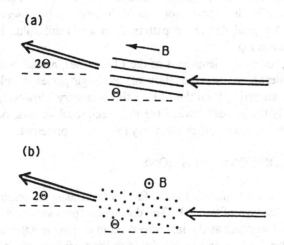

Fig. 1:   (a) Longitudinal and (b) transverse field geometries for small-angle neutron diffraction by the flux lattice. Θ is the Bragg angle, exaggerated for clarity.

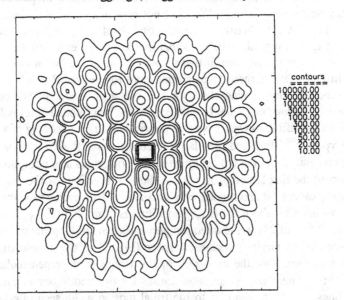

Fig 2.:   Contour plot (logarithmic scale) of a neutron diffraction pattern from the flux lattice in Nb at $B = 0.1$ T, $T = 1.6$K. The incident beam spot at the centre has been masked.

In Fig. 1(a), the neutron beam is incident at Θ to the magnetic flux lines: i.e. the neutron beam and magnetic field are almost parallel and the Bragg planes run through the flux lines into the page. In case (b), the Bragg planes have the same orientation relative to the neutron beam, but the neutrons are incident perpendicular to the flux lines. Usually, the first arrangement is to be preferred, because *all* the various Bragg planes can be brought into the diffracting condition by *small* (~2Θ) tilts of the field - up, down or out of the page. Alternatively, the mosaic spread of the FLL, if it is comparable with 2Θ, will allow all the diffraction spots about the main beam to be observed simultaneously without any tilt. A typical diffraction pattern from a triangular flux lattice under these conditions is shown in Fig. 2.

In case (b), for most orientations of the FLL planes, no diffraction is observed. If the FLL is disordered, or the orientation of its Bragg planes is unknown, it may be difficult to detect scattering from the FLL in this geometry. However, if it is desired to measure accurately the angles between the FLL reciprocal vectors, or the alignment of FLL planes with crystal planes, then geometry (b) is to be preferred.

3.2. RESOLUTION CONSIDERATIONS

It is of some importance to understand the factors controlling the resolution of the SANS setup for various features of the FLL. The diffraction spots shown in Fig. 2 will vary in shape, position and intensity as the instrument setup or sample angle are changed or the field is altered. These variations may be related to the perfection of the flux lattice, and to the distribution of values of wavelength and incidence angle of the neutron beam. If the latter are known, then in suitable circumstances the degree of perfection of the flux lattice may be deduced [18,19]. We shall not give here a detailed discussion, but point out some important qualitative considerations. Firstly, the incident beam will have a wavelength spread, typically ~10%, so that the Bragg angles Θ and 2Θ are likely to have a similar fractional spread, even if there is no spread of the flux lattice parameter in the sample. In fact, *both* of these factors contribute to the *radial* width of a spot- i.e. the width in the direction joining the spot and the main beam. In addition, the angular divergence of the incident beam (controlled by the input collimation) will tend to give a finite size to the diffraction spots, so that the *resolution in d-spacing* of a SANS machine is poor (typically >10%), and only large fractional spreads $\delta d/d$ may be measured. Due to the beam collimation, the typical *tangential* resolution is also poor, so that the angular orientation of the flux lattice planes about the *B* - direction (or the distribution of such orientations) cannot be determined very accurately. However, the width of the *rocking curve* of the intensity about the Bragg condition is typically $\prec 1°$ (~1/50 radian), which gives a resolution of 2% in the direction perpendicular to the plane of the detector. This may be seen rather nicely in reciprocal space (Fig. 3). If the mosaic spread of the flux lattice is b radians, then the extension of the Bragg spot perpendicular to the Ewald sphere is $b\tau_{hk}$. There are two possible causes for this width: one is non-straightness of the flux lines, which corresponds to the usual mosaic width seen in crystals. The other arises from a kind of "finite crystal" effect, arising if the correlation length of the flux lines along B is finite - for instance due to pinning of flux lines - which destroys long-range order [20]. Since $d = 2\pi/\tau_{hk}$, we may estimate the correlation length $L = 2\pi/b\tau_{hk}$. It should be emphasised that the width of the *rocking curve* is determined by a combination of the sample mosaic spread and the instrument resolution parameters [18], and since non-straightness may also be present, it may only be possible to set a *lower limit* on L. Nevertheless, with $d = 1000$ Å, and a rocking curve width of 1°, one is sensitive to values of L up to 5μ or even more.

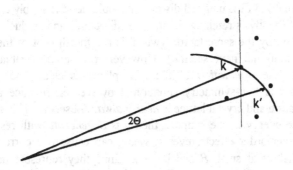

Fig. 3:   Scattering geometry in **k** - space for arrangement (a) of Fig. 1. A horizontal spot may be brought onto the Ewald sphere by a rotation of Θ about the vertical axis. The extension of the Bragg spot in the direction perpendicular to the Ewald sphere is related to the width of the mosaic spread of the FLL crystal about this axis.

It is also useful to note from Fig.3 that in this geometry, one has all the spots close to the Ewald sphere. Hence if the rocking curve is wide enough (as in Fig. 2), the diffraction pattern seen in the detector is just an image of the reciprocal lattice of the FLL. If the field is aligned in the geometry of Fig. 1(b), then the reciprocal lattice is rotated by 90°, and only two, if any, of the spots lie close to the Ewald sphere. However, the best resolution direction remains perpendicular to the detector, and hence (as mentioned earlier) one can measure the *orientation* of the spots very accurately (but with less resolution for L).

4.   Flux Lattice Structure, Orientation, Pinning and Melting

4.1.   EFFECTS OF CRYSTAL SYMMETRY

The Ginzburg-Landau (GL) and London approaches both predict that for an isotropic material the lowest-energy flux line structure is a regular triangular lattice. In practice, triangular lattices are usually observed, but often they are distorted or aligned in such a way as to indicate an interaction with the underlying crystal lattice [21,22]. However, if the underlying lattice has threefold or higher rotational symmetry about the field direction, then *no* such interaction is to be expected within the standard GL or London theories. This is because, within such theories, the crystal anisotropy appears as an anisotropic effective mass. If the field is applied in a high-symmetry direction, the mass - a second rank tensor - *has* to be *isotropic* - just like the resistivity in the plane perpendicular to the field. Furthermore, even when the symmetry is lower, so that the mass (and therefore magnetic penetration depth) are not equal for e.g. the *x* and *y*-axes, the situation can still be restored to symmetrical by an anisotropic scale transformation, so long as the field is applied along a principal axis of the effective mass tensor. This is useful in telling us what the FLL unit cell distortion should be: it is simply created from a regular triangular FLL by stretching it in one direction and reducing it in the perpendicular direction by the same factor, (which is the fourth root of the mass ratio: see [23], and Eq. 5 in the next section). However, the *rescaled* situation is now symmetrical, so that the *orientation* of the FLL planes is clearly not theoretically determined, even in this low-symmetry situation. However, in practice a correlation between FLL orientation and crystal lattice is *nearly always* observed. This should not be too surprising: if the energy in the simplest theory is *degenerate* with respect to FLL orientation, then higher-order effects, even if weak, can cause this correlation. All of these effects are smallest at small B and large κ, since they represent non-local (i.e. Cooper pair dimension) or *flux line core* effects added to the London model [24], which are equivalent to higher gradient terms added to the GL theory [25,26]. Core effects can cause a distortion and alignment of a triangular FLL [21,22,24,25,26], or even stabilise (the normally slightly higher energy) square FLL [4,5,21,22,24]. In High-T_c materials, there are also d-wave effects: a d-wave core has a 4-leaved clover shape, and in theory this can result in a distorted triangular or square FLL, and an alignment of the FLL with the d-wavefunction, and hence with the crystal lattice [27]. Finally, we should mention that extrinsic causes, such as pinning by crystal defects, twins, or the directions of cracks and surfaces, can also cause FLL alignment, which may therefore not always be due to fundamental causes [28].

4.2. EFFECTS OF CRYSTAL ANISOTROPY

In High-T_c materials, the superconductivity mainly resides in planes which are made up of 1, 2 or more closely coupled CuO_2 layers, with weak coupling to the next superconducting plane. This layered structure leads to anisotropic magnetic properties. To a first approximation, High-T_c materials are isotropic within the CuO_2 planes, and have a much higher effective mass perpendicular to the planes. For these high-κ superconductors, London theory with uniaxial anisotropy is a suitable initial framework. For a field B in a general direction, this leads to the prediction of a distorted-hexagonal FLL [23]. The reciprocal lattice vectors are expected to lie on an ellipse of axial ratio:

$$(minor \, / \, major) = (\cos^2 \theta + \gamma^{-2} \sin^2 \theta)^{1/2} \quad , \tag{5}$$

where $\gamma^2 = m_c/m_{ab}$ is the effective mass ratio and θ is the angle between the field B and the c direction (which is perpendicular to the CuO_2 ab planes). When the field is not along a symmetry direction, i.e. for any value of θ other than $0°$ and $90°$, the FLL *orientation* is also predicted by this theory: it has the shortest reciprocal lattice vector perpendicular to the plane containing B and c. However, the orientation-dependent term in the free energy is weak and falls off with increasing field [23], so may be overcome by other, particularly d-wave or defect pinning, effects.

At low fields and large values of θ there are other effects: the screening supercurrents flow nearly parallel to the CuO_2 planes, so that the *local* magnetic fields are no longer parallel to the flux line direction, which is B. This leads to an anisotropic interaction between flux lines, which can be *attractive* [29,30], and which would cause chains of flux lines to run parallel to the plane containing B and c, rather than form a distorted hexagonal lattice. Finally, there have been suggestions that a lattice of *straight* lines may be unstable with respect to distortions in a material of sufficient anisotropy when the field is not parallel to c [31].

4.3. EFFECTS OF LAYERED CRYSTAL STRUCTURE: "PANCAKE" VORTICES

It has become clear that an anisotropic continuum model is insufficient to describe the properties of the most anisotropic superconducting materials. This is because at low temperatures the superconducting coherence length perpendicular to the CuO_2 layers becomes much smaller than the interlayer spacing. Instead, we should regard materials such as BSCCO, with large values of γ, as consisting of weakly coupled superconducting layers. Flux lines can then be represented as a *stack* of "pancake vortices" [32] with each pancake having the current flow and magnetic fields due to a point vortex in a *single* superconducting layer. Pancakes in adjacent layers interact weakly via the "Josephson" tunneling energy, and by magnetic coupling, but as we shall see, the interlayer interaction can be weak enough for flux lines to fall apart.

If a magnetic field is applied parallel to the CuO_2 planes, flux line cores will form preferentially between the planes: this can give "intrinsic pinning" by the layered structure of the material. However, the average magnetic penetration depth for this orientation is so long that FLL diffraction intensities would be prohibitively weak and observations extremely difficult.

4.4. EFFECTS OF PINNING

Of extreme theoretical and practical importance is the phenomenon of *pinning* of flux lines by inhomogeneities in the superconductor. An unpinned FLL will move under the influence of the Lorentz force when a current flows and cause resistive losses. However, it was shown some years ago by Larkin and Ovchinnikov [20] that pinning, no matter how weak, would destroy the long range order of the flux lattice. In view of this, it is perhaps surprising that flux lines in real materials *do* give rise to observable diffraction spots. It now appears [33] that sufficiently weak pinning does not destroy the *topological* order of the FLL (i.e. no dislocations are created by the disorder) and the pinning causes only *algebraic* - rather than exponential - decay of the positional order parameter at long distances. This should give rise to broadened Bragg peaks with a power-law rather than Gaussian shape. With the finite resolution so far obtained in SANS experiments, these have appeared just like ideal FLL peaks. However, the resulting structure is pinned and has several features in common with the disordered *vortex glass* [34] which is believed to form with heavy pinning or at high fields (see [35] and section 6.2). This structure has therefore been named a *Bragg Glass* [33], and this is a more correct term than *flux line lattice* in any system with finite pinning. So far the detailed predictions of Bragg Glass theory have not been confirmed by SANS.

Another kind of pinning is so-called correlated disorder: this comes in two varieties in High T_c materials. Firstly in YBCO-type materials there are twin planes, which can act not just as point pinning sites, but can pin flux lines along their lengths. The effects of this on flux lattice orientation are discussed in section 5.1. Also of interest are *columnar defects*, which can be created as long amorphous damage tracks by irradiation with fast heavy ions. If the flux lines run parallel to these defects, they can be pinned very strongly along their length, and the resulting structure is known as a *Bose Glass* (from the similarity of the theory to that for the freezing of a liquid of bosons under the influence of disorder [36]). More recently, it has become clear that with random spacing of defects, the interactions of flux lines with each other will tend to prevent every columnar defect from being occupied by a flux line. The result is a *Weak Bose Glass* [37], with correlations between flux line positions that may be investigated by SANS.

4.5. EFFECTS OF THERMAL EXCITATIONS

We have described the flux line lattice as if it were a static crystal structure; however, like all crystals, its components will vibrate as the temperature is increased, and eventually may melt. Thermal excitations of the FLL are larger in High-T_c materials than most low T_c ones: not just because of the higher temperatures, but also because of the long penetration depths (which weaken the nearest neighbour forces between flux lines), and the large anisotropy, which reduces the flux line rigidity. Indeed it has been claimed that in the ideal case [38] the flux lines *never* form a solid, except at T=0. This view is not generally accepted, and appears to conflict with a sharp (probably melting) transition observed by magnetic measurements in BSCCO at least [39] and probably by observation of a latent heat of FLL melting in YBCO [40].

An approximate expression for the melting temperature as a function of field may be obtained using the *Lindemann criterion*, which states that a solid melts when the amplitude of vibration of the atoms is a sufficiently large fraction c_L of the atomic spacing (c_L is the Lindemann number).

Experimentally, it appears that c_L has the value 0.15-0.2 for the FLL [41]. Explicit calculations (see e.g. [42]) give the melting line $B_m(T)$ near T_c:

$$B_m(T) \approx 5.6 \frac{c_L^4}{Gi} H_{c2}(0)[1 - T/T_c]^2 \qquad (6)$$

Where Gi, the Ginzburg number, is a measure of the strength of thermal fluctuations. It depends on the ratio of kT_c to the zero-temperature condensation energy in a coherence volume:

$$Gi = \frac{1}{2} \left[\frac{kT_c}{\mu_0 H_c^2(0)(\xi_{ab}^2 \xi_c)} \right]^2 . \qquad (7)$$

From standard relationships [42], it may be shown that Gi is proportional to $(\lambda \kappa \gamma T_c)^2$, which shows how all the materials parameters of High T_c materials conspire to make thermal fluctuations *much* more important than in low T_c superconductors.

Eq. 6. predicts that the FLL in High T_c materials melts at fields and temperatures well below those that suppress the superconducting wavefunction, particularly when the anisotropy γ is high. The phenomenon is of some importance, because a flux liquid is far more difficult to "pin" than a solid. Flux lines that can move will cause resistivity, hence FLL melting represents an outer boundary to the region in which High-T_c materials are likely to be practically useful. To show that melting of the FLL really occurs requires *microscopic* methods of observation of the flux lattice structure, such as μSR [43] and SANS [6,7], although the phenomenon was first suspected from anomalies in transport measurements: see e.g. [44,45]. SANS allows us to detect the FLL structure and melting *directly*.

5.   Experimental Results on Flux Lattice Structures in YBCO

5.1. FIELD PARALLEL TO THE C-AXIS

$YBa_2Cu_3O_{7-\delta}$ (YBCO) has a tetragonal unit cell when oxygen-deficient and a weakly orthorhombic cell when the CuO "chains" in the crystal structure are well-oxygenated. When a tetragonal crystal formed at high temperatures is cooled in oxygen, it generally splits into domains which have either of two orientations of the orthorhombic structure - which have their **a** and **b** axes interchanged. These domains are separated by *twin planes*, which run at approximately 45° to the **a** and **b** axes, and which can be as closely spaced as 1000 Å. Neutron diffraction by flux lines in twinned YBCO with the field perpendicular to the CuO_2 planes gives a square pattern with four strong peaks (but noticeable intensity spread out between them) [46,47,48,49]. Typical results are shown in Fig. 4.

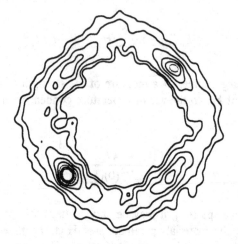

Fig. 4:   Contour plot of a neutron diffraction pattern from flux lines in twinned YBCO at T=4K, with $B = 2$ T applied parallel to c. The scattering from the sample has been removed by subtracting data taken in the normal state. Also, the incident beam spot at the centre has been masked. The a/b axes of the orthorhombic domains are horizontal and vertical.

A square pattern does *not* necessarily imply that the flux lines form a *square* lattice. Indeed, the most probable explanation of the data in Fig. 4 is that the flux lines form two 1-dimensionally ordered domains, with flux-line plane orientations controlled by pinning to the two possible sets of twin-planes, [47,48]. The strong diffraction spots in Fig. 4 correspond to these directions, and the weaker diffraction to less well aligned FLL planes. In another sample, presumably with wider-spaced twin planes, extra diffraction spots could be detected [50]. These results were correctly interpreted as due to four distorted triangular lattices, present in different parts of the sample. The superposition of their diffraction patterns is represented in Fig. 5.

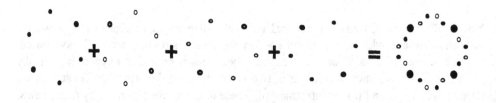

Fig. 5:   Representation of the square diffraction pattern observed in twinned YBCO [48] as a superposition of patterns due to four distorted triangular lattices. Each pattern has two of its spots at 45°, arising from flux lattice planes pinned to twin planes.

The alignment of these lattices is clearly due to twin planes, and the distortion of the hexagons from a regular shape is not (as was claimed in [50]) due to d-wave effects, but to anisotropy in the ab plane [28,51]. We shall now show this with new results from an untwinned crystal.

An "untwinned" YBCO crystal, with a large fraction of the crystal in one orthorhombic domain orientation, can be prepared by cooling it from high temperatures under uniaxial stress [52]. A typical FLL diffraction pattern [53] taken at fairly low field is shown in Fig.6: it clearly represents a polycrystalline flux lattice. However, all the spots lie on a single ellipse, so they represent various orientations of triangular FLL, distorted by the a-b anisotropy. The sign of this anisotropy is consistent with easier supercurrent flow parallel to the chains in the b-direction, which reduces λ_L and the flux lattice parameter in the a-direction, and hence increases the a* dimension of the ellipse.

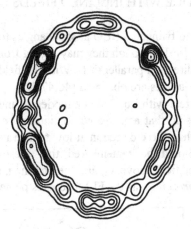

Fig. 6:   Contour plot of a neutron diffraction pattern from the flux lattice in *un*twinned YBCO at $B = 0.51$ T // to c. The a-axis of the crystal is vertical. The multitude of spots is due to four orientations of hexagonal lattice represented in Fig.7, all distorted by the a-b anisotropy.

Detailed analysis of the data in Fig. 6 [53] shows that there are four flux lattice orientations present: the diffraction patterns of these are represented in Fig. 7.

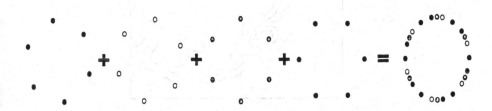

Fig. 7:   Representation of the four reciprocal lattices contributing to the diffraction pattern shown in Fig. 6. Two of these arise from FLLs pinned to twin planes, and the other two are aligned with the crystal axes.

It is clear that two of these are equivalent to the distorted hexagons aligned with twin planes observed in [50] - and that the distortions are due to a-b anisotropy. The *alignment with twin planes* carries the implication that our crystal is not *completely* twin-free, although the domains of the "wrong" orientation must be a very small fraction of the volume, otherwise we would observe another set of spots aligned on an ellipse at right angles. The other two FLL orientations that we observe are aligned with the crystal axes: they may even be aligned by cracks or the crystal surfaces. It should be mentioned that these results and those in [50] are in disagreement with the interpretation of STM observations taken at a rather higher field [54]. Those authors interpret their rather disordered patterns of flux line cores, observed at the surface of a twinned crystal, as *square* lattices, distorted by a-b anisotropy.

5.2. FLUX LINE LATTICE WITH INCLINED FIELDS

When the field is applied to twinned YBCO at small angles to **c** then the flux lines remain pinned to the twin planes [46], although they may zigzag from one twin plane to the next to maintain the *average* direction parallel to **B**. With the field at a sufficiently large angle to **c**, distorted hexagonal lattices are observed [46,47,48,49] and the angular dependence of the distortion is consistent with Eq. 5. and a moderate anisotropy ($\gamma \sim 5$). Even in this case, it is perhaps surprising that anisotropic continuum London theory works so well, since the coherence length in the **c**-direction at low temperatures is smaller than the layer spacing. However, it should be remembered that these experiments are generally performed by applying the field above T_c, and then cooling, so that the FLL observed at low temperatures is that *frozen in*, as the FLL becomes pinned not far below T_c.

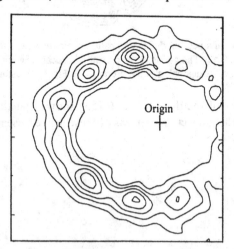

Fig. 8:   Contour plot of a neutron diffraction pattern from the FLL in twinned YBCO at $B = 1.5$ T applied at an angle $\theta = 51°$ to **c**. The sample and trapped field together were rotated about a vertical axis to bring the spots on the left of the origin close to the Bragg condition. A total of twelve spots from two orientations of FLL would be seen in a full pattern.

In general, the flux lattice *orientation* observed is *not* that given by anisotropic London theory [23]. A most interesting situation is when the field is applied at large and equal angles to both sets of twin planes; typical results are shown in Fig. 8. In this case, two distorted hexagonal lattices are formed [48,49], and their orientations are controlled by pinning of one set of flux line planes to the *1-dimensional* defects that are the *intersections* of the twin planes and CuO_2 planes.

When the field is applied at large angles to the **c** direction in an *untwinned* crystal, it is possible to favour fewer FLL orientations. In Fig. 9, we show diffraction patterns for an angle $\theta = 33°$ (rotated about **a**). This was chosen so that the a-c anisotropy would just *undo* the a-b anisotropy and give an almost exactly regular hexagonal pattern. However, the preferred orientation depends on the value of the field. At the moment we do not know the cause of this effect. There is clearly more work to do, in order to be sure we can disentangle sample-dependent effects from fundamental phenomena such as d-wave superconductivity.

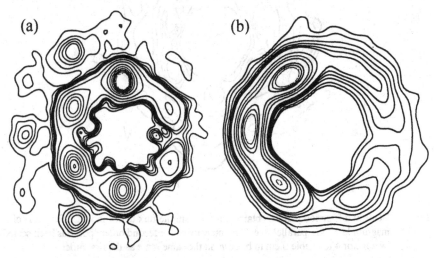

(a)   (b)

Fig. 9:   Contour plot of a neutron diffraction pattern from the flux lattice in *un*twinned YBCO with the field applied at an angle $\theta = 33°$ to **c**. (a) $B = 0.2$ T; (b) $B = 3.8$ T

6.   Experimental Results on Flux Lattice Structures in BSCCO-2212

6.1.   FLUX LINE LATTICE AT LOW FIELDS

At low fields, the situation in BSCCO is much simpler than in YBCO: there are no twin planes and the ab plane is almost isotropic. A hexagonal diffraction pattern is observed - even in suitable cases with weak 2nd-order features (Fig. 10). In all the specimens we have investigated, the FLL tends to be aligned so that a pair of spots lies along the **b**-

direction, which is vertical in the figure. It is likely that this orientation is controlled by the interaction of flux lines with defects on the surface [55], which can give an alignment through the bulk [56]. However, on cooling the sample to low temperatures in a field at increasing angle θ to the c-axis, the neutron diffraction signal disappears [57], so that no distorted hexagonal pattern is observed in this material. It is likely that the instability noted earlier [31] leads to a new FLL, perhaps with non-straight lines, so that the diffraction signal is too weak to be observed.

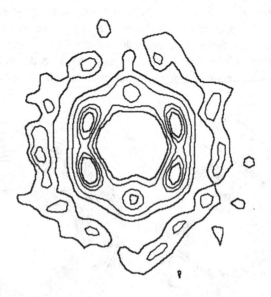

Fig. 10:   Neutron diffraction pattern obtained at 5K from the flux lattice in BSCCO with B of magnitude 20 mT parallel to c. The intensity of the second order spots has been scaled up by a factor 4 to enable them to be seen on the same scale as the first order.

6.2. FLUX LATTICE DECOMPOSITION

In BSCCO with the field parallel to c, it is found that the FLL signal at low temperatures *disappears* as increasing values of B are trapped in the material [6]. It is believed that this is due to a "crossover" in the pinning of the vortices at a field B_{cr}, at which the intra-(superconducting) -plane interactions between pancakes becomes stronger than the interplane interactions. Below B_{cr}, the vortices are pinned as flux lines, but above the crossover, they are pinned as pancakes, so that the flux lines dislocate or decompose, and the diffraction signal drops to zero. There are corresponding changes in magnetic hysteresis [58] and μSR signals [43].

In materials with moderate anisotropy, it is believed that this crossover should occur at a field $B_{cr} = B_{2d}$ given by:

$$B_{2d} = \Phi_0/(\gamma s)^2 \quad , \tag{8}$$

where s is the interlayer spacing. However, if γ is sufficiently large, then over most of the temperature range, Josephson coupling becomes negligible compared with magnetic coupling between pancakes, and B_{cr} is given by [59]:

$$B_{cr} = \Phi_0/\lambda_{ab}^2 \quad , \tag{9}$$

where λ_{ab} is the value of the penetration depth for supercurrents flowing in the ab planes. Experimental evidence for this formula is given by μSR data [59]. It should be noted that for YBCO, Eq. 8 applies, which predicts that B_{2d} is ~ 100 T, whereas for BSCCO B_{cr} ranges from 30 to 60 mT, depending on doping. Hence this decomposition phenomenon, which occurs at very low fields in BSCCO, should be at inaccessible fields in YBCO. Nonetheless, there is evidence from transport measurements [35] that the vortex structure in YBCO becomes more disordered and "glassy" at high but accessible fields.

7.   Flux Lattice melting

7.1. FLUX LATTICE MELTING IN BSCCO-2212

As described in section 4.5, thermal fluctuations in flux line positions are very important in High-T_c materials, particularly when the anisotropy is high. It is therefore not surprising that FLL melting was first observed by SANS in the highly anisotropic BSCCO.

In this material, it is found [6] that on increasing temperature at a constant low field, the FLL diffraction signal disappears well below T_c, so that the lattice structure is destroyed by thermal fluctuations even though the superconducting wavefunction remains. From these measurements, we can also say something about the *flux liquid* above the melting transition. There is no evidence in this state for a *ring* of diffracted intensity, as would be expected for a simple flux line liquid. This implies an absence of any long-range correlation of the flux line or pancake structure along the field direction, consistent with melting either into a liquid of highly curved flux lines or (more likely) decoupling of the lines into a liquid of 2D pancakes. The phase diagram deduced from the neutron diffraction, muon and magnetisation measurements on BSCCO is shown in Fig. 11. It is clear that there is interest in the form of the melting line above B_{cr} where the SANS signal rapidly goes to zero. At still higher fields, μSR measurements suggest a *2-dimensional melting* of pancakes in independent layers [60].

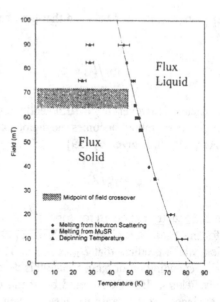

Fig. 11:   *B-T* phase diagram for a sample of "overdoped" BSCCO, which has a Tc of 84K. This
shows the melting line as determined by SANS and μSR measurements, and also the
crossover field [6]. It will be noted that a three-dimensional FLL exists in only a tiny
corner of the *B-T* plane: on the scale of the diagram, B_{c2} is an essentially vertical line.

7.2. FLUX LATTICE MELTING IN YBCO

YBCO is experimentally more difficult: twin planes provide strong pinning, which

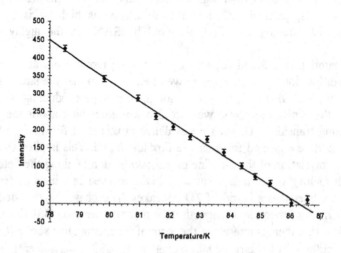

Fig. 12:   FLL diffracted intensity versus temperature at fixed *B* in a twinned YBCO sample with
B = 4.0 T in the orientation corresponding to Fig. 8.

destroys [61] the sharp resistive transition due to FLL melting which can be seen in small detwinned crystals of YBCO [45]. Until recently, the only YBCO crystals of sufficient size for SANS by the FLL were twinned; however, we have also recently carried out measurements on an untwinned [52] crystal. We see in Fig. 12. that in the twinned crystal [7,62], the intensity goes steadily to zero at the melting transition, indicating a 2nd order phase transition (with an approximately mean-field variation of the FLL order parameter!).

In the untwinned crystal [63] and Fig.13, the intensity falls rather more rapidly at FLL melting, possibly indicating a smeared first order transition. In both cases, the intensity above the melting transition is *zero*. Even though YBCO is *much* less anisotropic than BSCCO, it appears that in the liquid state the pancakes are decoupled as soon as the FLL melts.

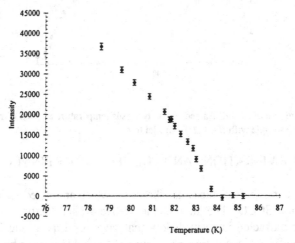

Fig. 13:   FLL diffracted intensity versus temperature at fixed B in an untwinned YBCO sample with $B = 5.0$ T applied parallel to c.

8.   Other information from Diffraction

8.1.   TEMPERATURE-DEPENDENCE OF PENETRATION DEPTH

It is clear from Eqs. 2 and 3 that the total diffracted intensity from a static flux lattice formed in a field much less than H_{c2} is a simple measure of the temperature-dependent penetration depth. The condition of a static flux lattice is most easily obtained in YBCO, and if a low field is employed then FLL vibrations, melting and the proximity of H_{c2} will only become important very close to T_c. Typical results obtained with the field parallel to c are shown in Fig. 14. The data has been plotted so that the ordinate is proportional to $\lambda^{-2}(T)$, i.e. the superfluid density. There is a clear tendency to a linear dependence at low temperatures, similar to that seen by microwave measurements of the same quantity [64].

This is nowadays accepted as evidence for nodes in the superconducting gap, probably arising from a pairing symmetry close to d-wave. If this is the case, then the detailed temperature- and field-dependence of the results should be affected by the depairing effects discovered in [11].

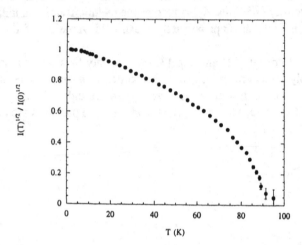

Fig. 14:   Square root of FLL diffracted intensity versus temperature at fixed B in an untwinned YBCO sample with $B = 0.2$ T parallel to c.

8.2.   FLUX LINE VIBRATIONS AND THE EFFECTS OF PINNING

There are two other effects of flux-line vibrations, which are expected in SANS results – particularly from BSCCO. The first is the Debye-Waller factor, which represents a reduction of the diffracted intensity below that given by Eqs. 2 and 3. At the melting temperature, the effect of the lattice vibrations $<u^2>$ should be given by:

$$I / I(<u^2> = 0) = \exp(-q^2 <u>^2 / 2) = \exp(-2\pi^2 c_L^2) \approx \exp(-0.44) = 0.64 \qquad (10)$$

for the first-order diffracted spots. Eq. 10 shows that this should be a noticeable effect. However, unlike ordinary crystals, the diffracted intensity from a static FLL is also temperature-dependent, as described in Section 8.1, so the DW factor may prove difficult to measure. There is some SANS evidence [60,65] and Fig. 15, for another effect of thermal excitation on the flux lattice. In the field region where flux lattice decomposition is occurring, the FLL actually becomes more *perfect* on *increasing* the temperature. This is believed to be due to depinning of individual pancakes, which realign with their neighbours. It can be argued that this process is an *entropic* effect: a depinned pancake has a higher energy, but more entropy in the broad potential well formed by the FLL. Thus the flux lattice diffracted intensity rises before finally falling to zero on melting. This effect is in qualitative agreement with recent theory [66].

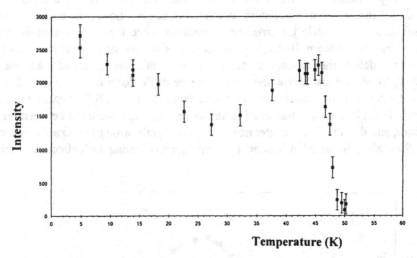

Fig. 15: FLL diffracted intensity versus temperature at $B = 82.5$ mT in BSCCO. This shows the *increase* in lattice perfection by thermal depinning, just before melting occurs.

Thermal fluctuations of flux line positions can be greatly reduced by columnar defects, discussed in Section 4.4. This allows the effect of the temperature-dependence of the penetration depth to be seen almost in isolation. An example of results [63] from μSR and SANS is shown in Fig. 16. The similarity to the results on YBCO (Fig. 14) is notable.

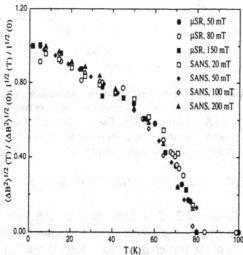

Fig. 16: RMS μSR linewidth versus temperature for various fields applied to a columnar-defected BSCCO sample, compared with the temperature-dependence of the normalised square root of the SANS intensity. With static flux lines, both should vary as $\lambda^{-2}(T)$.

The Q-dependence of the SANS intensity from a columnar defected sample can be used to find the degree of correlation between flux line positions. The columnar defects themselves are completely uncorrelated in position, since they are created in random positions by a broad beam. Hence, if there is one flux line per defect, and each flux line is pinned on a defect, the *intensity* scattered from each flux line should add, and the total SANS signal should just follow the Q-dependence of the form factor given in Eq. 3. If this is divided out, one obtains the pair correlation function $S(Q)$. Typical results are shown in Fig. 17. It is clear that at all fields, there is strong correlation between flux line positions, and that the flux line interactions are sufficiently strong to make the system the *weak Bose glass* discussed in Section 4.4. Simulations showing such effects are reported in [37,67].

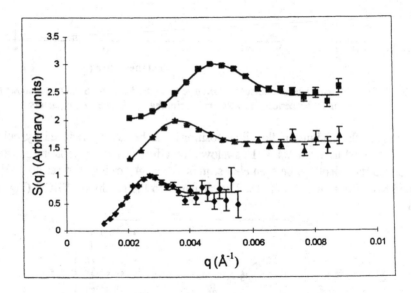

Fig. 17: $S(Q)$ versus Q for three different fields: $\frac{1}{2}B_\Phi$ (diamonds), B_Φ (triangles), and $2B_\Phi$ (squares). The vertical scale is arbitrary and the upper two graphs have been displaced vertically by 1.0 and 2.0 for clarity. The sample was BSCCO-2212 with $B_\Phi = 0.1$ T. $S(Q)$ was obtained by multiplying the observed intensity by Q^5, to remove approximately the London form factor and resolution effects.

8.3.   TRANSVERSE FIELD COMPONENTS IN YBCO

As described in Section 2.2, flux lines passing obliquely through an anisotropic superconductor give rise to transverse field components which can be detected by *polarisation analysis* of the scattered neutrons. Recent results [68] are shown in Fig. 18. They are in qualitative agreement with anisotropic London theory [13]

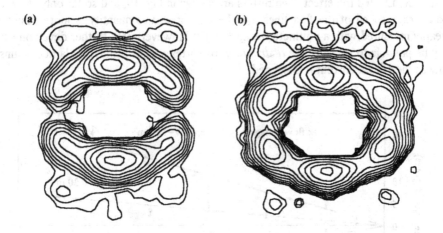

Fig. 18: Scattered intensity from the FLL formed in untwinned YBCO with an induction of 0.5 T applied at 45 to c: (a) Intensity of spin-flipped neutrons (b) Non-flipped. The experiment was performed on IN15 at the ILL.

8.4. FLUX LINE MOTION

There is a great deal of interest in the effects of flux line motion on flux structures and melting [e.g. 69,70]. Flux line motion, induced by a large enough transport current, averages and *reduces* the effect of pinning and is expected to create a *moving Bragg Glass* [70,71]. At the time of writing, there has been no confirmation of these ideas by SANS in High T_c materials; there is some indirect evidence from decoration measurements in $NbSe_2$ [72]. Further effects of current flow on flux lattice perfection are discussed in Section 10.2.

Many years ago, it was demonstrated that it is possible to *measure* flux line *motion* by SANS [73]. In these experiments, the change in Bragg angle for diffraction from the moving lattice was observed. It is also possible to measure the change $\Delta \varepsilon$ in *energy* of the neutrons on diffraction: this may easily be calculated as the energy change due to reflection from a moving mirror:

$$\Delta \varepsilon = h v_{ff} / d \quad , \qquad (11)$$

where v_{ff} is the speed of the flux lattice planes of spacing d. The speed may be related to the induction B and driving electric field E via:

$$v_{ff} = E / B \qquad (12)$$

For typical values of v_{ff} ($= E/B$) of < 1 ms^{-1}, this is a very small energy and can only be detected by the neutron spin-echo technique. A recent experiment in a low Tc sample [74] demonstrated the effect. The results are shown in Fig. 19, and so far only show that the *average* speed of the flux lines may be measured; this quantity is more easily and cheaply measured with a voltmeter using Eq. 12! However, with further development, it may be possible to measure the *distribution* of vortex speeds as the flux lattice begins to flow as the critical current is exceeded.

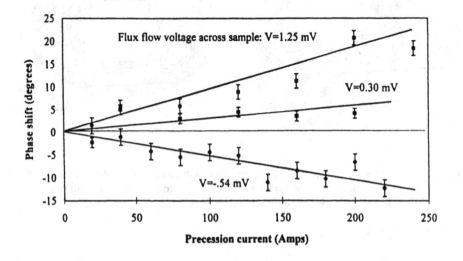

Fig. 19:   Spin-echo phase shift versus neutron spin precession field for various values of E in an Nb-Ta sample. The slope of the graph is proportional to the average neutron energy change and hence to the flux lattice velocity. Experiment performed on IN11 at ILL.

9.   Neutron depolarisation

9.1.   MEASUREMENTS OF AVERAGE INTERNAL FIELDS

Three-dimensional neutron depolarisation measurements may be used in two different ways. Firstly, one may use the fact that the transmitted beam has its overall polarisation *rotated*: this may be detected by 3-d analysis of the transmitted neutrons, which is carried out in any case to measure the depolarisation. With the incident neutron spin aligned perpendicular to the field in the sample region, the spin rotation gives a measure of the integral of the magnetic field along the neutron path. If the value of this integral is too large, so that the neutron velocity spread causes too much depolarisation, the average field in an unmagnetised sample may be compensated by an equal and opposite field in the neutron path outside the sample region. In this case, any additional rotation is proportional to the *magnetisation* of the sample if the neutron beam and field are exactly aligned [14]. In general, one also has to worry about the effects of *stray fields* outside the sample. However, if the sample magnetisation is perpendicular to the beam, the stray fields may be removed from the beam by connecting a magnetic yoke to the ends of the sample [see e.g. 17]. The rotation technique may be used in *tomographic* mode by restricting the neutron beam to a small aperture: a positional resolution of 0.2 mm is possible [14,75], without excessive loss of intensity. By this means, demagnetising effects of flat plates and remanent trapped fields have been measured as a function of position in High T_c samples.

9.2. DEPOLARISATION MEASUREMENTS

As described in Section 2.3, depolarisation measurements can give information about variations in field magnitude inside a sample. However, Eq. 4 tells us that we must *model* the field distribution inside the sample in order to obtain detailed information, because the depolarisation depends on the *product* of field variation $<B_n^2>$ and correlation length ζ. In particulate magnetic materials, microscopy may be able to provide the value of ζ, so that definite conclusions may be drawn [14]. In superconductors, one must assume the dimensions of "flux bundles" or the value of $<B_n^2>$ in order to make further progress. The most definite results that have been obtained so far are in refs. [15,16,17,75].

9.3. ADVANTAGES AND DISADVANTAGES OF 3-D ND

In principle, 3DND gives a large amount of information: a 3 x 3 "depolarisation matrix" relating each component of the input polarisation to the output polarisation. Also, since it uses the *straight through* beam, rather than scattered neutrons, it is not so limited in intensity as many polarised neutron techniques. Depolarisation is sensitive to tilts of flux lines through much larger angles than SANS would be able to measure, and can be used to detect variations in magnetic field over larger distances than SANS (either via the dependence of the results on ζ, or by the scanning tomographic technique). However, the need to *model* the results has meant that very little *completely new* information has been obtained on High T_c materials – and occasionally [76] the interpretation has been extremely speculative and probably wrong! 3DND is a technique that still has to prove its utility in superconductivity research.

10. Comparison with other superconductors

10.1. BOROCARBIDES

As mentioned in the introduction, FLL's in several other superconducting materials have been investigated in recent years. One interesting group are the borocarbides [4,5], of general formula RNi_2B_2C, with R = Y or heavy rare earths, which can have T_c's well above 10K and can also combine superconductivity with antiferromagnetic ordering. When *square* FLL's were observed in one of these materials, it was at first suspected that this departure from the usual shape was associated with magnetism [77]. However, more recent work has shown similar effects in non-magnetic variants, and demonstrated that there are phase transitions between triangular lattices at low fields and square at high fields, and also changes in FLL orientation [e.g.5]. From the agreement with theory [24, 26] there seems little doubt that these transitions are due to the non-local effects discussed in section 4.1.

10.2. NIOBIUM AND NIOBIUM SELENIDE

In the early years of SANS from the FLL, niobium was extensively investigated; however there have been two recent investigations. Firstly, it was claimed from the field-dependence of diffraction spot widths that FLL melting occurs in Nb, well below H_{c2} [78]. This could only be the case if the Lindemann number for the Nb FLL was much lower than high T_c materials, since the much lower T_c, κ and anisotropy of Nb makes thermal fluctuations of the FLL very much smaller in the low T_c material. Strong arguments were expressed by others that the broadening of the spots was due to distortion of the FLL by pinning [79] and more recent work by μSR [80] and SANS [81] confirms this view. This latter work also showed how pinning theory [20] may be tested by SANS.

There has also been in recent years some elegant work [82,83] on the flux lattice in NbSe2, relating the perfection of the flux lattice to the critical current, via Larkin-Ovchinnikov theory [20]. This work has also shown that the flux lattice perfection can be improved by passing a current. A similar observation was made in Nb many years before [84].

10.3. EXOTIC SUPERCONDUCTORS

The flux lattice in CeRu2 has attracted some attention [85]: this material is extremely magnetically reversible at low fields, and irreversible at high fields – the exact opposite of High T_c materials. It is as yet unclear whether this is due to unconventional superconductivity or the conventional increase in critical current near H_{c2}. The Heavy-Fermion superconductor UPt3 shows phase transitions in the mixed state, which have been widely interpreted as signs of non s-state pairing [86]. Investigation by SANS is extremely difficult because of the weak intensities caused by the long penetration depth. However, there have been heroic experiments [87] to look in the flux lattice signal for correlates of these phase transitions. Finally, we mention Sr2RuO4, which is certainly a non s-wave superconductor, and may be p-wave [88,89]. Recent SANS measurements [90] show that the flux lattice is *square*: this may be due to similar effects to those seen in the borocarbides, although there are no signs of square to hexagonal phase transitions. However the flux lattice structure may also be intimately related to the pairing state [91].

11. Summary and outlook

SANS measurements, along with μSR [43,80] have established microscopic evidence for flux lattice melting in High T_c materials, and given us a great deal of information about flux lattice structures, pinning and melting. There are still many areas to cover, since SANS in these materials with long penetration depths is intensity-limited, and defect-free crystals of sufficient size are not freely available. The detailed predictions of Bragg Glass theory [33] have not yet been tested. Polarisation analysis will allow extremely sensitive tests of whether anisotropic London theory is sufficient to describe tilted flux lattices. At rather higher fields than the 5T or so currently used, it may be possible to detect d-wave effects on the FLL structure in High-T_c materials, and *core effects* on the FLL of unconventional superconductors in general. The investigation of the columnar defect pinned state is still in its infancy [63]. A beginning has been made of the investigation of the effects of flux flow on vortex structure and motion, but there is still much to do in High-T_c materials. It would be nice to think that *inelastic* neutron measurements could be made of vortex thermal fluctuations in superconductors, but it is probable that the intensity of the signal is too small to measure. Even without this, we have plenty still to do with High T_c and other superconductors.

Finally, there are a couple of interesting technical developments on the horizon, which may prove useful in high-resolution SANS in general and FLL diffraction in particular. Both rely on *refraction* of neutrons. Firstly, there is the very recent demonstration [92] that *neutron lenses* can be constructed with a sufficiently short focal length to be useful in improving the optical setup in SANS from its present "pinhole camera" arrangement. Secondly, *prisms* can be useful in cancelling the effects of gravity, which bends and spreads the beam for slow neutrons with long flightpaths [93].

Acknowledgements

Much (but by no means all) of the High T_c SANS and μSR work reported here was carried out by collaborations in which the author has played a part. I acknowledge major contributions from Steve Lee, Bob Cubitt, Christof Aegerter, Don Paul, Mona Yethiraj and many others, and the help of staff at the neutron sources Risø, Oak Ridge and ILL, and at the muon sources PSI and ISIS. Without samples of sufficient size and quality from many crystal growers, particularly at Amsterdam-Leiden, ISTEC (Tokyo) and the University of Kyoto, we also would not be able to carry out the work.

References

1. D. Cribier, B. Jacrot, L. Madhav Rao.and B. Farnoux, *Phys.Lett.* 9 (1964) 106.
2. E.M. Forgan et al., *Nature* 343 (1990) 735.
3. U. Yaron, P.L. Gammel et al., *Phys. Rev. Lett.* 78 (1997) 3185, and refs. therein.
4. M.R. Eskildsen et al., *Phys.Rev.Lett.* 78 (1997) 1968.

5.   D.McK. Paul et *al.*, *Phys. Rev. Lett.* **80** (1998) 1517, and refs. therein.
6.   R. Cubitt, E.M. Forgan et *al.*, *Nature* **365** (1993) 407.
7.   M.T. Wylie, E.M. Forgan et *al.*, *Czech J. Phys.* **46** S3 (1996) 1569.
8.   D.K. Christen et *al.*, *Phys.Rev.* **B15** (1977) 4506.
9.   E.H. Brandt, *Phys.Rev.Lett.* **78** (1997) 2208.
10.   A. Yaouanc et *al.*, *Phys.Rev.* **B55** (1997) 11107.
11.   J.E. Sonier et *al.*, *Phys.Rev.* **B55** (1997) 11789.
12.   P. Böni and A. Furrer, *this volume*, chapter 1, section 6.
13.   S.L. Thiemann et *al.* *Phys.Rev.* **B39** (1989) 11406.
14   R. Rosman and M.Th. Rekveldt, *J.Magn.Magn.Mat.* **95** (1991) 319.
15   W. Roest and M.Th. Rekveldt, *Physica* **C252** (1995) 264.
16   H.W. Weber, *J.Low Temp.Phys.* **17** (1974) 49.
17   W. Roest and M.Th. Rekveldt, *Phys.Rev.* **B48** (1993) 6420.
18.   R.Cubitt, E.M.Forgan et *al.*, *Physica* **B180-1** (1992) 377.
19.   P. Harris, B. Lebech and J.S. Pedersen, *J.Appl.Cryst.* **28** (1995) 209.
20.   A.I. Larkin and Yu.N. Ovchinnikov, *J.Low Temp.Phys.* **34** (1979) 409.
21.   J. Schelten, in *Anisotropy Effects in Type-II Superconductors*, ed. H.W.Weber
      (Plenum, New York, 1977) p.113.
22.   D.K. Christen et *al.*, *Phys.Rev.* **B21** (1980) 102.
23.   L.J. Campbell, M.M. Doria and V.G. Kogan, *Phys. Rev.* **B38** (1988) 2439.
24.   V.G. Kogan et *al.*, *Phys.Rev.* **B55** (1997) 8693.
25.   D.K. Christen et *al.*, *Physica* **B135** (1985) 369.
26.   Y. De Wilde et *al.*, *Physica* **C282-7** (1997) 355.
27.   H. Won and K. Maki, *Phys.Rev.* **B53** (1996) 5927, and refs. therein.
28.   E.M. Forgan and S.L. Lee, *Phys.Rev.Lett.* **75** (1995) 1422.
29.   A.J. Buzdin and A. Yu. Simonov, *Physica* **B165-6** (1990) 1101.
30.   L.L. Daemen, L.J. Campbell and V.G. Kogan, *Phys.Rev.* **B46** (1992) 3631.
31.   A.M. Thompson and M.A. Moore, *Phys.Rev.* **B55** (1997) 3856, and refs. therein.
32.   J.R. Clem, *Phys.Rev.* **B43** (1991) 7837.
33.   T. Giamarchi and P. Le Doussal, *Phys.Rev.* **B52** (1995) 1242; *ibid.* **B55** (1997) 6577.
34.   D.S. Fisher, M.P.A. Fisher and D.A. Huse, *Phys.Rev.* **B43** (1991) 130.
35.   H. Safar et *al.*, *Phys.Rev.Lett.* **70** (1993) 3800.
36.   D.R. Nelson and V.M. Vinokur, *Phys.Rev.Lett.* **68** (1992) 2398.
37.   C. Wengel and U.C. Taüber, *Phys.Rev.Lett.* **78** (1997) 4845.
38.   M.A. Moore, *Phys.Rev.* **B55** (1997) 14136.
39.   E. Zeldov et *al.*, *Nature* **375** (1995) 373.
40.   A. Schilling et *al.*, *Phys.Rev.Lett.* **78** (1997) 4833.
41.   S.L. Lee et *al.*, *Phys.Rev.Lett.* **75** (1995) 922.
42.   G. Blatter et *al.*, *Rev.Mod.Phys.* **66** (1994) 1125, and refs. therein.
43.   S.L.Lee et *al.*, *Phys.Rev.Lett.* **71** (1993) 3862.
44.   P.L. Gammel et *al.*, *Phys.Rev.Lett.* **61** (1988) 1666.
45.   H. Safar et *al.*, *Phys.Rev.Lett.* **69** (1992) 824.
46.   E.M. Forgan et *al.*, *Physica* **C185-89** (1991) 247.
47.   E.M. Forgan et *al.*, *Physica Scripta* **T49** (1993) 143.
48.   M. Yethiraj et *al.*, *Phys.Rev.Lett.* **70** (1993) 857.
49.   B. Keimer et *al.*, *Science* **262** (1993) 83.
50.   B. Keimer et *al.*, *Phys.Rev.Lett.* **73** (1994) 3459.
51.   M.B. Walker and T. Timusk, *Phys.Rev.* **B52** (1995) 97.
52.   A.I. Rykov et *al.* in *Advances in Superconductivity VIII (Proc. ISS'95)*, eds.
      H. Hayakawa and Y. Enomoto (Springer, Tokyo, 1996) p. 341.

53.   S.T. Johnson et *al.*, *Phys.Rev.Lett.* (1998) submitted.
54.   I. Maggio-Aprile et *al.*, *Phys.Rev.Lett.* **75** (1995) 2754.
55.   H. Dai, J. Liu and C.M. Lieber, *Phys.Rev.Lett.* **72** (1994) 748.
56.   Z. Yao, S. Yoon, H. Dai, S. Fan and C.M. Lieber *Nature* **371** (1994) 777
57.   E.M. Forgan et *al.* in *Advances in Superconductivity VII (Proc. ISS'94)*, eds.
      K.Yamafuji and T.Morishita (Springer, Tokyo, 1995) p. 413.
58.   R. Cubitt, E.M.Forgan et *al.*, *Physica* **C235** (1994) 2583.
59.   C.M. Aegerter et *al.*, *Phys.Rev.* **B54** (1996) 15661.
60.   S.L. Lloyd et *al.*, to be published (1998).
61.   W.K. Kwok et *al.*, *Phys.Rev.Lett.* **69** (1992) 3370.
62.   C.M. Aegerter et *al.*, *Phys.Rev.* **B** (1998) in press.
63.   C.M. Aegerter et *al.*, to be published (1998).
64.   D.A. Bonn et *al.*, *Czech J. Phys.* **46** S6 (1996) 3195.
65.   E.M. Forgan, M.T. Wylie et *al.*, *Czech J. Phys.* **46** S3 (1996) 1571.
66.   D. Ertas and D.R. Nelson, *Physica* **C272** (1996) 79.
67.   U.C. Taüber and D.R. Nelson, *Phys.Rev.* **B52** (1995) 16106.
68.   P.G. Kealey et *al.*, to be published (1998).
69.   A.E. Koshelev and V.M. Vinokur, *Phys.Rev.Lett.* **73** (1994) 3580.
70.   S. Spencer and H.J. Jensen, *Phys.Rev.* **B55** (1997) 8473.
71.   T. Giamarchi and P. Le Doussal, *Phys.Rev.Lett.* **76** (1996) 3408.
72.   F. Pardo et *al.*, *Phys.Rev.Lett.* **78** (1997) 4633.
73.   J. Schelten, H. Ullmaier and G. Lippmann, *Phys.Rev.* **B12** (1971) 1772.
74.   E.M. Forgan, Ch. Simon et *al.*, to be published (1998).
75.   W. Roest and M.Th Rekveldt, *Physica* **C248** (1995) 213.
76.   M.L. Crow et *al.*, *J.Appl.Phys.*, **67** (1990) 4542.
77.   U. Yaron et *al.*, Nature **382** (1996) 236.
78.   J.W. Lynn et *al.*, *Phys.Rev.Lett.* **72** (1994) 3413; *ibid.* **74** (1995) 1698.
79.   E.M. Forgan et *al.*, *Phys.Rev.Lett.* **74** (1995) 1697.
80.   E.M. Forgan et *al.*, *Hyper.Inter.* **105** (1997) 61.
81.   P.L. Gammel et *al.*, *Phys.Rev.Lett.* **80** (1998) 833.
82.   U. Yaron et *al.*, *Phys.Rev.Lett.* **73** (1994) 2748.
83.   U. Yaron et *al.*, *Nature* **376** (1995) 753.
84.   P. Thorel et *al.*, *J.de Phys.* **34** (1973) 447.
85.   A. Huxley et *al.*, *Physica* **B224** (1996) 169.
86.   R. Joynt, *Phys.Rev.Lett.* **78** (1997) 3189, and refs. therein.
87.   U. Yaron et *al.*, *Phys.Rev.Lett.* **78** (1997) 3185, and refs. therein.
88.   T.M. Rice and M. Sigrist, *J.Phys.(Cond.Mat.)* **7** (1995) L643.
89.   A.P. Mackenzie et *al.*, *Phys.Rev.Lett.* **80** (1998) 181.
90.   T.M. Riseman et *al.*, to be published (1998).
91.   D.F Agterberg et *al.*, *Phys.Rev.Lett.* **78** (1997) 3374, and submitted (1998).
92.   M.R. Eskildsen, P.L. Gammel et *al.*, *Nature* **391** (1998) 563.
93.   E.M. Forgan and R. Cubitt, Neutron News (1998) in press.

Physics and Chemistry of Materials
with Low-Dimension Structures

Physics and Chemistry of Materials
with Low-Dimension Structures

9. L.J. de Jongh (ed.): *Magnetic Properties of Layered Transition Metal Compounds.* 1990
   ISBN 0-7923-0238-9
10. E. Doni, R. Girlanda, G. Pastori Parravicini and A. Quattropani (eds.): *Progress in Electron Properties of Solids.* Festschrift in Honour of Franco Bassani. 1989   ISBN 0-7923-0337-7
11. C. Schlenker (ed.): *Low-Dimensional Electronic Properties of Molybdenum Bronzes and Oxides.* 1989   ISBN 0-7923-0085-8
12. Not published.
13. H. Aoki, M. Tsukada, M. Schlüter and F. Lévy (eds.): *New Horizons in Low-Dimensional Electron Systems.* A Festschrift in Honour of Professor H. Kamimura. 1992
    ISBN 0-7923-1302-X
14. A. Aruchamy (ed.): *Photoelectrochemistry and Photovoltaics of Layered Semiconductors.* 1992   ISBN 0-7923-1556-1
15. T. Butz (ed.): *Nuclear Spectroscopy on Charge Density Wave Systems.* 1992
    ISBN 0-7923-1779-3
16. G. Benedek (ed.): *Surface Properties of Layered Structures.* 1992   ISBN 0-7923-1961-3
17. W. Müller-Warmuth and R. Schöllhorn (eds.): *Progress in Intercalation Research.* 1994
    ISBN 0-7923-2357-2
18. L.J. de Jongh (ed.): *Physics and Chemistry of Metal Cluster Compounds.* Model Systems for Small Metal Particles. 1994   ISBN 0-7923-2715-2
19. E.Y. Andrei (ed.): *Two-Dimensional Electron Systems.* On Helium and other Cryogenic Substrates. 1997   ISBN 0-7923-4738-2
20. A. Furrer: *Neutron Scattering in Layered Copper-Oxide Superconductors.* 1998
    ISBN 0-7923-5226-2

KLUWER ACADEMIC PUBLISHERS – DORDRECHT / BOSTON / LONDON